Characterization and Encapsulation of Natural Antioxidants: Interaction, Protection and Delivery

Characterization and Encapsulation of Natural Antioxidants: Interaction, Protection and Delivery

Editors

Li Liang
Hao Cheng

MDPI • Basel • Beijing • Wuhan • Barcelona • Belgrade • Manchester • Tokyo • Cluj • Tianjin

Editors
Li Liang
Jiangnan University
China

Hao Cheng
Jiangnan University
China

Editorial Office
MDPI
St. Alban-Anlage 66
4052 Basel, Switzerland

This is a reprint of articles from the Special Issue published online in the open access journal *Antioxidants* (ISSN 2076-3921) (available at: https://www.mdpi.com/journal/antioxidants/special_issues/Antioxidants_Characterization_Encapsulation).

For citation purposes, cite each article independently as indicated on the article page online and as indicated below:

LastName, A.A.; LastName, B.B.; LastName, C.C. Article Title. *Journal Name* **Year**, *Volume Number*, Page Range.

ISBN 978-3-0365-5455-6 (Hbk)
ISBN 978-3-0365-5456-3 (PDF)

© 2022 by the authors. Articles in this book are Open Access and distributed under the Creative Commons Attribution (CC BY) license, which allows users to download, copy and build upon published articles, as long as the author and publisher are properly credited, which ensures maximum dissemination and a wider impact of our publications.

The book as a whole is distributed by MDPI under the terms and conditions of the Creative Commons license CC BY-NC-ND.

Contents

About the Editors vii

Hao Cheng and Li Liang
Characterization and Encapsulation of Natural Antioxidants: Interaction, Protection, and Delivery
Reprinted from: *Antioxidants* **2022**, *11*, 1434, doi:10.3390/antiox11081434 1

Xin Yin, Kaiwen Chen, Hao Cheng, Xing Chen, Shuai Feng, Yuanda Song and Li Liang
Chemical Stability of Ascorbic Acid Integrated into Commercial Products: A Review on Bioactivity and Delivery Technology
Reprinted from: *Antioxidants* **2022**, *11*, 153, doi:10.3390/antiox11010153 5

Chenshan Shi, Miaomiao Liu, Hongfei Zhao, Zhaolin Lv, Lisong Liang and Bolin Zhang
A Novel Insight into Screening for Antioxidant Peptides from Hazelnut Protein: Based on the Properties of Amino Acid Residues
Reprinted from: *Antioxidants* **2022**, *11*, 127, doi:10.3390/antiox11010127 25

Jiumn-Yih Wu, Hsiou-Yu Ding, Tzi-Yuan Wang, Yu-Li Tsai, Huei-Ju Ting and Te-Sheng Chang
Improving Aqueous Solubility of Natural Antioxidant Mangiferin through Glycosylation by Maltogenic Amylase from *Parageobacillus galactosidasius* DSM 18751
Reprinted from: *Antioxidants* **2021**, *10*, 1817, doi:10.3390/antiox10111817 43

Lucija Mandić, Anja Sadžak, Ina Erceg, Goran Baranović and Suzana Šegota
The Fine-Tuned Release of Antioxidant from Superparamagnetic Nanocarriers under the Combination of Stationary and Alternating Magnetic Fields
Reprinted from: *Antioxidants* **2021**, *10*, 1212, doi:10.3390/antiox10081212 59

Wanwen Chen, Hao Cheng and Wenshui Xia
Construction of *Polygonatum sibiricum* Polysaccharide Functionalized Selenium Nanoparticles for the Enhancement of Stability and Antioxidant Activity
Reprinted from: *Antioxidants* **2022**, *11*, 240, doi:10.3390/antiox11020240 79

Muhammad Aslam Khan, Chufan Zhou, Pu Zheng, Meng Zhao and Li Liang
Improving Physicochemical Stability of Quercetin-Loaded Hollow Zein Particles with Chitosan/Pectin Complex Coating
Reprinted from: *Antioxidants* **2021**, *10*, 1476, doi:10.3390/antiox10091476 93

Ștefania Adelina Milea, Iuliana Aprodu, Elena Enachi, Vasilica Barbu, Gabriela Râpeanu, Gabriela Elena Bahrim and Nicoleta Stănciuc
Whey Protein Isolate-Xylose Maillard-Based Conjugates with Tailored Microencapsulation Capacity of Flavonoids from Yellow Onions Skins
Reprinted from: *Antioxidants* **2021**, *10*, 1708, doi:10.3390/antiox10111708 109

Xin Yin, Hao Cheng, Wusigale, Huanhuan Dong, Weining Huang and Li Liang
Resveratrol Stabilization and Loss by Sodium Caseinate, Whey and Soy Protein Isolates: Loading, Antioxidant Activity, Oxidability
Reprinted from: *Antioxidants* **2022**, *11*, 647, doi:10.3390/antiox11040647 123

Danilo Escobar-Avello, Javier Avendaño-Godoy, Jorge Santos, Julián Lozano-Castellón, Claudia Mardones, Dietrich von Baer, Javiana Luengo, Rosa M. Lamuela-Raventós, Anna Vallverdú-Queralt and Carolina Gómez-Gaete
Encapsulation of Phenolic Compounds from a Grape Cane Pilot-Plant Extract in Hydroxypropyl Beta-Cyclodextrin and Maltodextrin by Spray Drying
Reprinted from: *Antioxidants* **2021**, *10*, 1130, doi:10.3390/antiox10071130 145

Anna Valentino, Raffaele Conte, Ilenia De Luca, Francesca Di Cristo, Gianfranco Peluso, Michela Bosetti and Anna Calarco
Thermo-Responsive Gel Containing Hydroxytyrosol-Chitosan Nanoparticles (Hyt@tgel) Counteracts the Increase of Osteoarthritis Biomarkers in Human Chondrocytes
Reprinted from: *Antioxidants* **2022**, *11*, 1210, doi:10.3390/antiox11061210 163

Ting Liu, Zhipeng Gao, Weiming Zhong, Fuhua Fu, Gaoyang Li, Jiajing Guo and Yang Shan
Preparation, Characterization, and Antioxidant Activity of Nanoemulsions Incorporating Lemon Essential Oil
Reprinted from: *Antioxidants* **2022**, *11*, 650, doi:10.3390/antiox11040650 181

Maryam Mohammadi, Hamed Hamishehkar, Marjan Ghorbani, Rahim Shahvalizadeh, Mirian Pateiro and José M. Lorenzo
Engineering of Liposome Structure to Enhance Physicochemical Properties of *Spirulina plantensis* Protein Hydrolysate: Stability during Spray-Drying
Reprinted from: *Antioxidants* **2021**, *10*, 1953, doi:10.3390/antiox10121953 195

Karina Oliveira Lima, Meritaine da Rocha, Ailén Alemán, María Elvira López-Caballero, Clara A. Tovar, María Carmen Gómez-Guillén, Pilar Montero and Carlos Prentice
Yogurt Fortification by the Addition of Microencapsulated Stripped Weakfish (*Cynoscion guatucupa*) Protein Hydrolysate
Reprinted from: *Antioxidants* **2021**, *10*, 1567, doi:10.3390/antiox10101567 209

Tung Thanh Diep, Michelle Ji Yeon Yoo and Elaine Rush
Tamarillo Polyphenols Encapsulated-Cubosome: Formation, Characterization, Stability during Digestion and Application in Yoghurt
Reprinted from: *Antioxidants* **2022**, *11*, 520, doi:10.3390/antiox11030520 229

About the Editors

Li Liang

Li Liang is a professor in the State Key Laboratory of Food Science and Technology, School of Food Science and Technology at Jiangnan University. She received her B.S. in ocean chemistry at the Ocean University of Qingdao (Ocean University of China). After completing her Ph.D. in polymer chemistry and physics at Fudan University (China), she had her postdoctoral training in the Department of Food Science and Nutrition at Laval University (Canada). She won the Tanner Award of Institute of Food Technologists (IFT) in 2019. Her current research focuses on food chemistry involved in protein structure and function, protein encapsulation and immobilization, protection and delivery of bioactive components, and functional foods.

Hao Cheng

Hao Cheng is an associate professor in the State Key Laboratory of Food Science and Technology at Jiangnan University. He obtained his master's degree in food engineering at the School of Food Science and Technology, Jiangnan University in 2016. He studied in the faculty of agriculture, life and environmental sciences at the University of Alberta as a visiting scholar from September 2018 to October 2019. He obtained his Ph.D. in food science and technology at the School of Food Science and Technology, Jiangnan University in 2020. Since 2020, he has been an associate professor of the State Key Laboratory of Food Science and Technology at Jiangnan University. His research mainly focuses on protein structure–function relationships, biopolymer-based nutraceutical delivery systems, and development of functional foods.

Editorial

Characterization and Encapsulation of Natural Antioxidants: Interaction, Protection, and Delivery

Hao Cheng [1,2] and Li Liang [1,2,*]

[1] State Key Laboratory of Food Science and Technology, Jiangnan University, Wuxi 214122, China; haocheng@jiangnan.edu.cn
[2] School of Food Science and Technology, Jiangnan University, Wuxi 214122, China
* Correspondence: liliang@jiangnan.edu.cn

Natural antioxidants (e.g., polyphenols, carotenoids, vitamins, polyunsaturated fatty acids, peptides, and enzymes) are expected to inhibit the oxidative damage of proteins and lipids in food systems and reduce the risk of certain chronic diseases related to oxidative stress [1]. It is still a challenge to predict an antioxidant's activity based on its structural characteristics, especially for peptides [2]. Furthermore, most of natural antioxidants have poor aqueous solubility, high sensitivity to processing and storage conditions, and low bioaccessibility and bioavailability, which restrict their direct incorporation into commercial products and their related health benefits [3].

According to a Market Research Report of Global Industry Analysis Inc., the global market for food antioxidants, which is estimated at USD 1.2 billion in 2020, is projected to grow at a CAGR of 3.5% and reach a revised size of USD 1.6 billion by 2027, while natural food antioxidants are projected to record 3.8% CAGR and reach USD 1.1 billion by 2027 [4]. It is well known that encapsulation technology plays an important role in protecting and stabilizing natural antioxidants [5]. Inorganic materials, proteins, polysaccharides and lipids are potential carrier materials for antioxidants. Many of their structural and physicochemical properties facilitate the design and fabrication of delivery systems in a wide range of platforms, which can overcome the limitations of natural antioxidants used in food and pharmaceutical industries [6–8].

This Special Issue contains 13 research articles covering the prediction of antioxidant peptides and the improvement of the aqueous solubility, stability, bioactivity and release properties through enzymatic glycosylation and encapsulation technology in addition to a review about ascorbic acid, which describes the bioactivity and degradation mechanism of ascorbic acid, the strategies for improving its chemical stability, and the application of ascorbic acid in commercial products [9].

The antioxidant activity of peptides derived from protein is usually attributed to their amino acid composition, active amino acid position, molecular mass, and spatial structure. How to directly screen antioxidant peptides according to their physicochemical properties is essential to the efficiently manufacturing of peptides with high antioxidant activity. Shi et al. confirmed that the route of a molecular weight of 200 to 800 Da containing Tyr, Met, and Trp residues at the C-terminus and a grand mean hydropathy value between −2 and 1 could be used to screen antioxidant peptides. They found that peptides released from hazelnut protein falling within the route could effectively prevent the oxidation of linoleic acid and hazelnut oil [10].

Mangiferin, a C-glucosidic xanthone from *Mangifera indica* (mango) plant, exhibits many health-promoting activities. However, the poor hydrosolubility of mangiferin limits its application in the food industry. In order to improve the aqueous solubility of mangiferin, the enzymatic glycosylation of mangiferin was applied to produce more soluble mangiferin glucosides. The recombinant maltogenic amylase was produced by cloning a thermophile *Parageobacillus galactosidasius* DSM 18751 T (PgMA) into *Escherichia coli* BL21 (DE3) via the

expression plasmid pET-Duet-1. The obtained glucosyl-α-(1→6)-mangiferin and maltosyl-α-(1→6)-mangiferin from mangiferin by PgMA showed 5500-fold higher aqueous solubility than that of mangiferin. Both mangiferin glucosides showed similar DPPH free radical scavenging activities as mangiferin [11].

Inorganic nanoparticles, such as magnetic and selenium nanoparticles, can serve as antioxidants or delivery vehicles for bioactives. Mandić et al., synthesized mesoporous magnetite nanoparticles stabilized by PEG-4000 using a solvothermal method. The magnetite nanoparticles had a small size of around 10 nm and encapsulation efficiency of 20% for quercetin. The prolong quercetin release from PEG-coated magnetite nanoparticles could be realized by using combined stationary and alternating magnetic fields [12]. In another study, the *Polygonatum sibiricum* polysaccharide was used as a stabilizer of selenium nanoparticles in a simple redox system. The functionalization with the *Polygonatum sibiricum* polysaccharide effectively improved the free-radical-scavenging ability of selenium nanoparticles. The obtained polysaccharide-coated selenium nanoparticles showed a higher protective effect on PC-12 cells against H_2O_2-induced oxidative damage [13].

Protein particles with a hollow structure are preferred over solid ones due to their high loading capacity, sustained release, and low density. Khan et al. fabricated hollow zein particles through a sacrificial template method for the encapsulation of quercetin, obtaining a maximum encapsulation efficiency of 80% and loading capacity of 6.29%. The hollow zein particles were further coated with chitosan and pectin using a layer-by-layer technique, which improved the stability of hollow zein particles against heat treatment, pH variation, and salt. The obtained hollow zein composite particles could significantly improve the photostability and storage stability of quercetin [14]. Milea and coworkers prepared WPI-xylose Maillard-based conjugates for the encapsulation of flavonoids from yellow onion skins. The microcapsules formulated with WPI-xylose conjugates showed a high encapsulation efficiency of 90.53%. The incorporation of flavonoid-loaded microcapsules into nachos could improve the antioxidant activity of nachos [15]. Yin et al. investigated the interaction of whey protein isolate (WPI), sodium caseinate, and soy protein isolate with resveratrol, and they found that the protein species play an important role in loading and protecting antioxidants. WPI had the lowest loading capacity but showed the best protective effect for resveratrol against degradation during storage. The results highlight the importance of protein oxidability on stability [16].

Cyclodextrins with a hydrophobic interior and a hydrophilic outer surface can act as a carrier for phenolic compounds. Escobar-Avello et al. entrapped around 80% of the extracts from a grape cane in hydroxypropyl beta-cyclodextrin (HP-β-CD) through a spray-drying technique. The inclusion of grape cane extracts in HP-β-CD had no impact on their antioxidant activity [17]. Valentino and coworkers fabricated a thermo-responsive hydrogel with Pluronic F-127 and hyaluronic acid, which exhibited easy injectability and sol–gel transition as the temperature increased from room temperature to body temperature. The smart hydrogel was further used to encapsulate hydroxytyrosol-loaded chitosan nanoparticles for the treatment of osteoarthritis. The in vitro study demonstrated that the combination of hydrogels and antioxidant-loaded nanoparticles showed potential use in the treatment of the chronic inflammatory degenerative disease [18].

Lipid-based delivery systems, such as emulsions and liposomes, are effective carriers for enhancing the stability and bioavailability of bioactives [19]. Lemon essential oil with many biological activities is commonly used as a preservative or flavoring agent in the food industry. Liu and coworkers fabricated a stable nanoemulsion formulated with Tween-80 and Span-80 for the encapsulation of lemon essential oil. The optimal formulation was obtained by using a combination of single-factor experiments and response surface methodology. The lemon oil nanoemulsions exhibited improved antioxidant activity and stability during storage [20]. In another study, *Spirulina plantensis* protein hydrolysates were encapsulated into liposomes coated with chitosan. Around 90% of the initial antioxidant activity remained after storage at 4 °C for 30 days. The hydrolysate-loaded liposomes could be converted into powders by using a spray-dying technique, exhibiting all characteristics of peptide-loaded liquid formulations [21].

There are some examples of incorporating bioactive-loaded edible carriers into food products. Fish protein hydrolysates with high antioxidant and angiotensin-converting enzyme inhibitory capacity were microencapsulated into maltodextrin matrix using a spray-drying technique, which was further incorporated into yogurt products. Yogurts fortified with protein hydrolysates microcapsules have acceptable flavor and exhibited greater antioxidant and antihypertensive activities during a 7-day storage [22]. In another study, tamarillo polyphenols were successfully encapsulated in a cubosomal system. The addition of tamarillo polyphenol-loaded cubosome improved the physicochemical properties and nutritional values of yoghurts [23].

We would like to acknowledge all the authors who contributed to this Special Issue "Characterization and Encapsulation of Natural Antioxidants: Interaction, Protection, and Delivery". This Special Issue provides new insights into the importance of delivery carriers in the stabilization and improved performance of natural antioxidants. Despite the relevance of the topics covered in the papers published in this Special Issue, many aspects remain relatively limited, including the mechanisms of antioxidant–material interplay, the compatibility of antioxidant-loaded carriers with commercial products, and the evaluation of bioactivity in vivo models.

Funding: This research received no external funding.

Conflicts of Interest: The authors declare no conflict of interest.

References

1. Kasote, D.M.; Katyare, S.S.; Hegde, M.V.; Bae, H. Significance of Antioxidant Potential of Plants and its Relevance to Therapeutic Applications. *Int. J. Biol. Sci.* **2015**, *11*, 982–991. [CrossRef] [PubMed]
2. Wen, C.; Zhang, J.; Zhang, H.; Duan, Y.; Ma, H. Plant protein-derived antioxidant peptides: Isolation, identification, mechanism of action and application in food systems: A review. *Trends Food Sci. Technol.* **2020**, *105*, 308–322. [CrossRef]
3. McClements, D.J.; Li, F.; Xiao, H. The Nutraceutical Bioavailability Classification Scheme: Classifying Nutraceuticals According to Factors Limiting their Oral Bioavailability. *Annu. Rev. Food Sci. Technol.* **2015**, *6*, 299–327. [CrossRef]
4. Global Industry Analysts, Inc. Food Antioxidants. 2021. Available online: https://www.giiresearch.com/report/go906868-food-antioxidants.html (accessed on 14 July 2022).
5. Khalil, I.; Yehye, W.A.; Etxeberria, A.E.; Alhadi, A.A.; Dezfooli, S.M.; Julkapli, N.B.M.; Basirun, W.J.; Seyfoddin, A. Nanoantioxidants: Recent Trends in Antioxidant Delivery Applications. *Antioxidants* **2019**, *9*, 24. [CrossRef]
6. Wan, Z.-L.; Guo, J.; Yang, X.-Q. Plant protein-based delivery systems for bioactive ingredients in foods. *Food Funct.* **2015**, *6*, 2876–2889. [CrossRef]
7. Luther, D.C.; Huang, R.; Jeon, T.; Zhang, X.; Lee, Y.-W.; Nagaraj, H.; Rotello, V.M. Delivery of drugs, proteins, and nucleic acids using inorganic nanoparticles. *Adv. Drug Deliv. Rev.* **2020**, *156*, 188–213. [CrossRef]
8. Zhang, L.; McClements, D.J.; Wei, Z.; Wang, G.; Liu, X.; Liu, F. Delivery of synergistic polyphenol combinations using biopolymer-based systems: Advances in physicochemical properties, stability and bioavailability. *Crit. Rev. Food Sci. Nutr.* **2019**, *60*, 2083–2097. [CrossRef] [PubMed]
9. Yin, X.; Chen, K.; Cheng, H.; Chen, X.; Feng, S.; Song, Y.; Liang, L. Chemical Stability of Ascorbic Acid Integrated into Commercial Products: A Review on Bioactivity and Delivery Technology. *Antioxidants* **2022**, *11*, 153. [CrossRef]
10. Shi, C.; Liu, M.; Zhao, H.; Lv, Z.; Liang, L.; Zhang, B. A Novel Insight into Screening for Antioxidant Peptides from Hazelnut Protein: Based on the Properties of Amino Acid Residues. *Antioxidants* **2022**, *11*, 127. [CrossRef]
11. Wu, J.-Y.; Ding, H.-Y.; Wang, T.-Y.; Tsai, Y.-L.; Ting, H.-J.; Chang, T.-S. Improving Aqueous Solubility of Natural Antioxidant Mangiferin through Glycosylation by Maltogenic Amylase from *Parageobacillus galactosidasius* DSM 18751. *Antioxidants* **2021**, *10*, 1817. [CrossRef]
12. Mandić, L.; Sadžak, A.; Erceg, I.; Baranović, G.; Šegota, S. The Fine-Tuned Release of Antioxidant from Superparamagnetic Nanocarriers under the Combination of Stationary and Alternating Magnetic Fields. *Antioxidants* **2021**, *10*, 1212. [CrossRef] [PubMed]
13. Chen, W.; Cheng, H.; Xia, W. Construction of *Polygonatum sibiricum* Polysaccharide Functionalized Selenium Nanoparticles for the Enhancement of Stability and Antioxidant Activity. *Antioxidants* **2022**, *11*, 240. [CrossRef] [PubMed]
14. Khan, M.A.; Zhou, C.; Zheng, P.; Zhao, M.; Liang, L. Improving Physicochemical Stability of Quercetin-Loaded Hollow Zein Particles with Chitosan/Pectin Complex Coating. *Antioxidants* **2021**, *10*, 1476. [CrossRef] [PubMed]

15. Milea, A.; Aprodu, I.; Enachi, E.; Barbu, V.; Râpeanu, G.; Bahrim, G.E.; Stănciuc, N. Whey Protein Isolate-Xylose Maillard-Based Conjugates with Tailored Microencapsulation Capacity of Flavonoids from Yellow Onions Skins. *Antioxidants* **2021**, *10*, 1708. [CrossRef] [PubMed]
16. Yin, X.; Cheng, H.; Wusigale; Dong, H.; Huang, W.; Liang, L. Resveratrol Stabilization and Loss by Sodium Caseinate, Whey and Soy Protein Isolates: Loading, Antioxidant Activity, Oxidability. *Antioxidants* **2022**, *11*, 647. [CrossRef] [PubMed]
17. Escobar-Avello, D.; Avendaño-Godoy, J.; Santos, J.; Lozano-Castellón, J.; Mardones, C.; von Baer, D.; Luengo, J.; Lamuela-Raventós, R.; Vallverdú-Queralt, A.; Gómez-Gaete, C. Encapsulation of Phenolic Compounds from a Grape Cane Pilot-Plant Extract in Hydroxypropyl Beta-Cyclodextrin and Maltodextrin by Spray Drying. *Antioxidants* **2021**, *10*, 1130. [CrossRef]
18. Valentino, A.; Conte, R.; De Luca, I.; Di Cristo, F.; Peluso, G.; Bosetti, M.; Calarco, A. Thermo-Responsive Gel Containing Hydroxytyrosol-Chitosan Nanoparticles (Hyt@tgel) Counteracts the Increase of Osteoarthritis Biomarkers in Human Chondrocytes. *Antioxidants* **2022**, *11*, 1210. [CrossRef]
19. Rezhdo, O.; Speciner, L.; Carrier, R. Lipid-associated oral delivery: Mechanisms and analysis of oral absorption enhancement. *J. Control. Release* **2016**, *240*, 544–560. [CrossRef]
20. Liu, T.; Gao, Z.; Zhong, W.; Fu, F.; Li, G.; Guo, J.; Shan, Y. Preparation, Characterization, and Antioxidant Activity of Nanoemulsions Incorporating Lemon Essential Oil. *Antioxidants* **2022**, *11*, 650. [CrossRef]
21. Mohammadi, M.; Hamishehkar, H.; Ghorbani, M.; Shahvalizadeh, R.; Pateiro, M.; Lorenzo, J.M. Engineering of Liposome Structure to Enhance Physicochemical Properties of *Spirulina plantensis* Protein Hydrolysate: Stability during Spray-Drying. *Antioxidants* **2021**, *10*, 1953. [CrossRef]
22. Lima, K.O.; da Rocha, M.; Alemán, A.; López-Caballero, M.E.; Tovar, C.A.; Gómez-Guillén, M.C.; Montero, P.; Prentice, C. Yogurt Fortification by the Addition of Microencapsulated Stripped Weakfish (*Cynoscion guatucupa*) Protein Hydrolysate. *Antioxidants* **2021**, *10*, 1567. [CrossRef] [PubMed]
23. Diep, T.T.; Yoo, M.J.Y.; Rush, E. Tamarillo Polyphenols Encapsulated-Cubosome: Formation, Characterization, Stability during Digestion and Application in Yoghurt. *Antioxidants* **2022**, *11*, 520. [CrossRef] [PubMed]

Review

Chemical Stability of Ascorbic Acid Integrated into Commercial Products: A Review on Bioactivity and Delivery Technology

Xin Yin [1,2], Kaiwen Chen [1,2], Hao Cheng [1,2], Xing Chen [1,2], Shuai Feng [3], Yuanda Song [4] and Li Liang [1,2,*]

[1] State Key Laboratory of Food Science and Technology, Jiangnan University, Wuxi 214122, China; 17851312787@163.com (X.Y.); chenkaiwen1018@163.com (K.C.); haocheng@jiangnan.edu.cn (H.C.); xingchen@jiangnan.edu.cn (X.C.)
[2] School of Food Science and Technology, Jiangnan University, Wuxi 214122, China
[3] Luwei Pharmaceutical Group Co., Ltd., Shuangfeng Industrial Park, Zibo 255195, China; fengshuai@hlvitamin.com
[4] Colin Raledge Center for Microbial Lipids, School of Agricultural Engineering and Food Science, Shandong University of Technology, Zibo 255000, China; ysong@sdut.edu.cn
* Correspondence: liliang@jiangnan.edu.cn; Tel.: +86-(510)-8519-7367

Abstract: The L-enantiomer of ascorbic acid is commonly known as vitamin C. It is an indispensable nutrient and plays a key role in retaining the physiological process of humans and animals. L-gulonolactone oxidase, the key enzyme for the de novo synthesis of ascorbic acid, is lacking in some mammals including humans. The functionality of ascorbic acid has prompted the development of foods fortified with this vitamin. As a natural antioxidant, it is expected to protect the sensory and nutritional characteristics of the food. It is thus important to know the degradation of ascorbic acid in the food matrix and its interaction with coexisting components. The biggest challenge in the utilization of ascorbic acid is maintaining its stability and improving its delivery to the active site. The review also includes the current strategies for stabilizing ascorbic acid and the commercial applications of ascorbic acid.

Keywords: ascorbic acid; bioactivity; stability; delivery; application

1. Introduction

Ascorbic acid (L-enantiomer, Figure 1), commonly known as vitamin C, is composed of six carbons and related to the C_6 sugars. It is the aldono-1,4-lactone of a hexonic acid with an enediol group on carbons 2 and 3. As an essential micronutrient, ascorbic acid plays a vital role in maintaining normal metabolic processes and homeostasis within the human body. Although D-isoascorbic acid is the stereoisomer of ascorbic acid (Figure 1), such analogues hardly express the activity of ascorbic acid. Mammalian cells cannot synthesize ascorbic acid de novo due to the lack of L-gulono-1,4 lactone oxidase, which is an essential enzyme for the production of ascorbic acid [1]. Vegetables and fruits serve as natural sources of vitamin C intake, but only a limited number of plants are rich in vitamin C. Nowadays, ascorbic acid is industrially produced from D-glucose, and the procedure involves several complex chemical and biotechnological stages [2].

An ordinary diet of natural and synthetic ascorbic acid is the only way to maintain the physiological requirements. The well-known symptom of ascorbic acid deficiency is associated with connective tissue damage, such as scurvy, which is characterized by fragile tissues and poor wound healing [3]. The currently recommended dietary allowances (RDA) for ascorbic acid are 90 mg/day and 75 mg/day for men and women, respectively [4]. Researchers have found that the steady-state concentration of ascorbic acid in plasma is about 80 µmol/L, when sufficient fruits and vegetables are consumed every day. Oral dosing of ascorbic acid (1.25 g) can improve the concentration of ascorbic acid in plasma to 134.8 ± 20.6 µmol/L [5]. In order to maintain the ascorbic acid concentration required by the body, ascorbic acid-fortified dietary supplements or foods have attracted interest from consumers, in addition to fruits and vegetables found in nature.

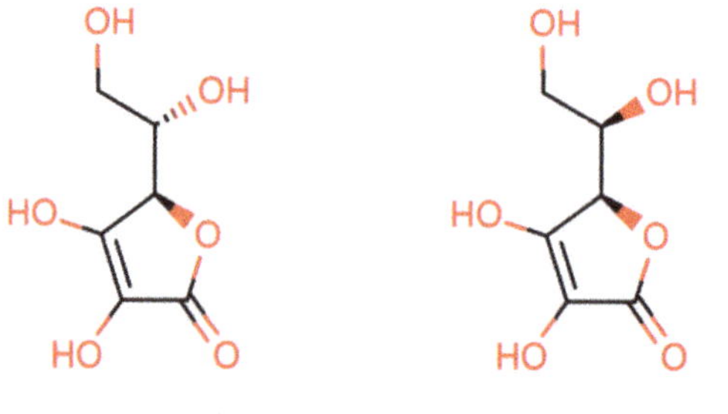

L-ascorbic acid D-isoascorbic acid

Figure 1. Structures of L-ascorbic acid and its stereoisomer.

The main challenge in the development of ascorbic acid products is its high instability and reactivity. Ascorbic acid is reversibly oxidized into dehydroascorbic acid (DHA) upon exposure to light, heat, transition metal ions and pH (alkaline condition), then DHA further irreversibly hydrolyzes to form 2,3-diketogulonic acid (Figure 2A). In recent decades, the strategy of shielding ascorbic acid from sensitive environments by encapsulating ascorbic acid within a layer of wall material has attracted much interest among researchers. A series of innovation delivery technologies have emerged, including microfluidic [6], melt extrusion [7], spray drying and chilling [8,9]. The particles prepared by these methods are usually on the microscale. Under certain conditions, nano-encapsulation of ascorbic acid can be realized through ion gelation of chitosan or complex coacervation with anionic polymers [10,11]. On the other hand, some bioactive compounds with low molecular weight can protect ascorbic acid by scavenging the pro-degradation factors of ascorbic acid in the solution [12].

A

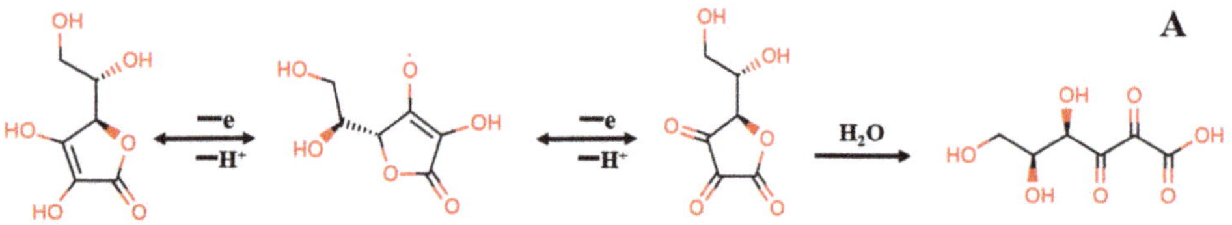

L-ascorbic acid monodehydroascorbic acid dehydroascorbic acid 2,3-diketogulonic acid

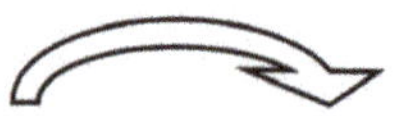

$$Fe^{3+} + AA \rightarrow Fe^{2+} + MDHA$$ **B**

$$Fe^{2+} + O_2 \rightarrow Fe^{3+} + O_2{\bullet}^-$$

$$Fe^{2+} + H_2O_2 \rightarrow Fe^{3+} + OH{\bullet} + OH^-$$

Figure 2. Degradation of L-ascorbic acid to dehydroascorbic acid and 2,3-diketogulonic acid (**A**) and pro-oxidant effects of ascorbic acid (**B**).

Stable ascorbic acid needs to be accurately delivered to the desired site, and released from the carrier at a desired rate and time. This is the basis for obtaining excellent bioavailability of ascorbic acid. For oral products, the expected release site of ascorbic acid is the small intestine rather than the stomach, since the absorption and metabolization of bioactive compounds mainly occur in the small intestine [13]. This is a challenge for the selection of suitable carrier and wall materials, which is related to the dissolution of the coating polymer in the gastrointestinal environment and its molecular weight. There is a hypothesis that the molecular weight of the used polymers is negatively related to the release of the encapsulated compounds [14]. The encapsulation efficiency of ascorbic acid in gelatin-coated microcapsules reached up to about 94%, but the release of ascorbic acid in the stomach was faster than that in the intestine [15]. However, only 30% of ascorbic acid in chitosan nanoparticles was released in a simulated gastric solution, while the release in simulated intestinal condition exceeded 75% [16].

Figure 3 shows the number of patents related to ascorbic acid from 1992 to 2021; there have been a stable and relatively high number of applications since 1997. From 2010 to 2013, the patents were even more than 1200 per year. It can be found that the importance of ascorbic acid has aroused widespread interest in the consumer market. Ascorbic acid itself or in conjunction with co-existing ingredients in the food matrix can express various physiological activities beneficial to health. This review mainly aims to provide a comprehensive summary about the strategies for stabilizing and controlling release of ascorbic acid in the past 20 years. The commercial products fortified with ascorbic acid are also summarized.

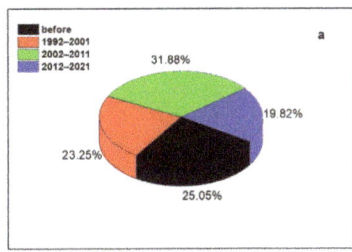
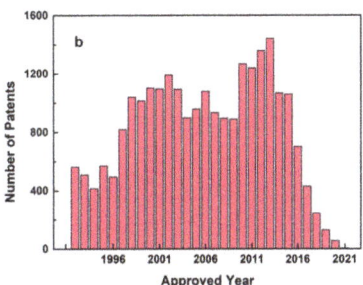

Figure 3. The proportion of patents related to ascorbic acid search terms before 2021 (**a**) and the number of ascorbic acid patents in each of the past 30 years (**b**).

2. Bioactivity of Ascorbic Acid

2.1. Antioxidant

The by-products of normal cell metabolism are reactive oxygen species (ROS), including superoxide radicals ($O_2 \cdot^-$), singlet oxygen ($^1O_2$), hydrogen peroxide (H_2O_2) and highly reactive hydroxyl radicals ($OH\cdot$). The adverse effect of ROS is that it can initiate a cascade of radicals, producing hydroxyl free radicals and other destructive species. These further induce protein and DNA damage, lipid peroxidation and finally lead to cell apoptosis [17]. The antioxidant defense system cannot fully eliminate the toxic ROS accumulated in the cells, that is, the so-called "oxidative stress" occurs [18]. In addition to enzymatic reactions, ROS can also be eliminated through non-enzymatic means such as antioxidants. Ascorbic acid is a free radical and other oxygen species scavenger, which can protect cells from oxidative damage caused by ROS. Antiradical capability commonly reflects the antioxidant ability, and ascorbic acid in foodstuff and bio-systems acts as antioxidant. As the most effective and natural antioxidant with the least side effects, ascorbic acid can inhibit various diseases caused by oxidative stress in the body, such as cancer, cardiovascular disease, aging and cataracts [15]. Studies have shown that the mortality from these diseases is inversely related to plasma concentration of ascorbic acid [19]. Ascorbic acid and its derivatives can reduce the level of lipid peroxidation in vivo due to aging [20]. In the absence of transition metals, ascorbic acid can reduce the frequency of mutations induced by H_2O_2 in human cells [21]. Compared with other polyphenols or flavonoid antioxidants, ascorbic acid terminates the free radical chain reaction through disproportionation reaction, and the reaction products such as DHA and 2,3-diketogulonic acid (Figure 2A) are non-damaging and non-radical products [22]. Another manifestation of antioxidant property is that ascorbic acid can form relatively stable ascorbic acid free radicals to donate single electrons [23]. As reported, antioxidants can also repair tryptophan free radicals produced by the one-electron oxidation of free tryptophan in lysozyme to maintain protein integrity [24].

Ascorbic acid is also used as an antioxidant to protect the sensory and nutritional properties of foods. As an anti-browning agent, it can inhibit the browning of vegetables and fruits caused by oxidation. The formation of quinones mediated by polyphenol oxidase causes the accumulation of H_2O_2, which in turn causes the browning of polyphenols mediated by peroxidase [25]. Ascorbic acid inhibits browning by reducing the o-quinone produced by polyphenol oxidase to the original diphenol through a process called "deactivation reaction" [26]. In addition to the regeneration mechanism of polyphenols, the protective effect is also attributed to the competitive inhibition of polyphenol oxidase activity by ascorbic acid. Meanwhile, addition of ascorbic acid causes a decrease in pH and is not conducive to the expression of polyphenol oxidase activity [27]. In meat products, ascorbic acid is widely used as a natural agent for color retention, which can inhibit lipid oxidation and maintain color stability [28]. Compared with other organic acids such as malic acid, citric acid and tartaric acid, ascorbic acid had the best protective effect on the quality of cured meat and was a suitable ingredient for cured meat products [29]. The surface of the pork sprayed with ascorbic acid and a mix of that and rosemary extract maintained good stability in color, water content and pH after frozen storage [30]. It is worth noting that this dietary source of ascorbic acid added in meat products is often overlooked. Norwegian researchers found that the content of ascorbic acid in sausages is 11–40 mg/mL, but ascorbic acid is usually ignored in the table of food ingredients because the added ascorbic acid is used as a color retention agent rather than a nutrient component. As a result, the actual ascorbic acid intake of Norwegian residents increased by 3–10% [31]. The ascorbic acid added to the edible polysaccharide film can eliminate or quench the free radicals generated by radiation. As a radiation inhibitor, ascorbic acid can inhibit the decrease in the viscosity of carrageenan caused by radiation and protect its rheological properties [32]. Ascorbic acid can inhibit food-borne pathogens in the early stage of biofilm formation due to its anti-quorum sensing activity and inhibition of extracellular polymer production. The efficacy of ascorbic acid is related to its concentration and the strain. For *Escherichia coli* and *Staphylococcus aureus*, the inhibitory effect of ascorbic acid at 25 mg/mL is the greatest, and lower concentrations of ascorbic acid are ineffective. For *Listeria monocytogenes*, ascorbic acid at 0.25 mg/mL shows an inhibitory effect [33].

2.2. Pro-Oxidant

Pro-oxidant activity is defined as the ability of antioxidants to reduce transition metal ions to a lower oxidation state, which refers to the Fenton reaction [34]. In the Fenton reaction, transition metal ions such as Fe^{3+} are reduced by ascorbic acid and then Fe^{2+} further react with oxygen and hydrogen peroxide to form highly active and destructive hydroxyl radicals (Figure 2B) [35]. Ascorbic acid does not always express antioxidant activity, and may be converted into a pro-oxidant and show toxic effects under certain conditions. The effect of ascorbic acid on the redox properties of bovine hemoglobin is dual with an antioxidant effect at the initial stage of the reaction. During the reaction process, ascorbic acid generates hydrogen peroxide under the mediation of oxygen or oxygenated hemoglobin. With the consumption of ascorbic acid, its own scavenging ability cannot balance the accumulated hydrogen peroxide, which leads to the formation of bilirubin and accelerates the oxidation of hemoglobin [36]. Studies have shown that ascorbic acid will transform from an antioxidant under physiological conditions to a pro-oxidant at higher concentrations. Researchers found that supplementation of 500 mg of ascorbic acid in the diet for 6 weeks increases the level of oxidative damage to peripheral blood lymphocytes, although this result is still controversial in the academic community [37,38]. Furthermore, the presence of transition metal ions in the system is also a key factor for ascorbic acid exerting pro-oxidant activity [39]. In mayonnaise, the added ascorbic acid works as a lipid antioxidant or pro-oxidant depending on the presence or absence of the fat-soluble antioxidant vitamin E. In the system containing vitamin E, the synergistically antioxidant effect of these two vitamins is stronger than the pro-oxidant effect of ascorbic acid. Without the addition of vitamin E, the hydrogen peroxide at the interface of the oil droplets promotes the lipid oxidation of lipoprotein particles in the mayonnaise, which in turn induces the oxidation of apolipoproteins and produces volatile odors [40]. In addition, dehydroascorbic acid may irreversibly degrade and produce highly reactive carbonyl intermediates, which can induce glycosylation of proteins. This is a non-enzymatic, non-specific reaction between carbonyl and amino groups, which is involved in a variety of age-related diseases [41]. It is worth noting that the pro-oxidation of ascorbic acid can induce the apoptosis of cancer cells, thereby exerting anti-cancer effects to a certain extent. As reported, the copper-dependent cellular redox state is an important factor in the cytotoxic effect of ascorbic acid on cancer cells. Ascorbic acid mobilizes nuclear copper to cause pro-oxidative cleavage of cellular DNA, and nuclear copper serves as a new molecular target for the toxic effects of cancer cells [42,43]. From this perspective, the pro-oxidation effect of ascorbic acid is beneficial.

Up to now, the mechanism and conditions that induce ascorbic acid to express pro-oxidant properties have not been clearly elucidated, and the definition of the conversion concentration between antioxidant and pro-oxidant is also unclear. Moreover, most of the reports on the promotion of oxidation of ascorbic acid are concentrated in vitro [42–44].

2.3. Co-Factors

Ascorbic acid can also be used as a co-factor for enzymes and other bioactive components to indirectly exert biological activities by acting as a free radical scavenger and electron transfer donor/acceptor to directly express its antioxidant properties. In the metabolic process of animals and plants, ascorbic acid does not directly participate in the catalytic cycle. As an enzyme co-factor, ascorbic acid exerts its indispensable function by regulating hydroxylation processes in multiple enzymatic reactions. For the active part of the enzyme with iron or copper, the role of ascorbic acid is to maintain the transition metal ions of these enzymes in a reduced form to exert their maximum physiological activity [45]. Ascorbic acid is a co-factor for non-heme iron α-ketoglutarate-dependent dioxygenases such as prolyl 4-hydroxylase with the role in the synthesis of collagen. As an electron donor, ascorbic acid can keep iron in the ferrous state, thereby maintaining the full activity of collagen hydroxylase. This promotes the hydroxylation of proline and lysine residues, allowing pro-collagen correct intracellular folding [46]. Ascorbic acid can also promote catecholamine synthesis by circulating tetrahydrobiopterin and enhance adrenal steroid production by increasing the expression of tyrosine hydroxylase [47]. As a co-factor, it helps dopamine β-hydroxylase convert dopamine to norepinephrine [48]. In addition, ascorbic acid can regulate cardiomyopathy and neurometabolic diseases. For example, as a co-factor for carnitine synthesis, it can shuttle fatty acids into the mitochondria and reduce oxidative stress [49]. In some clinical situations, as a co-factor for the biosynthesis of amidated opioid peptides, taking ascorbic acid can exhibit analgesic effects [50].

2.4. Synergistic Effect

As a natural antioxidant, ascorbic acid mostly exists in the form of coexistence with other components in nature. Combining it with other antioxidants may produce additive or even synergistic effects. Ascorbic acid and vitamin E, as chain-scission antioxidants, have an important inhibitory effect on the auto-oxidation of cell membrane polyunsaturated liposomes in vivo and the oxidation of lipids in vitro [51]. Studies have shown that the combination of 15% ascorbic acid and 1% α-tocopherol can significantly inhibit erythema and the formation of sunburn cells [52]. The synergy between α-tocopherol and ascorbic acid relies on the ability of ascorbic acid to regenerate α-tocopherol, and maintain the antioxidant capacity of α-tocopherol through circulation and inhibition of pro-oxidation [53]. The combined use of ascorbic acid and gallic acid is a promising strategy to prevent the formation of advanced glycation end products, showing the synergistic effect in the inhibition of amyloid cross-β-structure and protein carbonyl formation in fructose-induced BSA glycosylation samples [54]. Lycopene can inhibit inflammation and further stimulate the release of anti-inflammatory cytokine IL-10 when it combined with ascorbic acid and/or α-tocopherol [55]. Understanding the synergy between ascorbic acid and other bioactive compounds allows the antioxidant system of foods and drugs to be selected more specifically.

3. Sensitivity to Environment

3.1. Concentration and pH

Ascorbic acid is unstable in aqueous solutions, and its degradation has been considered the main cause of quality and color changes during food storage and processing. Stability analysis of ascorbic acid is a key point in its application. In addition to the interference of external factors, the concentration of ascorbic acid in the solution will also affect its stability. As reported, after storage at room temperature in the presence of light for 27 days, an aqueous solution of 1% concentration of ascorbic acid lost around 21% of its initial concentration, while the 10% ascorbic acid system only degraded about 8% [56]. In order to ensure sufficient content of ascorbic acid in the product, reinforcing the ascorbic acid content is a method commonly used in the food industry. The content of ascorbic acid in fortified milk decreased from 36.4 mg/L to 26.1 mg/L after sterilization, while the ascorbic acid content of normal milk dropped from 12.2 mg/L to 8.3 mg/L. Although the loss content of ascorbic acid increased with the addition of ascorbic acid, the loss efficiency decreased compared with the initial concentration of ascorbic acid. According to the stability and the degradation kinetics of ascorbic acid, a higher concentration of ascorbic acid has a lower degradation rate constant [57]. However, some studies have shown that excessive ascorbic acid (AH_2) is prone to auto-oxidation to produce dehydroascorbic acid anions, according to the following reaction [58]:

$$AH_2 \rightarrow AH^- + H$$
$$AH^- + O_2 \rightarrow \cdot A^- + O_2^- + H^+$$

Furthermore, the autoxidation of ascorbic acid depends on pH. Without transition metal catalysis, its spontaneous oxidation under neutral conditions is quite slow. At pH 7, 99.9% of ascorbic acid is in the form of ascorbate (AH^-), as well as a small amount of ascorbic acid (AH_2, ~0.1%) and ascorbate dianion (A^{2-}, 0.005%) [59].

$$AH_2 \xrightarrow{pK_a 4.2} AH^- \xrightarrow{pK_a 11.6} A^{2-}$$

$$A^{2-} + O_2 \rightarrow \cdot A^- + O_2^-$$

The amount of A^{2-} increases 10-fold for every unit increase in pH. Auto-oxidation occurs through A^{2-}, thus accelerating the oxidation rate of ascorbic acid and hydrolyzing to 2,3-diketogulonic acid in alkaline solution [60]. It is worthy to note that the pH value of the ascorbic acid solution changes with its degradation process. Ascorbic acid with an initial pH of alkaline would degrade to about 7.3, which was affected by degradation products generated from ascorbic acid once the reaction began [61]. In acidic solutions, the degradation products of ascorbic acid are also related to oxygen. Under aerobic conditions, dehydroascorbic acid is further degraded to form 2-furoic acid and 3-hydroxypyrone. The main degradation product in anaerobic environment is furfural, and the intermediate product does not involve dehydroascorbic acid [62]. However, in alkaline solutions, the main products of ascorbic acid are 2-methylfuran, 2,4-dimethylfuran, 2-acetyl-5-methylfuran and 2-methyl-2-cyclopentanone [61]. Compared with the hydrolytic pathway that directly cleaves the lactone ring of ascorbic acid under anaerobic conditions, the oxidoreductive pathway in which ascorbic acid forms dehydroascorbic acid in the presence of oxygen is more common in food systems [62]. Although the headspace can be filled with nitrogen, the dissolved oxygen in the solution is difficult to remove. Previous researchers found that there is a linear relationship between the first-order kinetic constant of ascorbic acid degradation in juice and the initial headspace oxygen concentration during storage at 22 °C [63].

3.2. Temperature

Ascorbic acid is severely degraded by heat, and the instability of ascorbic acid in thermal-processed foods impedes its application. The degradation of ascorbic acid involves complex oxidation and intermolecular rearrangement reactions, and is considered to be one of the main reasons for quality and color changes during food processing and storage [64]. From an analysis of the effects of temperature and pressure on the retention of ascorbic acid in processed juices, it revealed that the dominant factor determining the stability of ascorbic acid is the temperature, which directly affects the degradation rate of ascorbic acid [65]. Products containing ascorbic acid, such as fruit juice, need to undergo high-temperature pasteurization in order to guarantee safety and stability. The content of ascorbic acid in fresh orange juice ranges from 25–68 mg/100 mL. Studies have shown that the retention of ascorbic acid in the product after pasteurization (90 °C, 1 min) is about 82–92% [66,67]. It was observed that the maximum temperature of the ultra-high pressure homogenization treatment at 100, 200 and 300 MPa was 45 °C, 70 °C and 94 °C, respectively, and the continuous treatment time under the maximum temperature was 0.7 s or less. The loss of ascorbic acid in the ultra-high pressure treated juice is less than that in the traditionally heat-pasteurized juice [67]. As detected, the content of ascorbic acid in guava juice was around 42.2 ± 0.01 mg/mL. After 7 days of storage at 25 and 35 °C in the dark, ascorbic acid was degraded by 23.4% and 56.4%, respectively. Its degradation is significantly reduced at 4–10 °C [68]. The use of relatively mild temperature (75 °C) for heat treatment and a storage temperature below 25 °C is optimal for maintaining the ascorbic acid content of the product [69]. The degradation of ascorbic acid during storage and heat treatment follows first-order kinetics based on a classic dynamic model [70].

The degradation or oxidation products of ascorbic acid heated at 100 °C for 2 h include furfural, 2-furoic acid, 3-hydroxy-2-pyrone and an unknown compound. Among them, furfural is one of the main degradation products of ascorbic acid, which can polymerize or combine with amino acids to form brown melanoids, causing the browning of ascorbic acid-containing juice products [62]. Furthermore, thermally oxidized ascorbic acid was identified as a potential precursor of furan; it is a possible carcinogen usually produced in some heated food products [71]. Meanwhile, natural and synthetic antioxidants, such as chlorogenic acid, have a certain mitigation effect on the formation of furan induced by heated ascorbic acid, but the mitigation effect may decrease with the increase in heating time [72]. In addition, the thermal degradation process of ascorbic acid is also affected by pH, oxygen concentration, transition metal ions and oxidases, which is a complex system. Some researchers in food science believe that oxygen saturation decreases with increasing temperature and drops to 0 at 100 °C. However, according to the Tromans and Battino model, although 100–130 °C is the minimum oxygen solubility temperature, there is still dissolved oxygen in the system [73,74]. At temperatures above 100 °C, oxygen has a greater effect on ascorbic acid degradation than temperature. Therefore, removing all oxygen including dissolved oxygen is the best way to preserve ascorbic acid at high temperatures [75].

3.3. Light

In addition to heat-treatment sterilization, ultraviolet radiation (240 nm–300 nm) is a promising alternative and is gradually being used more for fruit juice sterilization [76]. The ultraviolet sterilization method includes the use of high-intensity pulsed ultraviolet radiation with wavelengths between 200 and 400 nm and a monochromatic ultraviolet system of which approximately 90% of the energy comes from a single wavelength [77].

However, ascorbic acid absorbs ultraviolet radiation in the wavelength range of 229–330 nm and undergoes degradation [78]. The formation of UV-induced free radicals may accelerate the loss of ascorbic acid. Ascorbic acid continues to degrade after UV treatment; higher initial UV dose values and storage temperature accelerate the degradation of ascorbic acid in the later stage [77]. In addition, the pH of the solution also affects the photo-degradation of ascorbic acid. Under alkaline conditions, AH^- produced by ionization of AH^2 is more prone to photo-degradation than AH^2 [79]. It is worth noting that the ingredients in products may absorb or scatter UV radiation, thereby affecting the degradation of ascorbic acid. Niacinamide, as a component of vitamin B-complex with vitamin C, acts as a photo-degradation accelerator to reduce the stability of ascorbic acid under UV-irradiation [79].

4. Strategies for Improving the Encapsulation and Delivery of Ascorbic Acid

4.1. Low-Molecular-Weight Stabilizer and Derivatives

Ascorbic acid can be protected by adding other antioxidants. Food is a system in which multiple ingredients coexist, and there may be preservation effects of certain antioxidants, which involve regeneration mechanisms [80]. Ascorbic acid and flavonoids can regenerate α-tocopherol by reacting with α-tocopheroxyl radical. The bond dissociation energy of coexisting antioxidants that play a regenerative effect is lower than or close to the O-H bond [81]. Similarly, ascorbic acid can also be regenerated by certain antioxidants. It is well known that the conversion between ascorbic acid and its degradation product dehydroascorbic acid is reversible. Tert-butyl hydroquinone (TBHQ), which is often used as an antioxidant in high-fat foods, has been found to accelerate the conversion of dehydroascorbic acid to ascorbic acid, thereby stabilizing ascorbic acid. This reaction follows the first-order kinetic model, and the regeneration efficiency is proportional to the reaction time [82]. Moreover, glutathione with the free sulfhydryl group acts as a nucleophile and reducing agent. In the ascorbic acid solution, glutathione reduces dehydroascorbic acid and inhibits the degradation of ascorbic acid. The degradation kinetic model of ascorbic acid gradually changes from first-order to zero-order with the increase in glutathione concentration [56]. Meanwhile, as an effective antioxidant, ferulic acid has a synergistic effect with ascorbic acid. The oxidation–reduction potential of ferulic acid (0.595) is significantly higher than that of ascorbic acid (0.282), thus the former protect effect on ascorbic acid is indirect. There is a hypothesis that ferulic acid preferentially reacts with pro-oxidant intermediates or acts as a sacrificial substrate [12]. As mentioned above, low-molecular-weight stabilizers can inhibit the degradation of ascorbic acid to a certain extent, but it is hard to mask the acidic taste of ascorbic acid.

Considering the long-term mechanism of antioxidation and the high stability requirements of commercial products, ascorbic acid derivatives are also widely used, in addition to adding antioxidants or preservatives to stabilize ascorbic acid. For example, 2-O-D-glucopyranosyl-L-ascorbic acid, the glycosylated ascorbic acid in which the hydroxyl group on the C_2 position is substituted by glucose residue, has excellent thermal stability and antioxidant properties [83]. Its application in anthocyanin-containing beverages can avoid the degradation of anthocyanins and maintain a high level of vitamin C content [84]. Ascorbate derivatives are also formed by introducing a phosphate group or combining sodium and magnesium salts at the C_2 position of ascorbic acid, showing better stability than ascorbic acid [85]. In addition to hydrophilic ascorbic acid derivatives, there are lipophilic-derivatives such as ascorbic acid 6-palmitate and tetra-isopalmitoyl ascorbic acid. However, these derivatives need to undergo some reactions in vivo to be converted into ascorbic acid and exert their physiological activities, and the high-cost is a limitation of their application into large-scale commercial products.

4.2. Construction of Carriers Based on Bio-Macromolecules

4.2.1. Chemical Interaction

Several technologies have been widely used to construct biomacromolecule-based carriers for ascorbic acid (Table 1), in order to shield the unfavorable environmental factors and improve the taste of the product. These processes involve physical and chemical interactions between carriers and ascorbic acid; chemical interactions mainly refer to covalent and non-covalent bonds.

Proteins are generally recognized as safe (GRAS) and have high nutritional value. The delivery systems based on proteins have received widespread attention in food field due to their biocompatibility, biodegradability and tunability. Ascorbic acid binds to β-lactoglobulin (β-LG) through ion contact to form a more stable conjugate than human serum albumin (HSA) and bovine serum albumin (BSA). β-LG, HSA and BSA can, respectively, bind about 50–60%, 40–55% and 35–50% of ascorbic acid, and the proteins can be used to deliver vitamin C in vitro [86]. Through a cationization reaction, the quaternary ammonium salt cationic group was attached to the soybean protein isolate (SPI) chain, which increases the solubility of the protein and favors the encapsulation of ascorbic acid [87]. However, the low loading capacity and carrier instability in the stomach and intestines are the main challenges that restrict proteins from being ideal delivery vehicles for ascorbic acid. Since the excellent hydrophilicity of ascorbic acid and its same charge as most proteins (isoelectric point, pI < 7) at physiological pH, their interactions such as hydrophobic interaction, electrostatic interaction, hydrogen bonds and van der Waals forces are usually weak or absent. This results in a low encapsulation efficiency and rapid release of ascorbic acid from protein nanoparticles in the aqueous solutions.

Chitosan is a cationic polysaccharide with excellent chelating and cross-linking properties and is widely used as a delivery vehicle in the food field. The formation of chitosan nanoparticles requires cross-linking with polyanions, such as tripolyphosphate (TPP). The amino groups of chitosan in the polymer backbone can interact with ascorbic acid to form a strong hydrogen bond, which captures and retains ascorbic acid on the polysaccharide [10,88]. The formed chitosan-ascorbic acid complexes have high singlet oxygen scavenging ability and then maintain the high antioxidant capacity of ascorbic acid. The nanoscale size and positive charge of the particles are very important for their adsorption on the mucosa, which is conducive to achieving a high uptake rate of the loaded ascorbic acid by the intestinal cells. Chitosan-ascorbic acid complex nanoparticles increase the residence time of ascorbic acid in the digestive tract of trout. Compared with protein nanoparticles, chitosan nanoparticles strengthen the interaction with ascorbic acid through electrostatic interaction, but the encapsulation efficiency is still relatively low [11]. This is related to the molecular weight and concentration of chitosan, the addition of ascorbic acid and the measurement method of the encapsulation. There are two views about the influence of chitosan molecular weight on the encapsulation efficiency of ascorbic acid. One is that high-molecular-weight chitosan has more surface charges to bind with more ascorbic acid molecules, thus the long backbone can capture more ascorbic acid. As the molecular weight of chitosan increased from 65 kDa to 110 kDa, the content of ascorbic acid loaded increased from 30% to 70%, respectively. With the further increase in chitosan molecular weight, the particle size increases but the overall surface area decreases, resulting in a decrease in the encapsulation efficiency of ascorbic acid [16]. Short fragments of low-molecular-weight chitosan are easier to protonate free amino groups, thereby complexing with ascorbic acid through electrostatic interactions. The average diameter of 55-kDa chitosan complex particles is 70.6 nm, and the loading efficiency of ascorbic acidic is about 66% [89].

4.2.2. Physical Barrier

In order to maintain the stability of ascorbic acid in food applications, ascorbic acid can be loaded into biomacromolecule-based delivery vehicles through physical encapsulation and adsorption. Compared with ascorbic acid nanoparticles chelated with protein and chitosan, the construction of physical barriers such as microcapsules based on protein and polysaccharide, solid lipids and liquid state multiple emulsions have a better loading capacity of ascorbic acid in the core and hence, this improves stability.

The process of encapsulating ascorbic acid in a core walled by polymers coating to isolate it from the external adverse factors is microencapsulation. The current preparation methods of microcapsules mainly include spray chilling, spray drying and complex coacervation. Among them, spray drying is one of the most common techniques due to its low cost, continuity and easy industrial scale production [90]. The selection of wall materials includes various proteins and polysaccharides, such as gum Arabic, maltodextrin, pectin, xyloglucan, sodium alginate. Gum Arabic and sodium alginate are low-cost and GRAS category polysaccharides, which are often used as food additives. The sodium alginate/gum Arabic microcapsules prepared by spray drying have an excellent loading capacity of ascorbic acid, which can reach more than 90%. Meanwhile, the thermal stability temperature of ascorbic acid is increased to 188 °C, which is higher than the temperature required for product preparation [91]. The xyloglucan extracted from *Hymenaea courbaril var. courbaril* seeds is a water-soluble polysaccharide containing gum Arabic, which is used as a thickener, stabilizer and crystallization inhibitor in the food industry. The spray-dried microcapsules can encapsulate around 96% of ascorbic acid. The system shows strong antioxidant activity and inhibits the formation of furan, an ascorbic acid degradation product, during the preheated process of products. After 60 days of storage at room temperature, the retention of ascorbic acid in the system is still around 90% [92]. However, the high viscosity of high-concentration polymers limits the granulation by spray-drying. To a certain extent, the loading capacity is related to the wall-to-core ratio and increases with the increase in the coating of wall materials [15]. Complex coacervation is the phase separation of at least two hydrocolloids from the initial solution, and then the coacervate phase is deposited around the suspended or emulsified bioactive compounds. One of the hydrocolloids is in the colloidal state. On the contrary to hydrophobic bioactive compounds, hydrophilic ascorbic acid needs to be emulsified before it is prepared [93]. Compared with spray drying, this method does not involve a heat treatment process and is more suitable for encapsulating thermally unstable ascorbic acid [94]. The microcapsules prepared with gelatin and pectin as wall materials improve the thermal stability of ascorbic acid, although the solubility of the microcapsules is relatively low [15]. The encapsulation efficiency of ascorbic acid using gelatin and acacia as wall materials is about 97% [93].

The systems based on lipids, such as solid lipid microcapsules and emulsions, can be obtained by high-pressure homogenization, microfluidics, and solvent evaporation. The solid lipid microcapsules prepared by polyglyceryl monostearate (PGMS) have the encapsulation capacity of ascorbic acid up to about 94%. The system can be added in to fortify milk, significantly inhibiting the Maillard reaction between milk proteins and ascorbic acid. Sensory analysis showed that there was no significant difference in most aspects between the control sample and the fortified sample encapsulated with ascorbic acid after 5 days of storage [95]. As reported, palm fat was used as wall material to fabricate the solid lipid microcapsules to encapsulate and protect ascorbic acid using a microfluidic technique. The internal phase was added with salt or chitosan to further improve the encapsulation efficiency of ascorbic acid. The two different mechanisms involve pore blockage and ascorbic acid chelation [6]. This system has better physical isolation performance than protein and/or polysaccharide solid microcapsules. However, the operation process includes thermal melting and ice bath cooling of liposomes. This method is limited to the laboratory scale and is difficult to industrialize. On the other hand, the storage stability of ascorbic acid in oil-containing systems may be affected by lipid oxidation and thermodynamic instability of emulsions, which is lower than that of carrier-stable protein and polysaccharide microcapsule systems [96,97].

The microcapsule system based on the physical barrier has a better loading capacity of ascorbic acid than complex nanoparticles (Table 1). In addition to the properties of the delivery carriers, it may also be related to the different measurement method of encapsulation efficiency. For delivery systems in micro-scale, the measurement conditions for encapsulation efficiency of ascorbic acid are gentler than those of protein and/or polysaccharide nanoparticles. The determination method includes separation by standing, ultrasonic and filter paper filtration [82–86]. Compared with the ultra-isolation method [79,80] used in the nanoparticle system, these methods reduce the release and diffusion of ascorbic acid during the measurement process.

Table 1. Different types of bio-macromolecule delivery vehicles of ascorbic acid.

Carrier	Material	Technology	Protective Effect	Encapsulation Efficiency	Reference
Microcapsules	Sodium alginate/gum Arabic	Spray drying	Thermal stability temperature of ascorbic acid is increased to 188 °C.	>90%	[91]
	Xyloglucan	Spray drying	After 60 days of storage at room temperature, the retention of ascorbic acid is around 90%.	~96%	[92]
	Gum Arabic/rice starch	Spray drying	The retention of ascorbic acid is around 81.3% after 90 days of storage at 21 °C.	~99.7%	[98]
	Gelatin/pectin	Complex coacervation	With low hygroscopicity and high thermal stability.	23.7% to 94.3%	[15]
	Gelatin/acacia	Complex coacervation	The retention of ascorbic acid is around 44% and 80% after 30 days of storage at 37 °C and 20 °C, respectively.	≥97%	[93]
Liposome	Palm fat/chitosan	Microfluidic technique	After 30 days, retained 98.58% and 97.62% of ascorbic acid at 4 °C and 20 °C, respectively.	~96.6%	[6]
	Polyglyceryl monostearate	Spray chilling	The system can inhibit the Maillard reaction between milk proteins and ascorbic acid.	~94.2%	[98]
	Milk fat globule membrane-derived phospholipids	Microfluidic technique	After 7 weeks at 4 °C and 25 °C, ascorbic acid in liposomes retained 67% and 30%, respectively.	~26%	[97]
W/O/W emulsions	Gelatin/tetraglycerin monolaurate condensed ricinoleic acid ester/decaglycerol monolaurate	Homogenization	The half-life for W/O/W emulsions containing 30% ascorbic acid at 4 °C was about 24 days.	≥90%	[99]
	Soybean oil/tetraglycerin condensed ricinoleic acid ester/gelatin	Homogenization and microchannel emulsification	The ascorbic acid exhibited 80% retention after 10 days storage at 4 °C.	>85%	[100]

4.2.3. Controlled Release of Ascorbic Acid

The challenge of ascorbic acid in food applications is not only to maintain its stability, but also to improve the effectiveness of delivery it to the active site. The release of bioactive compounds in the body is expected to occur in the intestine rather than the stomach, because the absorption mainly occurs in the small intestine. It was found that ascorbic acid in pomegranate juice was approximately 29% degraded during gastric digestion, which severely reduced the bioavailability of ascorbic acid. Additionally, the compounds that are transported by specialized processes are usually only absorbed in certain parts of the gastrointestinal tract. The absorption of riboflavin begins in the upper region of the small intestine, as does ascorbic acid [101]. Therefore, the changes in gastric and intestinal transit rates may affect the absorption efficiency of orally administered bioactive compounds. Additionally, the bioavailability of oral ascorbic acid is related to the following key steps: (1) Release of ascorbic acid in the gastrointestinal tract, and its solubility in gastrointestinal fluids. (2) Intestinal epithelial cells absorb ascorbic acid and undergo biochemical transformation. Studies have shown that a single high dose of ascorbic acid causes a temporary increase in plasma that is rapidly absorbed by the gastrointestinal tract and then quickly excreted in the urine [102]. A form of ascorbic acid that can be slowly released in the intestine is desired to maintain a constant level of ascorbic acid in plasma.

The release process of encapsulated ascorbic acid is as follows: absorption of solvent by the carrier, dissolution of the wall-coating, and the diffusion of inner core. The release of bioactive compounds in carriers depends on many factors, such as the selection of the wall material, the ratio of wall/core, the size of the carrier, the solubility of the bioactive compound, and the release conditions [15,97]. Ascorbic acid releases kinetics from ascorbate gummies, which were investigated using an in vitro simulated digestion model. The results show that the disintegration time of ascorbic acid candy was about 22 min, after which the functional ingredient ascorbic acid was gradually released, reaching 93.6% within 2 h. Notably, the components in gastric juice may have an effect on the release of ascorbic acid, with gastric juice containing 5% starch slowing the release of bioactive ascorbic acid in the gummies, but other dietary components had no significant effect on its release [103]. This may be related to the encapsulation of ascorbic acid in starch in the stomach. Compared with the afore-mentioned delivery vehicles based on polysaccharide and lipid, the protein carrier has poor stability in the stomach. The low pH of the gastric environment and the presence of pepsin cause the denaturation and degradation of the protein carrier, leading to the leakage of loaded bioactive compounds in the stomach before reaching the small intestine [104]. Gelatin/pectin microcapsules show a high loading capacity of ascorbic acid. However, due to the dissolution of gelatin coating in the gastric environment, the release of ascorbic acid in the gastric environment is faster than in the intestine [15]. Therefore, it is necessary to design a carrier that is relatively stable in the stomach and which can provide a sustained release of ascorbic acid in the intestine.

The small size and positive charge of the particles contribute to the high uptake rate by intestinal cells. The loading in chitosan nanoparticles effectively prolong the residence time of ascorbic acid in the intestine of rainbow trout [16]. Nanoparticles based on chitosan with low molecular weight have a higher delivery rate of ascorbic acid. The mechanism of ascorbic acid released from nanoparticles in the gastric environment and the intestinal environment are diffusion and erosion, respectively. Under the neutral conditions of the intestine, the ion exchange between chitosan and the release medium leads to the erosion of nanoparticles. The release rate of ascorbic acid increased from 30% in the stomach to more than 75% in the intestine [16]. As reported, the water-soluble derivative N,N,N-trimethylchitosan (TMC) as a carrier can efficiently transport hydrophilic molecules through mucosal epithelial tissues such as the oral cavity, nasal cavity, lungs and intestines [88]. Thus, the carriers based on chitosan coatings may be an effective strategy to achieve intestinal release of ascorbic acid. Based on the continuous deposition of positively charged chitosan and negatively charged sodium alginate on the surface of anionic nano-liposomes, a liposomal polyelectrolyte delivery system of ascorbic acid was prepared. The clinical results showed that the bioavailability of orally administered liposomal ascorbic acid was 1.77 times higher than that of non-liposomal ascorbic acid, with higher bioavailability [105]. The ability of the outer layer of chitosan to withstand the gastric environment is beneficial for maintaining the stability of the carrier structure. The excellent sealing properties of liposomes and better penetration with enterocyte phospholipid bilayers also contributed to the improved bioavailability of released ascorbic acid.

5. Commercial Application of Ascorbic Acid

Based on the afore-mentioned biological activity of ascorbic acid, ascorbic acid is mainly used as an antioxidant to inhibit food browning and as a dietary supplement for humans. Ascorbic acid is mainly used as an antioxidant to protect the senses of foods. As is well known, polyphenol oxidase catalyzes the enzymatic browning of phenol substrates to yield dark-colored melanin. Browning affects product sensory qualities and reduces consumer acceptance. Adding xyloglucan microcapsules containing ascorbic acid to baked foods such as tilapia fish burgers can significantly inhibit the browning that occurs during the preparation process and maintain the sensory qualities of the product [93]. The chitosan/tripolyphosphate nano-aggregates containing ascorbic acid enhanced the inhibition of mushroom slices browning induced by tyrosinase [106]. Acute heat stress during transport is known to predispose rainbow trout quality to deterioration, with negative effects on the histological, physicochemical and microbiological quality of fillets. Treatment with added ascorbic acid partially mitigated damage caused by acute heat stress. It can maintain tissue structure, delay protein oxidation and then prolong the shelf life of fish fillets to about 2 days [107]. In addition, as a nutritional supplement, ascorbic acid plays an important role as a co-factor in many biological processes. Unfortunately, fishes lack L-gluconolactone oxidase and cannot biosynthesize ascorbic acid by themselves, which is not conducive to the growth of their bone matrix and connective tissue. Lack of ascorbic acid can cause reduced wound-healing capacity and bone deformities in fish [108]. At present, in the aquaculture area, ascorbic acid is widely added to fish diets. Based on the healthcare function of ascorbic acid, it is also vital in nutrition fortification products. As an important source of protein supplementation, dairy products are popular beverages all over the world. At present, milk and soymilk have been fortified with ascorbic acid, including ascorbate and ascorbic acid isomers, to improve the iron absorption in the small intestine [109,110].

Food fortification can improve micronutrient malnutrition. It is worth noting that a category of foods tailored according to the necessary nutrients for a healthy life and their specific concentrations and ratios are called designer foods, also known as health foods, and are sought after and recognized by consumers. Such products often contain a variety of bioactive compounds. By adding calcium and antioxidants such as vitamins E and C to low-fat chicken patties, a high-quality product with high-quality animal protein, fat, multivitamins and minerals can be prepared. Ascorbic acid not only acted as a nutritional additive, but also maintained better color and flavor of chicken patties, and inhibited the formation of nitrosamines in the meat [111]. The addition of sodium ascorbate and vitamin A to pig feed can significantly improve the growth performance, antioxidant capacity and immune function of weaned piglets. Meanwhile, as an antioxidant, sodium ascorbate can delay the degradation of vitamin A [112]. A cornstarch-based baking premix was developed by addition of vitamin B, vitamin C and digestible iron, zinc, selenium and iodine. Although the added ascorbic acid in the baked bread degraded due to high temperature, it strengthened the structure of the bread and was benefit to product quality [113]. Meanwhile, it was found that the combination of butylated hydroxytoluene and ascorbic acid significantly inhibited the oxidation and isomerization of vitamin A in skim milk powder during thermally accelerated storage [104].

There are two major aspects in the current development of ascorbic acid-fortified products. On the one hand, the natural ascorbic acid is directly added, in order to use its antioxidant activity to maintain the sensory appearance of the product during the shelf life. The cost is low, but the retention activity of the final product is low. Another aspect is the addition of ascorbic acid derivatives, which is to ensure that sufficient physiological activity can be expressed after ingestion of the product. However, the cost of ascorbic acid derivatives is high, and they need to be converted before they can exert their functional properties. The related products of ascorbic acid and its main derivatives in the food fields in recent years are summarized in Table 2. Although various delivery technologies are available, they are still in the developmental stage of industrial transformation and have not been widely used. Combined with the above analysis of delivery strategies, these may be limited by the cost of wall materials, and the complexity of the process, which is not suitable for large-scale industrial production. Therefore, researchers still need to explore low-cost, simple, and high-yield encapsulation techniques of ascorbic acid for industrial application.

Table 2. Commercial products fortified with ascorbic acid and its derivatives.

Ascorbic Acid or Ascorbate	Product	Property of Added Bioactives	Challenges of Application	References
L-ascorbic acid	Liqueur chocolate, milk fortification, edible coating, juice, meat patties	With antioxidant properties and a series of physiological activities such as iron metabolism, it can eliminate bacterial biofilms and the cost is low.	Poor stability, sour taste.	[31,33,82,100,114–116]
L-ascorbic acid sodium	Fish feed, formulae and weaning foods, cured hams	With antioxidant properties and the cost is low.	Poor stability, and compared with ascorbic acid, sodium ascorbate has a potential anti-nutritional effect on protein after high-temperature baking.	[117,118]
2-O-D-glucopyranosyl-L-ascorbic acid	Berry beverage, black rice baking products, cured meat products, aquatic products	With anti-oxidation and stability, it avoids the degradation of anthocyanins caused by ascorbic acid and releases ascorbic acid under the catalysis of enzymes in vivo.	High cost and lowyield in industrial production.	[84,119,120]
L-ascorbic acid palmitic acid ester	Formula milk, heme iron-fortified bakery product, frying oil, nutritional powders	It is a lipophilicity L-ascorbic acid esters derivatives with antioxidant properties and can be converted into ascorbic acid by esterase.	The thermal stability is poor, and chemically modified products often contain mixtures.	[121–124]

6. Conclusions

In this review, the bioactivity and stability of ascorbic acid are introduced. There are many strategies for improving the bioavailability of ascorbic acid, and the influence of delivery systems on the stability and release properties of ascorbic acid is discussed. Besides the addition of low-molecular-weight antioxidants and preservatives, encapsulation technology is more and more widely used in the food field. The stabilization mechanism includes chemical chelation and physical barrier. Since the positively charged chitosan can interact with ascorbic acid through electrostatic interaction and hydrogen bond, it is the most superior carrier material. The complex system is mostly in the form of nano-sized particles. On the other hand, biomacromolecules can construct microcapsules with coatings through a series of technologies, such as spray drying, microfluidic technique and complex coacervation. The physical barrier restricts ascorbic acid within the inner core, reducing the contact between it and the external environment. The two mechanisms have their own limitations. For example, chemically complexed nanoparticles are beneficial to the absorption by mucosal membranes, but the encapsulation efficiency of ascorbic acid is low and ascorbic acid is accessible to solvent. The coating of the microcapsules can effectively shield the inner ascorbic acid from external environment, but the operation is more complicated and requires the assistance of a variety of equipment. In addition, the larger size and poor water-solubility of the microcapsules limit the absorption in the body to a certain extent. Therefore, understanding the pro-degradation factors of ascorbic acid and its properties are conducive to the targeted design of delivery systems. Adopting low-cost methods to design an effective fortification strategy to improve the stability of ascorbic acid during processing and storage is still the focus and challenge for researchers.

Author Contributions: X.Y. wrote the first manuscript draft. K.C. contributed to the first draft. H.C. and X.C. contributed to the revision. S.F. and Y.S. reviewed the manuscript. L.L. critically reviewed the manuscript for important intellectual contents. All authors have read and agreed to the published version of the manuscript.

Funding: This research received no external funding.

Conflicts of Interest: The authors declare that Luwei Pharmaceutical Group Co. had no role in the design of the study; in the collection, analyses, or interpretation of data; in the writing of the manuscript, or in the decision to publish the results and there is no conflict of interest.

Abbreviations

RDA	recommended dietary allowances
DHA	dehydroascorbic acid
ROS	reactive oxygen species
$O_2 \cdot^-$	superoxide radicals
$^1O_2$	singlet oxygen
H_2O_2	hydrogen peroxide
OH·	reactive hydroxyl radicals
AH_2	ascorbic acid
AH^-	ascorbate
A^{2-}	ascorbate dianion
TBHQ	tert-butyl hydroquinone
GRAS	generally recognized as safe
β-LG	β-lactoglobulin
HAS	human serum albumin
BSA	bovine serum albumin
SPI	soy protein isolate
TPP	tripolyphosphate
PGMS	polyglyceryl monostearate
TMC	N,N,N-trimethylchitosan

References

1. Hasan, L.; Vgeli, P.; Neuenschwander, S.; Stoll, P.; Stranzinger, G. The L-gulono-gamma-lactone oxidase gene (GULO) which is a candidate for vitamin C deficiency in pigs maps to chromosome 14. *Anim. Genet.* **2015**, *30*, 309–312. [CrossRef] [PubMed]
2. Günter, P.; Hanspeter, H. *Industrial Production of L-ascorbic Acid (Vitamin C) and D-isoascorbic Acid*; Springer: Berlin/Heidelberg, Germany, 2014; Volume 143, pp. 143–188.
3. Bei, R. Effects of vitamin C on health: A review of evidence. *Front. Biosci.* **2013**, *18*, 1017–1029. [CrossRef] [PubMed]
4. Monsen, E.R. Dietary reference intakes fop the antioxidant nutrients: Vitamin C, vitamin E, selenium, and carotenoids. *J. Am. Diet. Assoc.* **2000**, *100*, 637–640. [CrossRef]
5. Padayatty, S.J.; Sun, H.; Wang, Y.H.; Riordan, H.D.; Hewitt, S.M.; Katz, A.; Wesley, R.A.; Levine, M. Vitamin C pharmacokinetics: Implications for oral and intravenous use. *Ann. Intern. Med.* **2004**, *140*, 533–537. [CrossRef] [PubMed]
6. Comunian, T.A.; Abbaspourrad, A.; Favaro-Trindade, C.S.; Weitz, D.A. Fabrication of solid lipid microcapsules containing ascorbic acid using a microfluidic technique. *Food Chem.* **2014**, *152*, 271–275. [CrossRef] [PubMed]
7. Chang, D.W.; Abbas, S.; Hayat, K.; Xia, S.Q.; Zhang, X.M.; Xie, M.Y.; Kim, J.M. Encapsulation of ascorbic acid in amorphous maltodextrin employing extrusion as affected by matrix/core ratio and water content. *Int. J. Food Sci. Technol.* **2010**, *45*, 1895–1901. [CrossRef]

8. Carvalho, J.D.D.; Oriani, V.B.; de Oliveira, G.M.; Hubinger, M.D. Characterization of ascorbic acid microencapsulated by the spray chilling technique using palm oil and fully hydrogenated palm oil. *Lwt-Food Sci. Technol.* **2019**, *101*, 306–314. [CrossRef]
9. Alvim, I.D.; Stein, M.A.; Koury, I.P.; Dantas, F.B.H.; Cruz, C. Comparison between the spray drying and spray chilling microparticles contain ascorbic acid in a baked product application. *Lwt-Food Sci. Technol.* **2016**, *65*, 689–694. [CrossRef]
10. De Britto, D.; de Moura, M.R.; Aouada, F.A.; Pinola, F.G.; Lundstedt, L.M.; Assis, O.B.G.; Mattoso, L.H.C. Entrapment characteristics of hydrosoluble vitamins loaded into chitosan and N,N,N-trimethyl chitosan nanoparticles. *Macromol. Res.* **2014**, *22*, 1261–1267. [CrossRef]
11. Jimenez-Fernandez, E.; Ruyra, A.; Roher, N.; Zuasti, E.; Infante, C.; Fernandez-Diaz, C. Nanoparticles as a novel delivery system for vitamin C administration in aquaculture. *Aquaculture* **2014**, *432*, 426–433. [CrossRef]
12. Lin, F.H.; Lin, J.Y.; Gupta, R.D.; Tournas, J.A.; Burch, J.A.; Selim, M.A.; Monteiro-Riviere, N.A.; Grichnik, J.M.; Zielinski, J.; Pinnell, S.R. Ferulic acid stabilizes a solution of vitamins C and E and doubles its photoprotection of skin. *J. Investig. Dermatol.* **2005**, *125*, 826–832. [CrossRef]
13. Chen, X.; Yin, O.Q.P.; Zuo, Z.; Chow, M.S.S. Pharmacokinetics and modeling of quercetin and metabolites. *Pharm. Res.* **2005**, *22*, 892–901. [CrossRef]
14. Avnesh, K.; Sudesh, K.Y.; Subhash, C. Biodegradable polymeric nanoparticles based drug delivery systems. *Colloids Surf. B Biointerfaces* **2010**, *75*, 1–18.
15. Da Cruz, M.C.R.; Perussello, C.A.; Masson, M.L. Microencapsulated ascorbic acid: Development, characterization, and release profile in simulated gastrointestinal fluids. *J. Food Process Eng.* **2018**, *41*, e12922. [CrossRef]
16. Alishahi, A.; Mirvaghefi, A.; Tehrani, M.R.; Farahmand, H.; Shojaosadati, S.A.; Dorkoosh, F.A.; Elsabee, M.Z. Shelf life and delivery enhancement of vitamin C using chitosan nanoparticles. *Food Chem.* **2011**, *126*, 935–940. [CrossRef]
17. Su, L.J.; Zhang, J.H.; Gomez, H.; Murugan, R.; Hong, X.; Xu, D.; Jiang, F.; Peng, Z.Y. Reactive Oxygen Species-Induced Lipid Peroxidation in Apoptosis, Autophagy, and Ferroptosis. *Oxid. Med. Cell. Longev.* **2019**, *2019*, 5080843. [CrossRef]
18. Dalle-Donne, I.; Rossi, R.; Giustarini, D.; Milzani, A.; Colombo, R. Protein carbonyl groups as biomarkers of oxidative stress. *Clin. Chim. Acta* **2003**, *329*, 23–38. [CrossRef]
19. Khaw, K.T.; Bingham, S.; Welch, A.; Luben, R.; Wareham, N.; Oakes, S.; Day, N. Relation between plasma ascorbic acid and mortality in men and women in EPIC-Norfolk prospective study: A prospective population study. *Lancet* **2001**, *357*, 657–663. [CrossRef]
20. Taniguchi, M.; Arai, N.; Kohno, K.; Ushio, S.; Fukuda, S. Anti-oxidative and anti-aging activities of 2-O-α-glucopyranosyl-L-ascorbic acid on human dermal fibroblasts. *Eur. J. Pharmacol.* **2012**, *674*, 126–131. [CrossRef]
21. Lutsenko, E.A.; Carcamo, J.M.; Golde, D.W. Vitamin C prevents DNA mutation induced by oxidative stress. *J. Biol. Chem.* **2002**, *277*, 16895–16899. [CrossRef] [PubMed]
22. Davey, M.W.; Van Montagu, M.; Inze, D.; Sanmartin, M.; Kanellis, A.; Smirnoff, N.; Benzie, I.J.J.; Strain, J.J.; Favell, D.; Fletcher, J. Plant L-ascorbic acid: Chemistry, function, metabolism, bioavailability and effects of processing. *J. Sci. Food Agric.* **2000**, *80*, 825–860. [CrossRef]
23. Zhao, R.N.; Yuan, Y.H.; Liu, F.Y.; Han, J.G.; Sheng, L.S. A computational investigation on the geometries, stabilities, antioxidant activity, and the substituent effects of the L-ascorbic acid and their derivatives. *Int. J. Quantum Chem.* **2013**, *113*, 2220–2227. [CrossRef]
24. Hony, B.M.; Butler, J. The repair of oxidized amino acids by antioxidants. *Biochim. Biophys. Acta* **1984**, *791*, 212–218. [CrossRef]
25. Sukalovic, V.H.T.; Veljovic-Jovanovic, S.; Maksimovic, J.D.; Maksimovic, V.; Pajic, Z. Characterisation of phenol oxidase and peroxidase from maize silk. *Plant Biol.* **2010**, *12*, 406–413. [CrossRef]
26. Altunkaya, A.; Gokmen, V. Effect of various inhibitors on enzymatic browning, antioxidant activity and total phenol content of fresh lettuce (*Lactuca sativa*). *Food Chem.* **2008**, *107*, 1173–1179. [CrossRef]
27. Landi, M.; Degl'Innocenti, E.; Guglielminetti, L.; Guidi, L. Role of ascorbic acid in the inhibition of polyphenol oxidase and the prevention of browning in different browning-sensitive *Lactuca sativa var. capitata* (L.) and *Eruca sativa* (Mill.) stored as fresh-cut produce. *J. Sci. Food Agric.* **2013**, *93*, 1814–1819. [CrossRef]
28. Caiyun, L.; Jie, L.; Shoulei, Y.; Qingzhang, W. Research progress on application of ascorbic acid in food. *Food Sci. Technol.* **2021**, *46*, 228–232.
29. Kim, T.K.; Hwang, K.E.; Lee, M.A.; Paik, H.D.; Kim, Y.B.; Choi, Y.S. Quality characteristics of pork loin cured with green nitrite source and some organic acids. *Meat Sci.* **2019**, *152*, 141–145. [CrossRef] [PubMed]
30. Perlo, F.; Fabre, R.; Bonato, P.; Jenko, C.; Tisocco, O.; Teira, G. Refrigerated storage of pork meat sprayed with rosemary extract and ascorbic acid. *Cienc. Rural* **2018**, *48*. [CrossRef]
31. Fredriksen, J.; Løken, E.B.; Borgejordet, Å.; Gjerdevik, K.; Nordbotten, A. Unexpected sources of vitamin C. *Food Chem.* **2009**, *113*, 832–834. [CrossRef]
32. Aliste, A.J.; Del Mastro, N.L. Ascorbic acid as radiation protector on polysaccharides used in food industry. *Colloids Surf. Physicochem. Eng. Asp.* **2004**, *249*, 131–133. [CrossRef]
33. Przekwas, J.; Wiktorczyk, N.; Budzynska, A.; Walecka-Zacharska, E.; Gospodarek-Komkowska, E. Ascorbic Acid Changes Growth of Food-Borne Pathogens in the Early Stage of Biofilm Formation. *Microorganisms* **2020**, *8*, 10. [CrossRef] [PubMed]

34. Batalova, V.N.; Slizhov, Y.G.; Chumakov, A.A. Parameters for Quantitative Evaluation of the Radical-Generating (Pro-Oxidant) Capacity of Metal Ions and the Radical-Scavenging Activity of Antioxidants Using Voltammetric Method. *J. Sib. Fed. Univ. Chem.* **2016**, *9*, 60–67. [CrossRef]
35. Rietjens, I.; Boersma, M.G.; de Haan, L.; Spenkelink, B.; Awad, H.M.; Cnubben, N.H.P.; van Zanden, J.J.; van der Woude, H.; Alink, G.M.; Koeman, J.H. The pro-oxidant chemistry of the natural antioxidants vitamin C, vitamin E, carotenoids and flavonoids. *Environ. Toxicol. Pharmacol.* **2002**, *11*, 321–333. [CrossRef]
36. Chen, G.; Chang, T.M.S. Dual effects include antioxidant and pro-oxidation of ascorbic acid on the redox properties of bovine hemoglobin. *Artif. Cells Nanomed. Biotechnol.* **2018**, *46*, 983–992. [CrossRef] [PubMed]
37. Chakraborthy, A.; Ramani, P.; Sherlin, H.J.; Premkumar, P.; Natesan, A. Antioxidant and pro-oxidant activity of Vitamin C in oral environment. *Indian J. Dent. Res. Off. Publ. Indian Soc. Dent. Res.* **2014**, *25*, 499–504. [CrossRef]
38. Podmore, I.D. Vitamin C exhibits pro-oxidant properties. *Nature* **1998**, *392*, 559. [CrossRef]
39. Ullah, M.F.; Khan, H.Y.; Zubair, H.; Shamim, U.; Hadi, S.M. The antioxidant ascorbic acid mobilizes nuclear copper leading to a prooxidant breakage of cellular DNA: Implications for chemotherapeutic action against cancer. *Cancer Chemother. Pharmacol.* **2011**, *67*, 103–110. [CrossRef] [PubMed]
40. Yang, S.; Verhoeff, A.A.; Merkx, D.W.H.; van Duynhoven, J.P.M.; Hohlbein, J. Quantitative Spatiotemporal Mapping of Lipid and Protein Oxidation in Mayonnaise. *Antioxidants* **2020**, *9*, 13. [CrossRef]
41. Scheffler, J.; Bork, K.; Bezold, V.; Rosenstock, P.; Gnanapragassam, V.S.; Horstkorte, R. Ascorbic acid leads to glycation and interferes with neurite outgrowth. *Exp. Gerontol.* **2019**, *117*, 25–30. [CrossRef]
42. Putchala, M.C.; Ramani, P.; Sherlin, H.J.; Premkumar, P.; Natesan, A. Ascorbic acid and its pro-oxidant activity as a therapy for tumours of oral cavity—A systematic review. *Arch. Oral Biol.* **2013**, *58*, 563–574. [CrossRef] [PubMed]
43. Shi, M.Y.; Xu, B.H.; Azakami, K.; Morikawa, T.; Watanabe, K.; Morimoto, K.; Komatsu, M.; Aoyama, K.; Takeuchi, T. Dual role of vitamin C in an oxygen-sensitive system: Discrepancy between DNA damage and dell death. *Free Radic. Res.* **2005**, *39*, 213–220. [CrossRef]
44. Kondakci, E.; Ozyurek, M.; Guclu, K.; Apak, R. Novel pro-oxidant activity assay for polyphenols, vitamins C and E using a modified CUPRAC method. *Talanta* **2013**, *115*, 583–589. [CrossRef]
45. Padh, H. Cellular functions of ascorbic acid. *Biochem. Cell Biol. Biochim. Biol. Cell.* **1990**, *68*, 1166–1173. [CrossRef] [PubMed]
46. Libby, P.; Aikawa, M. Vitamin C, collagen, and cracks in the plaque. *Circulation* **2002**, *105*, 1396–1398. [CrossRef]
47. Patak, P.; Willenberg, H.S.; Bornstein, S.R. Vitamin C is an important cofactor for both adrenal cortex and adrenal medulla. *Endocr. Res.* **2004**, *30*, 871–875. [CrossRef] [PubMed]
48. May, J.M.; Qu, Z.C.; Meredith, M.E. Mechanisms of ascorbic acid stimulation of norepinephrine synthesis in neuronal cells. *Biochem. Biophys. Res. Commun.* **2012**, *426*, 148–152. [CrossRef]
49. Pekala, J.; Patkowska-Sokola, B.; Bodkowski, R.; Jamroz, D.; Nowakowski, P.; Lochynski, S.; Librowski, T. L-Carnitine—Metabolic Functions and Meaning in Humans Life. *Curr. Drug Metab.* **2011**, *12*, 667–678. [CrossRef]
50. Carr, A.C.; McCall, C. The role of vitamin C in the treatment of pain: New insights. *J. Transl. Med.* **2017**, *15*, 77. [CrossRef] [PubMed]
51. Liu, D.H.; Shi, J.; Ibarra, A.C.; Kakuda, Y.; Xue, S.J. The scavenging capacity and synergistic effects of lycopene, vitamin E, vitamin C, and beta-carotene mixtures on the DPPH free radical. *Lwt-Food Sci. Technol.* **2008**, *41*, 1344–1349. [CrossRef]
52. Lin, J.Y.; Selim, M.A.; Shea, C.R.; Grichnik, J.M.; Omar, M.M.; Monteiro-Riviere, N.A.; Pinnell, S.R. UV photoprotection by combination topical antioxidants vitamin C and vitamin E. *J. Am. Acad. Dermatol.* **2003**, *48*, 866–874. [CrossRef]
53. Niki, E. Role of vitamin E as a lipid-soluble peroxyl radical scavenger: In vitro and in vivo evidence. *Free Radic. Biol. Med.* **2014**, *66*, 3–12. [CrossRef]
54. Adisakwattana, S.; Thilavech, T.; Sompong, W.; Pasukamonset, P. Interaction between ascorbic acid and gallic acid in a model of fructose-mediated protein glycation and oxidation. *Electron. J. Biotechnol.* **2017**, *27*, 32–36. [CrossRef]
55. Hazewindus, M.; Haenen, G.; Weseler, A.R.; Bast, A. The anti-inflammatory effect of lycopene complements the antioxidant action of ascorbic acid and alpha-tocopherol. *Food Chem.* **2012**, *132*, 954–958. [CrossRef]
56. Touitou, E.; Alkabes, M.; Memoli, A.; Alhaique, F. Glutathione stabilizes ascorbic acid in aqueous solution. *Int. J. Pharm.* **1996**, *133*, 85–88. [CrossRef]
57. Oey, I.; Verlinde, P.; Hendrickx, M.; Loey, A.V. Temperature and pressure stability of l-ascorbic acid and/or [6s] 5-methyltetrahydrofolic acid: A kinetic study. *Eur. Food Res. Technol.* **2006**, *223*, 71–77. [CrossRef]
58. Zou, M.Y.; Nie, S.P.; Yin, J.Y.; Xie, M.Y. Ascorbic acid induced degradation of polysaccharide from natural products: A review. *Int. J. Biol. Macromol.* **2020**, *151*, 483–491. [CrossRef]
59. Buettner, G.R. In the absence of catalytic metals ascorbate does not autoxidize at pH 7: Ascorbate as a test for catalytic metals. *J. Biochem. Biophys. Methods* **1988**, *16*, 27–40. [CrossRef]
60. Yang, S.; Buettner, G.R. Thermodynamic and kinetic considerations for the reaction of semiquinone radicals to form superoxide and hydrogen peroxide. *Free Radic. Biol. Med.* **2010**, *49*, 919–962.
61. Li, Y.; Yan, Y.; Yu, A.N.; Wang, K. Effects of reaction parameters on self-degradation of L-ascorbic acid and self-degradation kinetics. *Food Sci. Biotechnol.* **2016**, *25*, 97–104. [CrossRef]
62. Yuan, J.P.; Chen, F. Degradation of Ascorbic Acid in Aqueous Solution. *J. Agric. Food Chem.* **1998**, *46*, 5078–5082. [CrossRef]

63. Bree, I.V.; Baetens, J.M.; Samapundo, S.; Devlieghere, F.; Laleman, R.; Vandekinderen, I.; Noseda, B.; Xhaferi, R.; Baets, B.D.; Meulenaer, B.D. Modelling the degradation kinetics of vitamin C in fruit juice in relation to the initial headspace oxygen concentration. *Food Chem.* **2012**, *134*, 207–214. [CrossRef]
64. Du, J.; Cullen, J.J.; Buettner, G.R. Ascorbic acid: Chemistry, biology and the treatment of cancer. *Biochim. Et Biophys. Acta-Rev. Cancer* **2012**, *1826*, 443–457. [CrossRef] [PubMed]
65. Dhakal, S.; Balasubramaniam, V.M.; Ayvaz, H.; Rodriguez-Saona, L.E. Kinetic modeling of ascorbic acid degradation of pineapple juice subjected to combined pressure-thermal treatment. *J. Food Eng.* **2018**, *224*, 62–70. [CrossRef]
66. Sánchez-Moreno, C.; Plaza, L.; Elez-Martínez, P.; De, A.B.; Martín-Belloso, O.; Cano, M.P. Impact of high pressure and pulsed electric fields on bioactive compounds and antioxidant activity of orange juice in comparison with traditional thermal processing. *J. Agric. Food Chem.* **2005**, *53*, 4403–4409. [CrossRef]
67. Velázquez-Estrada, R.; Hernández-Herrero, M.; Rüfer, C.; Guamis-López, B.; Roig-Sagués, A. Influence of ultra high pressure homogenization processing on bioactive compounds and antioxidant activity of orange juice. *Innov. Food Sci. Emerg. Technol.* **2013**, *18*, 89–94. [CrossRef]
68. Siriwoharn, T.; Surawang, S. Protective effect of sweet basil extracts against vitamin C degradation in a model solution and in guava juice. *J Food Process. Preserv.* **2018**, *42*, 7. [CrossRef]
69. Akyildiz, A.; Mertoglu, T.S.; Agcam, E. Kinetic study for ascorbic acid degradation, hydroxymethylfurfural and furfural formations in Orange juice. *J. Food Compos. Anal.* **2021**, *102*, 103996. [CrossRef]
70. Peleg, M.; Normand, M.D.; Dixon, W.R.; Goulette, T.R. Modeling the degradation kinetics of ascorbic acid. *Crit. Rev. Food Sci. Nutr.* **2018**, *58*, 1478–1494. [CrossRef]
71. Seok, Y.J.; Her, J.Y.; Kim, Y.G.; Kim, M.Y.; Jeong, S.Y.; Kim, M.K.; Lee, J.Y.; Kim, C.I.; Yoon, H.J.; Lee, K.G. Furan in Thermally Processed Foods—A Review. *Toxicol. Res.* **2015**, *31*, 241–253. [CrossRef]
72. Mingyue, S.; Fan, Z.; Tao, H.; Jianhua, X.; Yuting, W.; Shaoping, N.; Mingyong, X. Comparative study of the effects of antioxidants on furan formation during thermal processing in model systems. *LWT-Food Sci. Technol.* **2017**, *75*, 286–292.
73. Zhang, S.W.; Tromans, D. Temperature and pressure dependent solubility of oxygen in water: A thermodynamic analysis. *Hydrometallurgy* **1998**, *48*, 327–342.
74. Battino, R. Oxygen and Ozone. *Solubility Data* **1981**, *46*, B1513–B1516.
75. Al Fata, N.; George, S.; Dlalah, N.; Renard, C.M.G.C. Influence of partial pressure of oxygen on ascorbic acid degradation at canning temperature. *Innov. Food Sci. Emerg. Technol.* **2018**, *49*, 215–221. [CrossRef]
76. Aguilar, K.; Garvin, A.; Lara-Sagahon, A.V.; Ibarz, A. Ascorbic acid degradation in aqueous solution during UV-Vis irradiation. *Food Chem.* **2019**, *297*, 124864.1–124864.6. [CrossRef] [PubMed]
77. Tikekar, R.V.; Anantheswaran, R.C.; Laborde, L.F. Ascorbic Acid Degradation in a Model Apple Juice System and in Apple Juice during Ultraviolet Processing and Storage. *J. Food Sci.* **2015**, *76*, H62–H71. [CrossRef]
78. Koutchma, T. Advances in Ultraviolet Light Technology for Non-thermal Processing of Liquid Foods. *Food & Bioprocess Technol.* **2009**, *2*, 138–155.
79. Ahmad, I.; Mobeen, M.F.; Sheraz, M.A.; Ahmed, S.; Anwar, Z.; Shaikh, R.S.; Hussain, I.; Ali, S.M. Photochemical interaction of ascorbic acid and nicotinamide in aqueous solution: A kinetic study. *J. Photochem. Photobiol. B* **2018**, *182*, 115–121. [CrossRef]
80. Doert, M.; Krüger, S.; Morlock, G.E.; Kroh, L.W. Synergistic effect of lecithins for tocopherols: Formation and antioxidant effect of the phosphatidylethanolamine—l-ascorbic acid condensate. *Eur. Food Res. Technol.* **2017**, *243*, 583–596. [CrossRef]
81. Pedrielli, P.; Skibsted, L.H. Antioxidant synergy and regeneration effect of quercetin, (−)-epicatechin, and (+)-catechin on alpha-tocopherol in homogeneous solutions of peroxidating methyl linoleate. *J. Agric. Food Chem.* **2002**, *50*, 7138–7144. [CrossRef] [PubMed]
82. Hang, Y.; Yhabc, D.; Mwabc, D.; Fyabc, D.; Yxabc, D.; Ygabc, D.; Ycabc, D.; Wyabc, D. Regenerative efficacy of tert-butyl hydroquinone (TBHQ) on dehydrogenated ascorbic acid and its corresponding application to liqueur chocolate. *Food Biosci.* **2021**, *42*, 101129.
83. Han, R.Z.; Liu, L.; Li, J.H.; Du, G.C.; Chen, J. Functions, applications and production of 2-O-D-glucopyranosyl-L-ascorbic acid. *Appl. Microbiol. Biotechnol.* **2012**, *95*, 313–320. [CrossRef] [PubMed]
84. Zhang, W.; Huang, Q.; Yang, R.; Zhao, W.; Hua, X. 2-O-D-glucopyranosyl-L-ascorbic acid: Properties, production, and potential application as a substitute for L-ascorbic acid. *J. Funct. Foods* **2021**, *82*, 104481. [CrossRef]
85. Andersen, F.A. Final report of the safety assessment of L-Ascorbic Acid, Calcium Ascorbate, Magnesium Ascorbate, Magnesium Ascorbyl Phosphate, Sodium Ascorbate, and Sodium Ascorbyl Phosphate as used in cosmetics. *Int. J. Toxicol.* **2005**, *24*, 51–111.
86. Chanphai, P.; Tajmir-Riahi, H.A. Conjugation of vitamin C with serum proteins: A potential application for vitamin delivery. *Int. J. Biol. Macromol.* **2019**, *137*, 966–972. [CrossRef]
87. Nesterenko, A.; Alric, I.; Silvestre, F.; Durrieu, V. Comparative study of encapsulation of vitamins with native and modified soy protein. *Food Hydrocoll.* **2014**, *38*, 172–179. [CrossRef]
88. Desai, K.; Park, H.J. Encapsulation of vitamin C in tripolyphosphate cross-linked chitosan microspheres by spray drying. *J. Microencapsul.* **2005**, *22*, 179–192. [CrossRef] [PubMed]
89. Yang, H.C.; Hon, M.H. The effect of the molecular weight of chitosan nanoparticles and its application on drug delivery. *Microchem. J.* **2009**, *92*, 87–91. [CrossRef]

90. Anandharamakrishnan, C.; Ishwarya, S.P. Spray drying for nanoencapsulation of food components. In *Spray Drying Techniques for Food Ingredient Encapsulation*; John Wiley & Sons: New York, NY, USA, 2015; pp. 180–197.
91. Barra, P.A.; Márquez, K.; Gil-Castell, O.; Mujica, J.; Ribes-Greus, A.; Faccini, M. Spray-Drying Performance and Thermal Stability of L-ascorbic Acid Microencapsulated with Sodium Alginate and Gum Arabic. *Molecules* **2019**, *24*, 2872. [CrossRef]
92. Farias, M.D.P.; Albuquerque, P.B.S.; Soares, P.A.G.; de Sa, D.; Vicente, A.A.; Carneiro-da-Cunha, M.G. Xyloglucan from *Hymenaea courbaril* var. *courbaril* seeds as encapsulating agent of L-ascorbic acid. *Int. J. Biol. Macromol.* **2018**, *107*, 1559–1566. [CrossRef]
93. Comunian, T.A.; Thomazini, M.; Alves, A.; Junior, F.E.D.M.; Balieiro, J.C.D.C.; Favaro-Trindade, C.S. Microencapsulation of ascorbic acid by complex coacervation: Protection and controlled release. *Food Res. Int.* **2013**, *52*, 373–379. [CrossRef]
94. Eghbal, N.; Choudhary, R. Complex coacervation: Encapsulation and controlled release of active agents in food systems. *LWT* **2018**, *90*, 254–264. [CrossRef]
95. Lee, J.B.; Ahn, J.; Lee, J.; Kwak, H.S. L-ascorbic acid microencapsulated with polyacylglycerol monostearate for milk fortification. *Biosci. Biotechnol. Biochem.* **2004**, *68*, 495–500. [CrossRef] [PubMed]
96. Rozman, B.; Gasperlin, M. Stability of vitamins C and E in topical microemulsions for combined antioxidant therapy. *Drug Deliv.* **2007**, *14*, 235–245. [CrossRef] [PubMed]
97. Farhang, B.; Kakuda, Y.; Corredig, M. Encapsulation of ascorbic acid in liposomes prepared with milk fat globule membrane-derived phospholipids. *Dairy Sci. Technol.* **2012**, *92*, 353–366. [CrossRef]
98. Trindade, M.A.; Grosso, C.R.F. The stability of ascorbic acid microencapsulated in granules of rice starch and in gum arabic. *J. Microencapsul.* **2000**, *17*, 169–176.
99. Khalid, N.; Kobayashi, I.; Neves, M.A.; Uemura, K.; Nakajima, M. Preparation and characterization of water-in-oil emulsions loaded with high concentration of L-ascorbic acid. *Lwt-Food Sci. Technol.* **2013**, *51*, 448–454. [CrossRef]
100. Khalid, N.; Kobayashi, I.; Neves, M.A.; Uemura, K.; Nakajima, M.; Nabetani, H. Monodisperse W/O/W emulsions encapsulating L-ascorbic acid: Insights on their formulation using microchannel emulsification and stability studies. *Colloids Surf. Physicochem. Eng. Asp.* **2014**, *458*, 69–77. [CrossRef]
101. Perez-Vicente, A.; Gil-Izquierdo, A.; Garcia-Viguera, C. In vitro gastrointestinal digestion study of pomegranate juice phenolic compounds, anthocyanins, and vitamin C. *J. Agric. Food. Chem.* **2002**, *50*, 2308–2312. [CrossRef]
102. De Lorenzo, A.; Andreoli, A.; Sinibaldi Salimei, P.; D'Orazio; Guidi, A.; Ghiselli, A. Determination of the blood ascorbic acid level after administration of slow-release vitamin C. *Clin. Ter.* **2001**, *152*, 87–90.
103. Xu, X.-F.; Zhong, H.Q.; Liu, W.; Xia, W.X.; Yang, J.G. A Kinetic Study on the In Vitro Simulated Digestion of Gummy Vitamin C and Calcium Candies. *Mod. Food Sci. Technol.* **2018**, *34*, 83–89.
104. Liu, G.; Huang, W.; Babii, O.; Gong, X.; Chen, L. Novel protein-lipid composite nanoparticles with inner aqueous compartment as delivery systems of hydrophilic nutraceutical compounds. *Nanoscale* **2018**, *10*, 10629–10640. [CrossRef]
105. Gopi, S.; Balakrishnan, P. Evaluation and clinical comparison studies on liposomal and non-liposomal ascorbic acid (vitamin C) and their enhanced bioavailability. *J. Liposome Res.* **2021**, *31*, 356–364. [CrossRef] [PubMed]
106. Ojeda, G.A.; Sgroppo, S.C.; Martin-Belloso, O.; Soliva Ortuny, R. Chitosan/tripolyphosphate nanoaggregates enhance the antibrowning effect of ascorbic acid on mushroom slices. *Postharvest Biol. Technol.* **2019**, *156*, 110934. [CrossRef]
107. Zhao, M.; You, X.; Wu, Y.; Wang, L.; Wu, W.; Shi, L.; Sun, W.; Xiong, G. Acute heat stress during transportation deteriorated the qualities of rainbow trout (*Oncorhynchus mykiss*) fillets during chilling storage and its relief attempt by ascorbic acid. *LWT*, 2021. (in press)
108. Lall, S.P.; Lewis-Mccrea, L.M. Role of nutrients in skeletal metabolism and pathology in fish—An overview. *Aquaculture* **2007**, *267*, 3–19. [CrossRef]
109. Tzu Ming, P.; Cheng-Lun, W. Soybean Milk to Which Vitamin C, Vitamin C Salt, or Vitamin C Stereoisomer Is Added. Patent 2013039122 A, 28 February 2013.
110. Fujii, K.; Yasuda, A.; Orikoshi, E.; Kimura, Y.; Chaen, H. Milk Constituent-Containing Food Reinforced in Vitamin C. Patent 2006320222 A, 30 November 2006.
111. Nitin, M.; Sharma, B.D.; Kumar, R.R.; Pavan, K.; Prakash Malav, O.; Kumar Verma, A. Fortification of low-fat chicken meat patties with calcium, vitamin E and vitamin C. *Nutr. Food Sci.* **2015**, *45*, 688–702.
112. Zhou, H.B.; Huang, X.Y.; Bi, Z.; Hu, Y.H.; Wang, F.Q.; Wang, X.X.; Wang, Y.Z.; Lu, Z.Q. Vitamin A with L-ascorbic acid sodium salt improves the growth performance, immune function and antioxidant capacity of weaned pigs. *Animal* **2021**, *15*, 7. [CrossRef]
113. Tsaloeva, M.R.; Dubtsov, G.G.; Bogdanov, A.R.; Pavlyuchkova, M.S. A vitamin-mineral premix for bakery products to be used in preventive diets. *Nutrition* **2013**, *3*, 29–31.
114. Kurzer, A.B.; Dunn, M.L.; Pike, O.A.; Eggett, D.L.; Jefferies, L.K. Antioxidant effects on retinyl palmitate stability and isomerization in nonfat dry milk during thermally accelerated storage. *Int. Dairy J.* **2014**, *35*, 111–115. [CrossRef]
115. Hwang, K.E.; Kim, H.W.; Song, D.H.; Kim, Y.J.; Ham, Y.K.; Choi, Y.S.; Lee, M.A.; Kim, C.J. Effect of Mugwort and Rosemary Either Singly, or Combination with Ascorbic Acid on Shelf Stability of Pork Patties. *J. Food Process. Preserv.* **2017**, *41*, e12994. [CrossRef]
116. Sripakdee, T.; Mahachai, R.; Chanthai, S. Phenolics and Ascorbic Acid Related to Antioxidant Activity of MaoFruit Juice and Their Thermal Stability Study (Review Article). *Orient. J. Chem.* **2017**, *33*, 74–86. [CrossRef]
117. Yoshitomi, B. Depletion of ascorbic acid derivatives in fish feed by the production process. *Fish. Sci.* **2004**, *70*, 1153–1156. [CrossRef]

118. Berardo, A.; De Maere, H.; Stavropoulou, D.A.; Rysman, T.; Leroy, F.; De Smet, S. Effect of sodium ascorbate and sodium nitrite on protein and lipid oxidation in dry fermented sausages. *Meat Sci.* **2016**, *121*, 359–364. [CrossRef] [PubMed]
119. Lee, B.J.; Hendricks, D.G.; Cornforth, D.P. A comparison of carnosine and ascorbic acid on color and lipid stability in a ground beef pattie model system. *Meat Sci.* **1999**, *51*, 245–253.
120. Itaru, Y.; Norio, M.; Toshio, M. Alpha-glycosyl-L-ascorbic acid, and its preparation and uses. Patent EP0398484 B1, 31 May 1995.
121. Bamidele, O.P.; Duodu, K.G.; Emmambux, M.N. Encapsulation and antioxidant activity of ascorbyl palmitate with maize starch during pasting. *Carbohydr. Polym.* **2017**, *166*, 202–208. [CrossRef]
122. Aleman, M.; Bou, R.; Tres, A.; Polo, J.; Codony, R.; Guardiola, F. Oxidative stability of a heme iron-fortified bakery product: Effectiveness of ascorbyl palmitate and co-spray-drying of heme iron with calcium caseinate. *Food Chem.* **2016**, *196*, 567–576. [CrossRef]
123. Lin, F.-Q.; Chen, Y.-H.; He, S. Application of L-ascorbyl palmitate in formula milk. *Mod. Food Sci. Technol.* **2010**, *26*, 1114–1116.
124. He, S.; Lin, F.-Q.; Chen, Y.-H. Effect of L-ascorbyl palmitate on the stability of frying oil. *Mod. Food Sci. Technol.* **2010**, *26*, 972–974.

Article

A Novel Insight into Screening for Antioxidant Peptides from Hazelnut Protein: Based on the Properties of Amino Acid Residues

Chenshan Shi [1], Miaomiao Liu [1], Hongfei Zhao [1], Zhaolin Lv [1], Lisong Liang [2,3,4,5,*] and Bolin Zhang [1,*]

[1] Beijing Key Laboratory of Forest Food Processing and Safety, College of Biological Sciences and Technology, Beijing Forestry University, Beijing 100083, China; shichenshan@bjfu.edu.cn (C.S.); 17853555780@163.com (M.L.); zhaohf518@163.com (H.Z.); zhaolinlv@bjfu.edu.cn (Z.L.)
[2] Key Laboratory of Tree Breeding and Cultivation of the State Forestry and Grassland Administration, Research Institute of Forestry, Chinese Academy of Forestry, Beijing 100091, China
[3] State Key Laboratory of Tree Genetics and Breeding, Chinese Academy of Forestry, Beijing 100091, China
[4] Hazelnut Engineering and Technical Research Center of the State Forestry and Grassland Administration, Beijing 100091, China
[5] National Innovation Alliance of Hazelnut Industry, Beijing 100091, China
* Correspondence: lianglisong@caf.ac.cn (L.L.); zhangbolin888@bjfu.edu.cn (B.Z.); Tel.: +86-010-6288-9634 (L.L.); +86-010-6233-8221 (B.Z.)

Abstract: This study used the properties of amino acid residues to screen antioxidant peptides from hazelnut protein. It was confirmed that the type and position of amino acid residues, grand average of hydropathy, and molecular weight of a peptide could be comprehensively applied to obtain desirable antioxidants after analyzing the information of synthesized dipeptides and BIOPEP database. As a result, six peptides, FSEY, QIESW, SEGFEW, IDLGTTY, GEGFFEM, and NLNQCQRYM were identified from hazelnut protein hydrolysates with higher antioxidant capacity than reduced Glutathione (GSH) against linoleic acid oxidation. The peptides having Tyr residue at C-terminal were found to prohibit the oxidation of linoleic acid better than others. Among them, peptide FSEY inhibited the rancidity of hazelnut oil very well in an oil-in-water emulsion. Additionally, quantum chemical parameters proved Tyr-residue to act as the active site of FSEY are responsible for its antioxidation. This is the first presentation of a novel approach to excavating desired antioxidant peptides against lipid oxidation from hazelnut protein via the properties of amino acid residues.

Keywords: antioxidant peptide; amino acid residues; hazelnut; BIOPEP database; DFT calculation

1. Introduction

Peptides, consisting of amino acid residues, are now popular as antioxidants owing to their advantages related to absorption and safety [1]. The antioxidant activity is usually attributed to such properties as the amino acid composition, active amino acid position, molecular mass, and spatial structure of the peptides [1]. Regarding peptides, active amino acid residues, such as tyrosine (Tyr), tryptophan (Trp), cysteine (Cys), methionine (Met), and histidine(His) may act as hydrogen donors; acidic amino acid residues, such as aspartic acid (Asp) and glutamic acid (Glu) can chelate metal ion; hydrophobic amino acid residues, such as alanine (Ala), valine (Val), Proline (Pro), Phenylalanine (Phe) and leucine (Leu) may help to improve the solubility of peptides in the lipid phase, and facilitate interactions between the peptide and lipid-free radicals, thus increasing antioxidant activity [2]. At the same time, active amino acids Cys, Met, Trp, and Tyr, as well as peptides which are designed based on these residues, have been confirmed to eliminate reactive oxygen species (ROS), reactive nitrogen species (RNS), as well as ABTS (2,2-azino-bis-3-ethylbenzothiazoline-6-sulfonic acid) and DPPH (2,2′-diphenyl-1-picrylhydrazyl) radicals in real peptides' system [1,3]. Thus, is it possible to directly screen antioxidant peptides according to the features of active amino acids? A survey of relevant literature by us has shown that few studies have been done in this aspect.

Currently, real-time updated protein databases, bioinformatics tools, and computer-aided mathematical models have become available for investigating the bioactive peptides. The databases BIOPEP, UniProt, and SwePep collect the physicochemical information of various protein-derived peptides for predicting their bioactivities from released protein sequences; the websites Peptide Ranker, GRAVY (Grand average of hydropathy), ProtParam, and Compute pI/Mw allow us easily to determine the possible physical and chemical properties of any peptide, such as pI (Isoelectric Point), Mw (Molecular weight) and potential active; use of molecular docking and quantum chemical calculation can help us to predict/or explain how a peptide works [2,4]. It is noted that in-silico peptide databases and software have been used to identify 10 novel bioactive DPP-IV, renin, and ACE inhibitory peptides from meat proteins [5], as well as tyrosinase inhibitory peptide FPY from walnut protein [6]. However, few reports have been presented to show the possibility of the in-silico tools in screening antioxidant peptides.

Hazelnut, belonging to the family Betulaceae, is an important food source both for oil and protein supplies, in which edible oil occupies 60% of total dry weight, and protein is 15% [7]. Easy lipid oxidation is the main challenge for direct consumption of hazelnut due to the presence of about 90% unsaturated fatty acids in nuts [7,8]. However, as a major by-product of woody oil production, hazelnut protein is usually treated as feed for animals or fertilizers for soil due to its potential allergenicity to humans [9]. To date, limited data on characterizing the bioactive components of hazelnut protein have been reported [10], especially the hazelnut-derived peptides involved in the inhibition of lipid oxidation. Therefore, studies must be carried out to identify if it is possible to separate antioxidant peptides against oil oxidation from hazelnut protein according to the properties of amino acid residues by in-silico tools.

In this study, a hybrid hazelnut (*Corylus. heterophylla* Fisch. × *Corylus. avellana* L., cv. Dawei) widely cultivated in the north part of China was selected to identify antioxidant peptides in terms of active amino acids (Met, Trp, and Tyr), Mw and GRAVY. The selection of protease was detected by searching expected cleavage sites from the database ExPASy ENZEMY. The peptides processed by the selected enzyme from hazelnut protein were sequenced, in which antioxidant peptides able to inhibit linoleic acid oxidation were identified by featuring the properties of active amino acid residues. Identified peptides were then used to protect hazelnut oil from oxidation in an oil-in-water emulsion system. Next, the geometries and active sites were visualized using quantum chemical computation to verify the significance of active amino acids in improving the antioxidant capacity of a biopeptide. The following are all results.

2. Materials and Methods

2.1. Materials

All reagents were of analytical grade. Peptides (22 dipeptides and 9 oligopeptides) were synthesized using a solid phase peptide synthesis method by Zhejiang Ontores Biotech Co., Ltd. (Hangzhou, China) with a purity of more than 90%. After cold-pressing for crude oil, defatted hazelnut flour (DHF) was obtained after a second solvent extraction to remove excess oil.

2.2. Experimental Analysis

2.2.1. Inhibition Activity Assay of Linoleic Acid Oxidation

The ability of amino acids and peptides to inhibit lipid oxidation was determined using a linoleic acid emulsified model. Briefly, amino acids or peptides were dissolved in 50 mM phosphate buffer (pH = 7.0) and then added to a linoleic acid emulsion. In previous studies, different incubation temperatures were designed for linoleic acid oxidation assessment at 20 °C [11], 40 °C [4,12], and 60 °C [13], respectively. However, most studies selected 40 °C for linoleic acid oxidation assay. Subsequently, the reaction mixture was then incubated in tubes at 40 °C in dark for 48 h. The degree of linoleic acid oxidation was measured at 500 nm after mixing with 30% ammonium thiocyanate (dissolved in water) and ferrous chloride solution (dissolved in 3.5% HCl). Effect on the oxidation of linoleic acid was described as inhibition rate (IR), which was calculated by Equation (1). GSH was used as control and the results were expressed as GSH equivalents.

$$IR\% = (1 - A_{sample}/A_{blank}) \times 100 \quad (1)$$

where A means absorbance value of samples (A_{sample}) and blank (A_{blank}) at 500 nm.

2.2.2. Superoxide Radical Scavenging Activity Assay

The superoxide radical scavenging activity was measured at room temperature by monitoring the inhibition effect of pyrogallol auto-oxidation described by Li et al. [14]. Sample solutions (0.1 mL) were incubated at 25 °C for 10 min after added 2.8 mL Tris HCl-EDTA buffer (0.1 M, pH 8.0). The optical density was measured at 325 nm every 10 s for 240 s after mixing with pyrogallol solution (3 mM). The slopes represented rates of pyrogallol auto-oxidation. Superoxide radical scavenging activity was calculated by Equation (2). GSH was used as the control and the results were expressed as GSH equivalents.

$$K = (V_c - V_s)/V_c \times 100 \quad (2)$$

where K means superoxide radical scavenging activity assay, %; V_c indicates auto-oxidation rate of pyrogallol; V_s. stands for oxidation rate of pyrogallol after adding the sample.

2.2.3. Metal Ion Chelation Activity Assay

Fe^{2+} chelation activities were determined according to the methods described by Zhang [15] with some modifications. First, 0.05 mL of 2 mM $FeCl_2$ was mixed with 0.1 mL of 5 mM ferrozine. After the addition of samples solutions, the final volume was increased up to 3 mL with ultrapure water. The absorbance change was measured at 562 nm after being incubated for 10 min at room temperature. The metal ion chelation activity was calculated by Equation (3).

$$\text{Metal ion chelation activity}\% = (1 - A_{sample}/A_{blank}) \times 100 \quad (3)$$

where A means absorbance value of samples (A_{sample}) and blank (A_{blank}) at 562 nm.

2.3. Synthesis of Dipeptides

A total of 22 dipeptides were synthesized according to amino acid activities and properties. The active amino acid residues located at the N-terminus are represented by the letter X; acidic amino acid (D, Asp), basic amino acid (H, His), and hydrophobic amino acid (P, Pro) residues located at the C-terminus were named the dipeptides of XD, XH, and XP, respectively. The dipeptides of XX consist of active amino acid residues located at both C-terminal and N-terminal. To evaluate the antioxidant capacity of all synthesized dipeptides containing active amino acids, four dipeptides without active amino acid, that is, IR with ID 8215, KP with ID 8218, AH and KD obtained from the BIOPEP database (http://www.uwm.edu.pl/biochemia/, accessed on 1 June 2020), were also artificially made the controls.

2.4. BIOPEP Database Analysis

The database BIOPEP as a bioinformatics tool allows the detection of biologically active fragments within protein sequences. BIOPEP categorizes proteins as potential sources of bioactive fragments [16–18] and is used to characterize food-derived peptides [19]. According to the keyword "antioxidative", 294 peptides were found, in which 86 biopeptides related to lipid oxidation were selected. Additionally, their molecular weight, GRAVY value (Calculated by gravy-calculator from http://www.gravy-calculator.de, calculated on 20 May 2020), and location of active amino acids were analyzed.

2.5. Selection of Protease

The cleavage specificity of an enzyme plays a determining role in proteolytic peptide release [20], so the proteolytic sites of the enzymes are accessible from ExPASy (https://enzyme.expasy.org/, accessed on 13 August 2020). According to the properties of antioxidant peptides collected from the BIOPEP database and dipeptides in our study, a desirable protease was selected depending on the cleavage specificity.

2.6. Preparation of Hazelnut Protein Hydrolysates

Hazelnut protein was isolated as Tatar et al. reported [21]. Briefly, DHF dissolved in distilled water was adjusted to a pH of 8.0 with 1% NaOH solution, and stirred for 30 min. Then, the supernatant was collected by centrifugation (9391× g, 20 min) and adjusted to a pH of 4.5 by 1% HCl solution for precipitation. Next, the precipitates were collected by centrifugation (3000× g, 20 min) and freeze-dried for powder harvest after being neutralized to pH 7.0 by a 0.2% NaOH solution.

Hazelnut protein hydrolysates were prepared according to the method described by Liu et al. with some modification [22]. Briefly, 1 g freeze-dried protein, after being mixed with 50 mL distilled water, was denatured at 90 °C for 15 min. Selected proteases were added to mixtures at room temperature. The enzyme amount, reaction time, optimal pH, and reaction temperature were depended on different proteases. For alkaline and neutral proteinases, these parameters are 10,000 U/g protein, 120 min, pH 8.5, 54 °C [22], and 17,000 U/g protein, 120 min, pH 7.0, 44 °C [23], respectively. Obtained hazelnut hydrolysates, after being heated at 100 °C for 10 min to inactivate enzymes, were readjusted to neutral pHs (1% HCl solution for hydrolysates of alkaline protease). Next, the hydrolysates, after being centrifuged at 3910× g for 15 min to remove a little sediment, were freeze-dried for peptide preparation. The lyophilized peptide powders were stored at −20 °C prior to use. The purity of hazelnut hydrolysates was measured by Folin-phenol protein quantitative assay [22].

2.7. Analysis of Peptide Sequence

First, 1 mg of the lyophilized hydrolysates powder, after mixed with 1 mL ultrapure water, was transferred to a 3 KD ultrafiltration tube. The mixture was centrifuged at 12,000× g at 4 °C for 10 min and repeated twice. Then the disulfide bond of hydrolysates was treated by reductive alkylation with 10 mM DTT (DL-Dithiothreitol) and 20 mM IAA (Iodoacetamide) before LC-MS/MS (Liquid chromatography-tandem mass spectrometry, Thermo Fisher Scientific, Waltham, MA USA) analysis. The peptides, after being desalted and freeze-dried, were resuspended in 2 to 20 µL of 0.1% formic acid for sequencing. The hazelnut peptides were sequenced by Beijing Bio-Tech Pack Technology Company Ltd., (Beijing, China).

2.8. DFT Calculations

DFT method (Density Functional Theory method) with B3LYP/6-311G(d,p) was used to optimize the geometries of the molecules. The optimized stable conformation was confirmed to be real minima by frequency calculation (no imaginary frequency). Additionally, HOMO, E-gap (expressed by $E_{LUMO-HOMO} = E_{LUMO} − E_{HOMO}$) [24], and Fukui functions were applied to predict reactivity sites. Fukui functions ($f^−$, f^+, f^0) were calculated using Equations (4)–(6) [25]. q from Equations (4)–(6) represents the atomic charge at the rth atomic site with the neutral (N), anionic (N + 1), and cationic (N-1) chemical species [26]. All calculations were performed using the Gaussian 09, Revision D.01 program package (Gaussian, Inc., Wallingford, CT, USA) [27].

$$f^+ = q(N + 1) − q(N) \quad \text{for nucleophilic attack} \quad (4)$$

$$f^- = q(N) - q(N-1) \quad \text{for electrophilic attack} \tag{5}$$

$$f^0 = (q(N+1) - q(N-1))/2 \quad \text{for radical attack} \tag{6}$$

2.9. Inhibition of Hazelnut Oil Oxidation Assay

The oil-in-water emulsion was prepared according to a previous study. To obtain oil without any antioxidants, hazelnut crude oil was stripped by silica gel, activated charcoal, and sucrose [28]. The aqueous phase of the emulsion was prepared by dispersing 0.5 wt% Tween 20 in 10 mM phosphate buffer at pH 7.0 followed by stirring at room temperature for 20 min to ensure complete dispersion. Hazelnut oil-in-water emulsions were prepared by homogenizing 10 wt% oil phases with 90 wt% aqueous phases at ambient temperature using a high-speed blender for 2 min, followed by ultrasonic vibration for 20 min. Peptides or TBHQ (tert-Butylhydroquinone, as positive control) were added to this emulsion with a final concentration of 0.02% to test the inhibition ability of peptides against oil. Sodium azide (NaN$_3$, 0.02% (w/w)) was used as an antibacterial agent.

Incubation temperature for the oil-in-water emulsion system was generally set at 37 °C [29] or 40 °C [30]. In this work, a longer incubation time was required for analyzing the possible oxidative products of oil-in-water emulsion compared to that of linoleic acid oxidation, so an incubation temperature of 37 °C was selected. During incubation at 37 °C for 14 days, the levels of lipid hydroperoxides as primary lipid oxidation products were monitored at regular intervals to assess the degree of oxidation of hazelnut oil. In brief, 0.3 mL of emulsions were breakdown by 1.5 mL of isooctane/2-propanol (3:1, v/v), vortexed, and then centrifuged at 1000× g for 2 min. Next, 200 µL of organic solvent phases were collected and mixed with 2.8 mL of methanol/1-butanol (2:1, v/v), followed by 15 µL ferrous iron solution (prepared by mixing 0.132 M BaCl2 and 0.144 M FeSO$_4$, dissolved in 0.4 M HCl), and 15 µL ammonium thiocyanate solution (3.94 M, dissolved in water). The absorbance was measured at 510 nm after 20 min. Lipid hydroperoxides (µmol/g oil) were calculated using a standard curve prepared by cumene hydroperoxide (0, 20, 40, 80, 160, 300, and 400 µM) [31].

2.10. Statistical Analysis

IBM SPSS 26.0 software (IBM Corporation. Armonk, NY, USA) was used for statistical analysis between groups. The data are expressed as the mean ± standard deviation (n = 3).

3. Results

3.1. Antioxidant Activity of Amino Acids and Dipeptides

It has been reported that GSH [32] with cysteine residue showed excellent $O_2^{\bullet-}$ scavenging ability than peptides with other residues. Our study confirmed that dipeptides with cysteine residue showed stronger activities than others due to the strong ability of Cys in scavenging $O_2^{\bullet-}$ (shown in Figure 1a). It was seen that the $O_2^{\bullet-}$ scavenging activities of these synthesized dipeptides ranked in the order CP(>GSH) > CD > CH > WC > YC > MC > IR > WH > AH > YH > MH > KP > MD > WD > KD > WY > MW > WP > YP > YD > MP > MY. However, the dipeptides with W, M, or Y residues seem to be inactive in scavenging $O_2^{\bullet-}$. Additionally, these synthesized dipeptides did not show any ability to chelate Fe^{2+} (data not shown).

Furthermore, as shown in Figure 1b, the amino acids Cys, Met, Trp, and Tyr were confirmed to have significant inhibition activity against linoleic acid oxidation after analyzing the antioxidant activities of 20 amino acids. The amino acid Cys showed stronger activity than GSH, followed by Tyr, Met, and Trp. The GE values of Cys, Met, Trp, and Tyr were 1.22 ± 0.12, 0.07 ± 0.01, 0.03 ± 0.01, and 0.07 ± 0.01 mmol/mmol, respectively. The remining amino acids did not show antioxidant activities. Moreover, 22 dipeptides were synthesized based on the inhibition activity of the 4 amino acids towards the oxidation of linoleic acid. It was observed that among these synthesized dipeptides, the peptide WY had the best antioxidant capacity with the GE value 80.29 ± 0.68 mmol/mmol, followed by peptides MY, MW, YH, MH, MC, WC(GSH), YC, MD, WD, CP, YD, CH, CD, YP, MP,

WH, KD, WP and AH(0.12 ± 0.04 mmol/mmol) (IR and KP showed no inhibitory activity). Clearly, the more active amino acid residues existed, the stronger the inhibitory activity of dipeptides exhibited. Interestingly, Cys was observed to show an excellent antioxidation, but the dipeptides with Cys residue did not have more activity against linoleic acid oxidation than the dipeptides with Met, Trp, and Tyr residues. In addition, the dipeptides IR and KP did not show any activity in inhibiting linoleic acid oxidation even though they were reported to scavenge oxygen radicals [33]. Thus, data from our study indicated that the dipeptides containing Met, Trp, or Tyr residues should be selected as potential inhibitors of stopping linoleic acid oxidation.

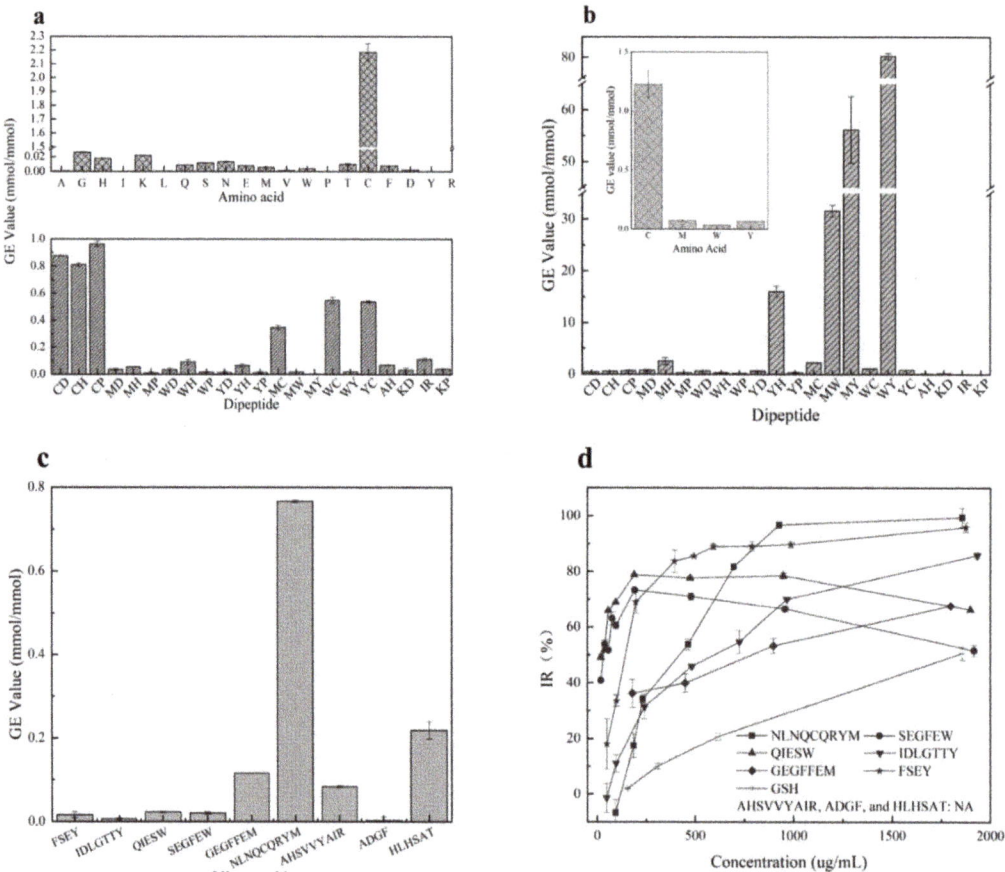

Figure 1. Antioxidant activities of amino acids and peptides. (**a**) means activity of amino acids and dipeptides against $O_2^{\bullet-}$; (**b**) stands for activities of amino acids and dipeptides against linoleic acid; (**c**) shows effect of oligopeptides on the oxidation of $O_2^{\bullet-}$; (**d**) indicates effect of oligopeptides on the oxidation of linoleic acid, AHSVVYAIR, ADGF, HLHSAT were not presented for the poor capacity (NA: No activity was observed at concentration less than 2000 μg/mL.).

3.2. BIOPEP Database Analysis

To evaluate the possible inhibition of antioxidants towards oil oxidation, bioactive peptides are collected from the BIOPEP database and various publications (they are listed in Table S1). Further analysis showed that the peptides whose molecular weight is from 200 to 800 plus a GRAVY value of −2 to 1 as the major proportion of these biopeptides had a possibility of stopping oil oxidation (as presented in Figure S1a,b). Interestingly, it is noted that almost three-quarters of the potential antioxidant peptides containing Tyr, Trp, Cys, or Met residues and mainly located at the C-terminus or N-terminus, especially Tyr residue (as presented in Figure S1c). Thus, it could be presumed that the peptides have a molecular weight of 200 to 800, GRAVY value of −2 to 1, and Tyr, Trp, Cys, or Met residues at N- or C- terminus should be selected as effective candidates for antioxidants.

3.3. Selection of Protease

To find a targeted peptide that consists of active amino acid residues at the N- or C- terminus, the database ExPASy ENZYME integrating available information about proteolytic sites and enzymes was used to select appropriate protease. The use of ExPASy ENZYME allows us easily to determine the cleavage site between all pairs of amino acids in the N- or C-terminal [34]. According to the preferential cleavage sites searched from ExPASy ENZYME (https://enzyme.expasy.org/enzyme-search-ec.html, accessed on 13 August 2020) and from a review [35], the alkaline proteinase (EC number: 3.4.21.62), chymotrypsin (EC number: 3.4.21.1) and pepsin A (EC number: 3.4.23.1) are likely to hydrolyze proteins to the peptides with Tyr, Trp, or Met as C-terminus or N-terminus (Seen in Table 1). Among the three proteases, alkaline hydrolysates exhibited the highest inhibition effect on linoleic acid oxidation [36]. To further reveal the preference of the proteases for producing Tyr, Trp, and Met residues, alkaline and neutral proteinases were chosen to "really" hydrolyze hazelnut protein. It was seen from Table 1 that alkaline proteinase hydrolysates exhibited a high inhibition rate of 95.11 ± 0.17% after 1 mL of the hydrolysates were added to linoleic acid emulsions incubated at 40 °C for 48 h. However, the inhibition rate of linoleic acid by neutral hydrolysates was 81.44 ± 1.94%. In addition, protein hydrolysates hydrolyzed by alkaline + neutral proteinase showed an in-between inhibition rate (83.35 ± 1.02%). Similar to the result reported by Ngamsuk [37], that alkaline proteinase was found to give high activity hydrolysate compared to neutrase and mix.

Table 1. Preferential cleavage sites of several proteinases and inhibition rate of hazelnut protein hydrolysates against oxidation of linoleic acid.

EC Number	Name	Preferential Cleavage Sites	Inhibition Rate (IR, %)
3.4.21.1	Chymotrypsin	Cleaves, Tyr-∣-Xaa, Trp-∣-Xaa, Phe-∣-Xaa, and Leu-∣-Xaa	—[1]
3.4.21.62	Alcalase Novo/ *Bacillus subtilis* alkaline proteinase	Hydrolysis of proteins with broad specificity for peptide bonds, especially aromatic or hydrophobic amino acids. Cleaves, Glu-∣-Xaa, Met-∣-Xaa, Leu-∣-Xaa, Tyr-∣-Xaa, Lys-∣-Xaa, Trp-∣-Xaa, and Gln-∣-Xaa	95.11 ± 0.71
3.4.23.1	Pepsin A (pH = 1.3)	Preferential cleavage, hydrophobic, preferably aromatic. Cleaves, Phe-∣-Val, Gln-∣-His, Glu-∣-Ala, Ala-∣-Leu, Leu-∣-Tyr, Tyr-∣-Leu, Gly-∣-Phe, Phe-∣-Phe, and Phe-∣-Tyr	—[1]
3.4.24.27	Bacillolysin/ *Bacillus subtilis* neutral proteinase	Cleaves, Xaa-∣-Leu> Xaa-∣-Phe	81.44 ± 1.94

Note: 1 means not analyzed.

Furthermore, the purity of hazelnut hydrolysates processed by alkaline proteinase was 73.66 ± 2.50%. It was clear from our data that the proteinase alkaline could be appropriately selected for processing the antioxidant peptides from hazelnut protein. The alkaline protease hydrolysates were freeze-dried for further study.

3.4. Screening of Antioxidant Peptides

Peptides with molecular weights less than 3 kD, which produced from hazelnut protein were sequenced for the screening of potential antioxidants. Furthermore, the peptides which have Tyr, Trp, and Met residues, whose molecule weight is less than 800 and with GRAVY value of −2 to 1 are desirable for us, as sorted out in Table 2. It was seen that seven peptides from hazelnut protein, designed from No.1 to No.7, were screened as potential antioxidants due to up to the desired requirements. Next, they were artificially prepared as follows regarding the determined amino acid sequences. The five peptides FSEY, QIESW, SEGFEW, IDLGTTY, and GEGFFEM were artificially made based on the active amino acids of their C-terminal. The peptide AHSVVYAIR (designed as No.6) was synthesized in terms of Tyr-containing residue in its middle position. The peptide NLNQCQRYM (named as No.7) was synthesized because of the existence of Tyr, Cys, and Met residues. Besides this, two reported peptides HLHSAT and ADGF from hazelnut protein were artificially synthesized according to their ability to scavenge ABTS and DPPH radical [22,23]. Thus, the four peptides NLNQCQRYM, AHSVVYAIR, HLHSAT, and ADGF were artificially prepared as the control, and they were designed to clarify the feasibility of amino acid residues in screening antioxidant peptides.

Table 2. Sequences of synthetic peptides containing Tyr, Trp, Cys, and Met residues.

Number	Sequence	Length	Mw (Da)	GRAVY	Features
1	FSEY	4	544.55	−0.70	
2	QIESW	5	661.70	−0.84	
3	SEGFEW	6	753.75	−1.05	With Tyr, Trp, or Met residue at the C-terminal.
4	IDLGTTY	7	781.85	0.24	
5	GEGFFEM	7	815.88	−0.04	
6	AHSVVYAIR	9	1015.16	0.74	With Tyr residue in the sequence.
7	NLNQCQRYM	9	1169.33	−1.29	With Cys, Tyr, and Met residues in the sequence.
8	HLHSAT	6	664.71	−0.38	With ABTS and DPPH radical scavenging ability.
9	ADGF	4	408.40	0.18	

Notes: A = Alanine, R = Arginine, N = Asparagine, D = Aspartic Acid, C = Cysteine, E = Glutamic Acid, Q = Glutamine, G = Glycine, H = Histidine, I = Isoleucine, L = Leucine, K = Lysine, M = Methionine, F = Phenylalanine, P = Proline, S = Serine, T = Threonine, W = Tryptophan, Y = Tyrosine, V = Valine. HLHSAT and ADGF are antioxidant peptides obtained from hazelnut in other studies [22,23].

As predicated by us, six synthesized peptides showed a significant impact on retarding the oxidation of linoleic acid (Figure 1d). Peptides FSEY and NLNQCQRYM showed the best antioxidant activity, followed by QIESW, SEGFEW, IDLGTTY, and GEGFFEM. However, the inhibition of these synthesized peptides against linoleic acid oxidation functioned in a dose-dependent manner. Peptide NLNQCQRYM with a concentration higher than 900 µg/mL performed excellent activity in stopping linoleic acid from oxidation, while inhibition rates of the peptides QIESW and SEGFEW which contain Trp residue were less than 80% when their concentrations exceeded 200 µg/mL. It was noted that peptide AHSVVYAIR showed a poor capacity, and its IR% was only 50 even when its concentration was elevated to 5000 µg/mL (Figure 1 not shown). Our work indicated that peptides HLHSAT and ADGF, which have been reported to have ABTS and DPPH radical scavenging capacity [22,23] did not show any inhibition against the oxidation of linoleic acid. Moreover, the peptides that contain Cys residue showed perfect $O_2^{\bullet -}$ scavenging activity compared to others, as these dipeptides did (see Figure 1c). Clearly, data from the artificially synthesized peptides support our assumption that the occurrence of active amino acid residues plays a crucial role in promoting the antioxidant capacity of a peptide. Featuring the properties of the amino acid residues should be a simple and feasible tool for the quick selection of desirable antioxidant peptides from hazelnut protein. Besides this, despite the excellent impact on retarding linoleic acid oxidation of FSEY and NLNQCQRYM, FSEY was selected for further study as the peptide falls within a molecular weight of 200 to 800 and GRAVY value of −2 to 1, as well as containing Tyr residue at the C-terminal.

3.5. DFT Calculation of Peptides

3.5.1. Frontier Molecular Orbital Energy of Peptides

To convince the antioxidant performance of four active amino acids, 22 synthesized dipeptides were prepared, and their quantum chemical parameters were obtained by DFT calculations [38]. As shown in Table 2, the frontier molecular orbital energy of each dipeptide, expressed by E_{HOMO} (Energy of highest occupied molecular orbital) and E_{LUMO} (Energy of lowest unoccupied molecular orbital), represents the active level of

22 synthetics [26]. Theoretically, a higher E_{HOMO} means more unstable electrons, which are more likely to scavenge free radicals as hydrogen donors. It was seen from Table 3 that the free radical-scavenging ability of these dipeptides, according to their calculated E_{HOMO}, ranked from high to low in the order WH > WP > WY > WC > WD > MW > MH > YH > MY > YP > MP > CH > MD > MC > YD > IR > YC > KP > KD > AH > CP > CD. Moreover, a lower energy gap (E-gap) represents a higher chemical reactivity [39]. It means that the antioxidative activity of 22 dipeptides, based on their energy gap, ranked from strong to poor in the order WH > WC > WY > WP > WD > MW > MC > MH > MD > MY > YH > YC > YD > YP > CH > KD > CD > IR > AH > MP > KP > CP. To evaluate the real antioxidation of the 22 peptides, their inhibition towards linoleic acid oxidation was examined (see Figure 1b and Table 3). As a result, the best-synthesized dipeptide which retards the oxidation of linoleic acid was WY, followed by peptides MY, MW, YH, MH, MC, WC, YC, MD, WD, CP, YD, CH, CD, YP, MP, WH, KD, WP, AH, IR, and KP. It was noted that the dipeptides including KP, KD, IR, and AH with low E_{HOMO} and high E-gap values had a low antioxidant capacity, whereas the dipeptides like WC, WY, MW, MC, MY, MH, and YH which have high E_{HOMO} and low E-gap values showed a higher inhibition towards the oxidation of linoleic acid (see Table 3). Although several dipeptides had a poor activity in inhibiting the oxidation of linoleic acid, most of the synthesized dipeptides showed good antioxidant activity, especially these dipeptides containing W, M, or Y residue. It means that there is a corresponding relationship between the antioxidant activity of a dipeptide and its frontier molecular orbital energy.

Table 3. Frontier molecular orbital energy(eV) and GE (mM/mM) values of the dipeptides against the oxidation of linoleic acid.

	Dipeptide	E_{HOMO}	E_{LUMO}	E-Gap	GE Value
With GE value higher than 1	WC	−5.61	−0.73	4.87	1.13 ± 0.04
	WY	−5.59	−0.61	4.98	80.28 ± 0.68
	MW	−5.91	−0.75	5.15	31.59 ± 1.13
	MC	−6.21	−0.88	5.33	2.24 ± 0.03
	MY	−6.08	−0.59	5.49	56.17 ± 6.44
	MH	−5.91	−0.52	5.39	2.56 ± 0.68
	YH	−5.99	−0.34	5.65	16.06 ± 1.09
With GE value between 0.5~1	WD	−5.75	−0.61	5.13	0.70 ± 0.02
	MD	−6.17	−0.73	5.45	0.75 ± 0.21
	YC	−6.29	−0.63	5.66	0.85 ± 0.04
	YD	−6.26	−0.60	5.66	0.64 ± 0.08
	CH	−6.17	−0.47	5.70	0.57 ± 0.04
With GE value less than 0.5	WH	−4.77	−0.29	4.48	0.33 ± 0.03
	WP	−5.59	−0.52	5.07	0.20 ± 0.00
	YP	−6.14	−0.47	5.67	0.36 ± 0.12
	KD	−6.42	−0.68	5.75	0.22 ± 0.03
	CD	−6.63	−0.76	5.88	0.46 ± 0.01
	IR	−6.27	−0.16	5.94	0.00 ± 0.00
	AH	−6.54	−0.34	6.00	0.12 ± 0.04
	MP	−6.16	−0.14	6.02	0.35 ± 0.04
	KP	−6.34	−0.05	6.29	0.00 ± 0.00
	CP	−6.57	−0.18	6.39	0.70 ± 0.05

To clarify the possible active site of each peptide, we constructed the dimensional structures of 22 synthesized dipeptides as well as FSEY. It is well known that HOMO always acts as the active site of any organic compound [40]. As shown in Figures 2 and 3a, the HOMOs of all dipeptides as well as FSEY are located at their active amino acid residues, that is, C, W, M, or Y. It indicates that these amino acids will firstly lose their electrons once when interacting with free radicals [41]. Clearly, data from the analysis of the HOMO site and linoleic acid oxidation inhibition confirmed the significance of active amino acid residues in determining the activity of antioxidant peptides. The results from Hougland also et al. supported our findings [42].

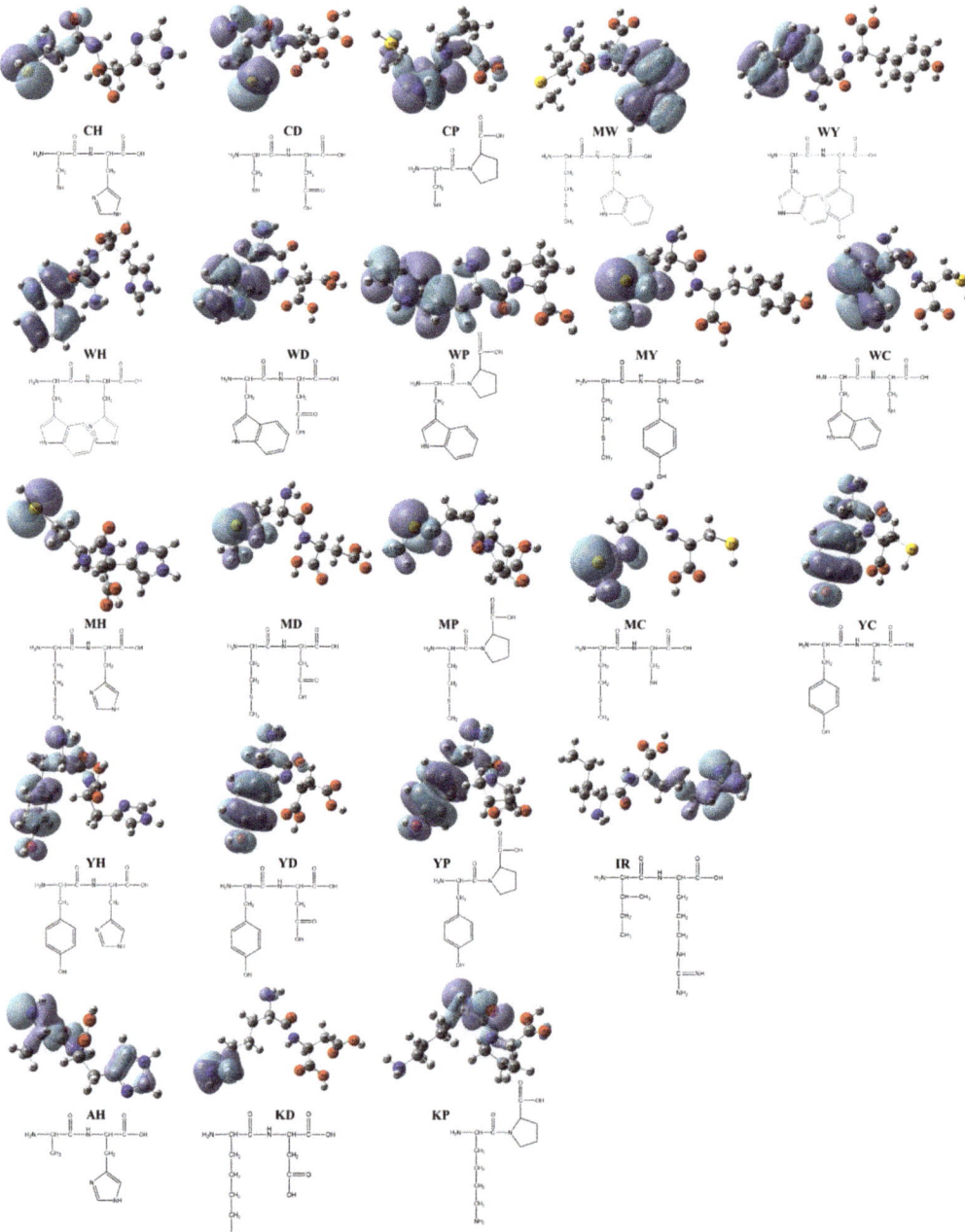

Figure 2. HOMO distribution of the dipeptides. Note: red ball represents the oxygen atom; blue ball represents the nitrogen atom; dark gray ball represents the carbon atom; yellow ball represents the sulfur atom; light gray ball represents the hydrogen atom.

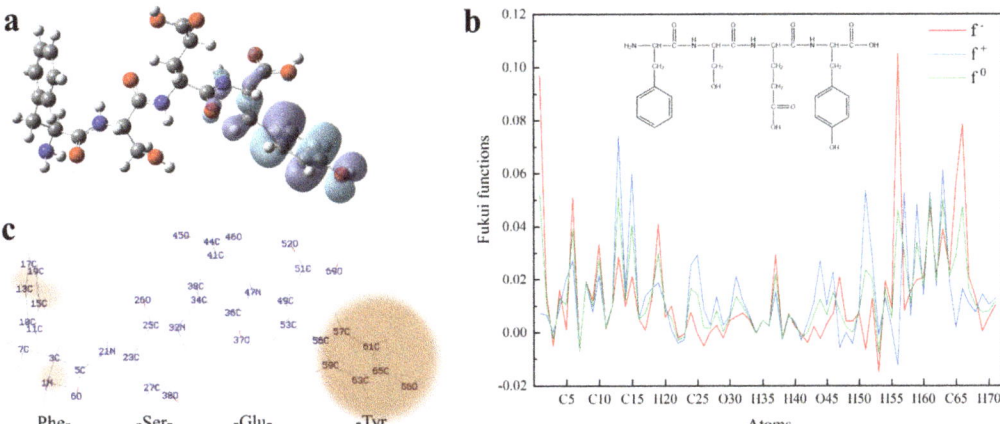

Figure 3. Quantum chemical parameters of FSEY. (**a**) shows HOMO distribution of peptide FSEY; (**b**) stands for Fukui functions; (**c**) means the predicting sites of FSEY more prone to a nucleophilic, electrophilic or radical attack. Note: red ball represents the oxygen atom; blue ball represents the nitrogen atom; dark gray ball represents the carbon atom; light gray ball represents the hydrogen atom.

3.5.2. Fukui Function

To predict which atom would be most susceptible to a nucleophilic or electrophilic attack, peptide FSEY was selected to investigate the tendency by Fukui functions. Fukui function is a local reactivity parameter, which is widely used for molecular reactivity analysis, indicating the tendency of a molecule to lose or gain an electron thus predicting which atom in the molecule would be more prone to a nucleophilic or electrophilic attack. When a molecule prefers to accept an electron, the Fukui function is f^+, it is the index of nucleophilic attack. While when a molecule has a tendency to lose an electron, the Fukui function is f^- and is also termed as the index of electrophilic attack [22]. In our study, the individual atomic charges were calculated by natural population analysis (NPA) with B3LYP/6-311G (d, p) basis set. For all atomic sites of peptide FSEY, their Fukui functions (f^-, f^+, f^0) were presented in Figure 3b,c, respectively.

Blue, red, and green colors in Figure 3b represent nucleophilic, electrophilic, and radical attacks, respectively. It is found that nucleophilic, electrophilic, and radical attacks of peptide FSEY are located at N_1, C_{13}, C_{15}, C_{51}, C_{56}, C_{57}, C_{59}, C_{61}, C_{63}, C_{65}, and O_{66} atoms, especially at the atoms of Tyr residue. These results further highlighted a fact that Tyr residue should act as a biological activity site, as shown in Figure 3c. Moreover, the molecular reactivity site of peptide FSEY, as indicated by Fukui functions, totally corresponded to that of its HOMO.

3.6. Effects of Tyrosine Residue's Location on the Antioxidant Activity of Peptides

To clarify how the position of tyrosine residue governs the antioxidant activity of a peptide against free radicals, three peptides FSEY, FYSE, and YFSE were selected to examine their inhibitions towards the oxidation of linoleic acid (see Figure 4a). The three peptides have the same composition of amino acids but have different locations for tyrosine residue. After incubated with linoleic acid at 40 °C for 48 h, the peptide FSEY showed the strongest ability in stopping the oxidation of the fat acid compared to others. The peptide YFSE had the lowest antioxidant activity. It is concluded that if tyrosine residue is located at the C-terminal, the Tyr-containing peptide should have a stronger activity against linoleic acid radical, as shown in Figure 4a.

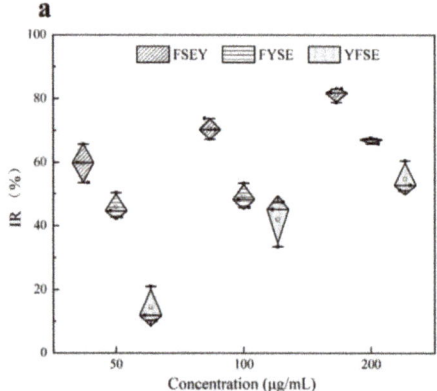 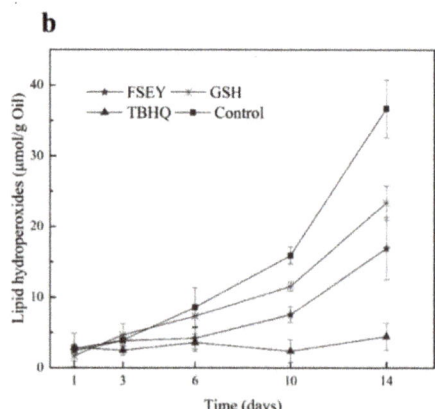

Figure 4. Antioxidant activities of peptides. (**a**) stands for activities of FSEY, FYSE, and YFSE on the oxidation of linoleic acid; (**b**) represents the effects of the antioxidants on oil oxidation, GSH and TBHQ as positive controls.

3.7. Application of Hazelnut Peptide FSEY in Inhibiting Lipid Oxidation

To validate the antioxidant ability of selected peptides in a real-emulsion, 0.02% hazelnut-original peptide FSEY was added to a hazelnut oil-in-water emulsion system for the evaluation of antioxidant activity against oil rancidity. It was seen that the hazelnut-derived peptide FSEY inhibited the rancidity of oil very well by analyzing hydroperoxides on days 1, 3, 6, 10, and 14 (see Figure 4b). Furthermore, the antioxidant activity of peptide FSEY was compared with that of TBHQ, which is a commercial additive for protecting the oil from rancidity, as well as GSH, which is an antioxidant peptide (see Figure 4b). After incubation at 37 °C for 14 days, hydroperoxides of the emulsion system was 36.69 μmoL/g oil without antioxidant, whereas that was 16.94 μmoL/g oil, 22.35 μmoL/g oil, and 4.44 μmoL/g oil in the presence of FSEY, GSH, and TBHQ, respectively. It was clear that peptide FSEY showed a higher ability than GSH in controlling lipid oxidation, but lower activity than TBHQ. These results indicated that hazelnut-original peptide FSEY could be used as an antioxidant in the emulsion system for delaying the rancidity of oil. In addition, due to a weak ability in $O_2^{\bullet-}$ scavenging and Fe^{2+} chelation, we speculate that the peptide FSEY act as a radical scavenger by contributing phenolic hydrogen atom to peroxyl radical.

4. Discussion

Generally, antioxidant peptides against oil oxidation should act as one or more roles, that is, containing hydrophobic amino acids which expose more active sites to terminate lipid chain reaction; having free radical scavenging agents (such as $O_2^{\bullet-}$ and peroxyl radical) or as metal ions chelating agents; and possessing strong lipase-inhibitory activities [2]. Obviously, the efficiency of antioxidation peptides in an emulsion system depends greatly on their ability to present more active sites, scavenge superoxide radicals or peroxyl radicals, and chelate metal ions. In addition, GSH was selected as a positive control for the outstanding antioxidant properties. Our study was pictured in Figure 5. As is shown, we used chemical experiments and physical properties of biopeptides as well as DFT calculations to verify which properties of the amino acid residues could be used to screen antioxidant peptides.

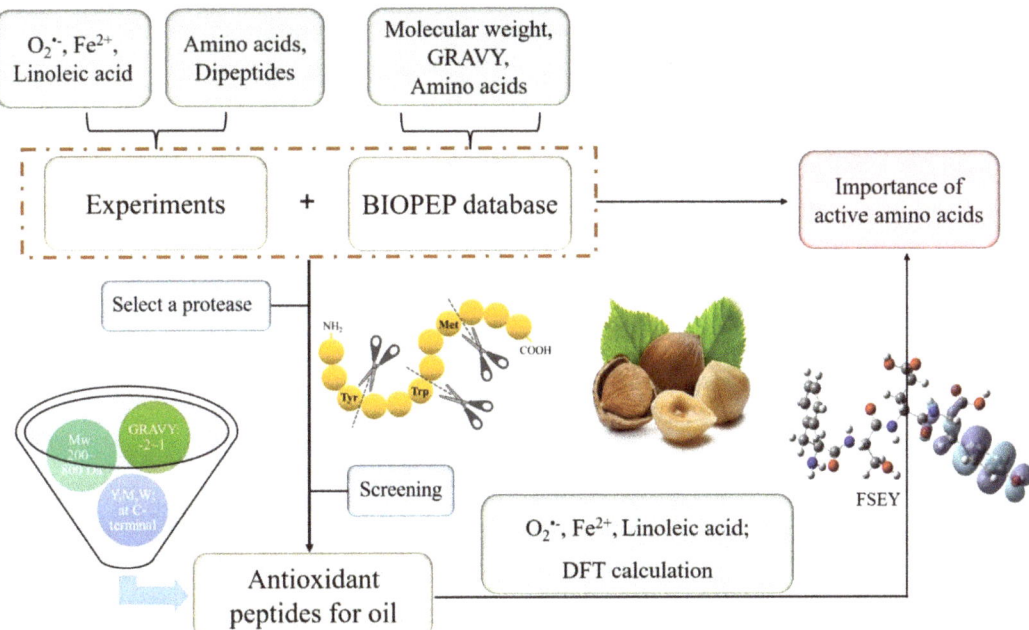

Figure 5. Screening antioxidant peptides depending on properties of amino acid residues.

4.1. Chemical Test: We Found Four Key Amino Acid Residues

The radicals scavenging capacity of a peptide depends greatly upon its amino acid residues, especially upon Tyr, Trp, Met, and Cys. To illustrate this point, various tests firstly were carried out to confirm that Tyr, Trp, and Met and dipeptides containing these residues have excellent antioxidant capacities against the oxidation of linolic acid. It was observed that the absence of these residues caused the dipeptides to lose their activities (see Figure 1b). These amino acids, which are crucial in scavenging free radicals, have been reported by several research works. Amino acids Tyr, Trp, Cys, and Met as well as peptides containing these amino acids showed activities against ABTS radicals and oxygen radicals (ORAC, oxygen radical absorbance capacity) [43], active against ROS or RNS [3], and effectively against AAPH-induced peroxyl radicals [41].

4.2. DFT Calculation: Tyr, Trp, Met, and Cys Are Active Sites

Theoretically, quantum chemical computations can gain prediction of behaviors of organic compounds, such as their structural features and chemical reactivity, and therefore, help to analyze the relationship between the biological potencies and the type of compounds [38]. The distributions of HOMO correspond to the active sites of the peptides able to scavenge free radicals [26]. By our DFT calculations, HOMOs of the tested peptides are located at their active amino acids, that is, Cys, Trp, Met, or Tyr (see Figure 2). The HOMOs of some peptides including EAAY, PMRGGGYHY, PMRGGYHY, PMRGYHY, PMRYHY, and YHY have been reported to be concentrated on the phenolic hydroxyl structure in Tyr [41]. The peptides PVETVR, QEPLLR, RDPEER, and LDDDGRL have the HOMOs of guanidyl in Arg, and the active sites of peptides KELEEK, DAAGRLQE, and GFAGDDAPRA are located at Lys-Glu, Gly, and Asp [24,39]. Clearly, data from HOMOs addressed that the residues Cys, Trp, Met, or Tyr are key components responsible for the antioxidant activity of the tested peptides in our study.

Generally, a high E_{HOMO} or a low E-gap value means flexible chemical reactivity and could be used to predict the antioxidant activity of each peptide [24]. As predicted in our study, seven synthetic dipeptides, having a higher E_{HOMO} and a lower E-gap value, showed a good ability to inhibit the oxidation of linoleic acid. It is found the seven antioxidant dipeptides possess the active residues Tyr, Trp, or Met. The presence of Tyr, Trp, and Met significantly enhanced the antioxidant activity of these dipeptides compared to other tested peptides (see Table 3). In a similar study, Wang et al. used E_{HOMO} and E-gap to predicate the antioxidant activity of five peptides with only one exception [39]. Experiments conducted by Wu et al. also indicated that E_{HOMO} and E-gap were feasible to describe the antioxidant behaviors of a set of man-made peptides, which were designed from the parent peptide "PMRGGGGYHY" [41]. Consistent with other studies reported [44], the presence of active residues Tyr, Trp, or Met as well as high E_{HOMO} and low E-gap should be the characteristics of a peptide responsible for inhibiting the oxidation of linoleic acid. Amino acids, Tyr and Trp, act as active sites were also confirmed by Molecular docking. Wang et al. found that Trp1 and Tyr4 in peptide WLSYPMNPATGH could form hydrogen bonds with DPPH, which means responsible of Trp and Tyr in scavenging DPPH free radical. These two emerging approaches are helpful in analyzing antioxidative products, meanwhile, DFT calculation is a useful tool in screening antioxidant peptides [45].

4.3. BIOPEP Database Analysis: Rules of Molecular Weight, GRAVY Value and Active Amino Acid Residue's Location

Few reports have focused on how a molecular weight, GRAVY value, and position of amino acid residues affect the capacity of a peptide against oil oxidation yet. In our study, a peptide in the BIOEPE database falling within a molecular weight of 200 to 800, GRAVY value of −2 to 1, and active amino acid residues at N- or C- terminus produced a strong inhibition towards lipid oxidation, as shown in Table S1 and Figure S1.

Regarding molecular weight or numbers of amino acid residues, numerous studies have indicated that peptides containing amino acid residues between 2 and 11 [46] or weighing less than 1000 Da [47] will exhibit good antioxidant ability. Peptides with more molecular weights (>2000 Da) easily decrease in their antioxidant activity due to the hiding of the active site [47]. It is well-known that the interfacial phase, the contact region between the oil phase and the aqueous phase, is the critical region in the system with regard to the development of lipid peroxidation [48]. Thus, in a given emulsion system, GRAVY value could not be ignored for antioxidant estimation, because a higher GRAVY means higher hydrophobicity. Various researches have proposed that peptides with higher hydrophobicity can protect linoleic acid from oxidation by donating protons to hydrophobic peroxy-radicals [49,50]. In our study, it is seen that peptides from BIOPEP database which able to stop lipid oxidation have GRAVY values ranging from −2 to 1 (see Table S1 and Figure S1b). Thus, a GRAVY value of −2 to 1 was proposed to be an ideal criterion for looking for antioxidant peptides from protein hydrolysate. A recent study has overviewed the roles of amino acid composition and sequence in conferring the antioxidant activities of peptides. The phenolic hydroxyl of Tyr, the indolyl of Trp, the thiol group of Cys, and the thioether of Met are regarded to act as hydrogen donors for free radicals [2,39,51]. Our study has found that the most of antioxidative peptides which are searched from the database BIOPEPE have the residues Tyr, Trp, Cys, or Met. Similarly, the antioxidant activities of peptides LGFEY and LGFYY were attributed to the presence of Tyr residues [52]. Concerning the inhibition of lipid oxidation, most peptides searched from the BIOPEPE database are observed to have active amino acid residues located at C- terminus. In our case, a linoleic acid oxidation system was designed to confirm the strong antioxidant activity of the peptide FSEY having Tyr residue located at the C-terminal (see Figure 4a). HOMO analysis from the synthesized peptides FYSE and YFSE also presents evidence that the reaction sites are all located at Tyr (see Figure S2). Similarly, studies done by Guo et al. [53] and Torkova et al. [54] indicated that peptides having Tyr residue located at the C-terminus strongly scavenged hydroxyl-radical, hydrogen-peroxide, and peroxyl radicals. While these peptides exhibited a better inhibition against ABTS cation radical when Tyr residue is located at the N-terminus [54]. Clearly, the types and positions of amino acid residues should be considered in searching for an antioxidant peptide from protein.

In conclusion, a wanted antioxidant peptide could be quickly screened by determining the types and location of amino acid residues as well as molecular weight and GRAVY, especially Tyr, Trp, Cys, and Met which act as H donors. Among the four amino acids, Tyr-containing peptides show a prominent antioxidant activity, especially when it is located at the C-terminus.

5. Conclusions

In our study, amino acids, Met, Tyr, Try, and peptides containing these active amino acids show antioxidant activity against linoleic acid radicals. These amino acid residues, as Tyr residue in peptide FSEY does, act as an active site for scavenging lipid free radicals. More meaningfully, the active amino acid residues located at C-terminal are more active than other positions. Compared to traditional technology for manufacturing bioactive peptides, our work presents a practical route able to successfully screen desirable high-activity antioxidant peptides from hazelnut protein hydrolysates by featuring the properties of amino acid residues. Our technical route consists of two steps. Firstly, peptides from hazelnut protein hydrolyzed by alkaline protease are sequenced; secondly, the peptides falling within a molecular weight of 200 to 800 and GRAVY value of −2 to 1 as well as containing Tyr, Met, Trp residues at C- terminus are selected for antioxidant candidates inhibiting the oxidation of the oil. Using this route successfully releases a peptide from hazelnut protein which inhibits oil oxidation very well. To our knowledge, it is the first attempt to prepare antioxidant peptides based on the properties of amino acid residues. Perhaps, the new findings out of our work will be beneficial for screening bioactive peptides from various protein resources with different purposes.

Supplementary Materials: The following are available online at https://www.mdpi.com/article/10.3390/antiox11010127/s1, Supplementary Materials to this article contains Table S1. Biopeptides collected from BIOPEP database and other publications, Figure S1. Analysis of the biopeptides harvested from BIOPEP database and other publications, Figure S2. The highest occupied molecular orbital (HOMO) of FYSE and YFSE.

Author Contributions: C.S., Writing—original draft, Methodology, Supervision. M.L., Investigation. H.Z., Funding acquisition. Z.L., Writing—review & editing. L.L. and B.Z., Supervision, Writing—review & editing. All authors have read and agreed to the published version of the manuscript.

Funding: This work was supported by the Fundamental Research Funds for the Central Universities (No. 2015ZCQ-SW-05).

Institutional Review Board Statement: Not applicable.

Informed Consent Statement: Not applicable.

Data Availability Statement: The data is contained within the article or supplementary material.

Acknowledgments: Authors acknowledge Fen Wang (College of Chemistry and Chemical Engineering, Taishan University, Taian, Shandong, China) for her great help in DFT calculations.

Conflicts of Interest: The authors declare no conflict of interest. The funders had no role in the design of the study; in the collection, analyses, or interpretation of data; in the writing of the manuscript, or in the decision to publish the results. The author Lisong Liang is part of the National Innovation Alliance of Hazelnut Industry, the industry had no role in the design of the study; in the collection, analyses, or interpretation of data; in the writing of the manuscript, or in the decision to publish the results.

References

1. Zheng, L. Structure-Activity Relationship and Directional Preparation of Antioxidant Peptide. Ph.D. Thesis, South China University of Technology, Guangzhou, China, 2015.
2. Wen, C.; Zhang, J.; Zhang, H.; Duan, Y.; Ma, H. Plant protein-derived antioxidant peptides: Isolation, identification, mechanism of action and application in food systems: A review. *Trends Food Sci. Technol.* **2020**, *105*, 308–322.
3. Matsui, R.; Honda, R.; Kanome, M.; Hagiwara, A.; Terashima, M. Designing antioxidant peptides based on the antioxidant properties of the amino acid side-chains. *Food Chem.* **2017**, *245*, 750.
4. Jie, Y.; Zhao, H.; Sun, X.; Lv, X.; Zhang, Z.; Zhang, B. Isolation of antioxidative peptide from the protein hydrolysate of Caragana ambigua seeds and its mechanism for retarding lipid auto-oxidation. *J. Sci. Food Agric.* **2019**, *99*, 3078–3085.
5. Lafarga, T.; O'Connor, P.; Hayes, M. Identification of novel dipeptidyl peptidase-IV and angiotensin-I-converting enzyme inhibitory peptides from meat proteins using in silico analysis. *Peptides* **2014**, *59*, 53–62.
6. Feng, Y.; Wang, Z.; Chen, J.; Li, H.; Wang, Y.; Ren, D.; Lu, J. Separation, identification, and molecular docking of tyrosinase inhibitory peptides from the hydrolysates of defatted walnut (*Juglans regia* L.) meal. *Food Chem.* **2021**, *353*, 129471.
7. Jiang, J.; Liang, L.; Ma, Q.; Zhao, T. Kernel nutrient composition and antioxidant ability of *Corylus* spp. in China. *Front. Plant Sci.* **2021**, *12*, 690966.
8. Moscetti, R.; Saeys, W.; Keresztes, J.C.; Goodarzi, M.; Cecchini, M.; Danilo, M.; Massantini, R. Hazelnut Quality Sorting Using High Dynamic Range Short-Wave Infrared Hyperspectral Imaging. *Food Bioprocess Technol.* **2015**, *8*, 1593–1604.
9. Zhong, Y. Hazelnut Protein Physical Modification Methods and the Physical and Chemical Properties Change Research. Master's Thesis, Shenyang Agricultural university, Shenyang, China, 2017.
10. Çağlar, A.F.; Çakır, B.; Gülseren, İ. LC-Q-TOF/MS based identification and in silico verification of ACE-inhibitory peptides in Giresun (*Turkey*) hazelnut cakes. *Eur. Food Res. Technol.* **2021**, *247*, 1189–1198.
11. Sabeena Farvin, K.H.; Andersen, L.L.; Nielsen, H.H.; Jacobsen, C.; Jakobsen, G.; Johansson, I.; Jessen, F. Antioxidant activity of Cod (*Gadus morhua*) protein hydrolysates: In vitro assays and evaluation in 5% fish oil-in-water emulsion. *Food Chem.* **2014**, *149*, 326–334.
12. Sakanaka, S.; Tachibana, Y.; Ishihara, N.; Raj Juneja, L. Antioxidant activity of egg-yolk protein hydrolysates in a linoleic acid oxidation system. *Food Chem.* **2004**, *86*, 99–103.
13. Alashi, A.M.; Blanchard, C.L.; Mailer, R.J.; Agboola, S.O.; Mawson, A.J.; He, R.; Girgih, A.; Aluko, R.E. Antioxidant properties of Australian canola meal protein hydrolysates. *Food Chem.* **2014**, *146*, 500–506.
14. Li, Y.; Bo, J.; Tao, Z.; Mu, W.; Jian, L. Antioxidant and free radical-scavenging activities of chickpea protein hydrolysate (CPH). *Food Chem.* **2008**, *106*, 444–450.
15. Zhang, T.; Li, Y.; Miao, M.; Jiang, B. Purification and characterisation of a new antioxidant peptide from chickpea (*Cicer arietium* L.) protein hydrolysates. *Food Chem.* **2011**, *128*, 28–33.
16. Minkiewicz, P.; Iwaniak, A.; Darewicz, M. BIOPEP-UWM database of bioactive peptides: Current opportunities. *Int. J. Mol. Sci.* **2019**, *20*, 5978.
17. Piotr, M.; Jerzy, D.; Małgorzata, D.; Anna, I.; Nałecz, M. Food Peptidomics. *Food Technol. Biotechnol.* **2008**, *46*, 1–10.
18. Piotr, M.; Jerzy, D.; Anna, I.; Marta, D.; Magorzata, D. BIOPEP Database and Other Programs for Processing Bioactive Peptide Sequences. *J. AOAC Int.* **2019**, *4*, 965–980.
19. Iwaniak, A.; Minkiewicz, P.; Darewicz, M.; Sieniawski, K.; Starowicz, P. BIOPEP database of sensory peptides and amino acids. *Food Res. Int.* **2016**, *85*, 155–161.
20. Chatterjee, A.; Kanawjia, S.K.; Khetra, Y.; Saini, P. Discordance between in silico & in vitroanalyses of ACE inhibitory & antioxidative peptides from mixed milk tryptic whey protein hydrolysate. *J. Food Sci. Technol.* **2015**, *52*, 5621–5630.
21. Tatar, F.; Tunç, M.T.; Kahyaoglu, T. Turkish Tombul hazelnut (*Corylus avellana* L.) protein concentrates: Functional and rheological properties. *J. Food Sci. Technol.* **2015**, *52*, 1024–1031.
22. Liu, C.; Ren, D.; Li, J.; Fang, L.; Wang, J.; Liu, J.; Min, W. Cytoprotective effect and purification of novel antioxidant peptides from hazelnut (*C. heterophylla* Fisch) protein hydrolysates. *J. Funct. Foods* **2018**, *42*, 203–215.
23. Chen, Y.; Lv, C.; Han, J.; Ji, Y.; Fu, Q.; Liu, L.; Xu, J. Separation, Purification and Sequence Analysis of Antioxidative Peptides from the Halnut Meal Protein Hydrolysate. *Sci. Technol. Food Ind.* **2018**, *39*, 114–118.
24. Wen, C.; Zhang, J.; Zhang, H.; Duan, Y.; Ma, H. Study on the structure–activity relationship of watermelon seed antioxidant peptides by using molecular simulations. *Food Chem.* **2021**, *364*, 130432.
25. Zacharias, A.O.; Varghese, A.; Akshaya, K.B.; Savitha, M.S.; George, L. DFT, spectroscopic studies, NBO, NLO and Fukui functional analysis of 1-(1-(2,4-difluorophenyl)-2-(1H-1,2,4-triazol-1-yl)ethylidene) thiosemicarbazide. *J. Mol. Struct.* **2018**, *1158*, 1–13. [CrossRef]
26. Sevvanthi, S.; Muthu, S.; Raja, M.; Aayisha, S.; Janani, S. PES, molecular structure, spectroscopic (FT-IR, FT-Raman), electronic (UV-Vis, HOMO-LUMO), quantum chemical and biological (docking) studies on a potent membrane permeable inhibitor: Dibenzoxepine derivative. *Heliyon* **2020**, *6*, e04724.
27. Frisch, M.J.; Trucks, G.W.; Schlegel, H.B.; Scuseria, G.E.; Robb, M.A.; Cheeseman, J.R.; Scalmani, G.; Barone, V.; Mennucci, B.; Petersson, G.A.; et al. *Gaussian 09, Revision D.01*; Gaussian, Inc.: Wallingford, CT, USA, 2013.
28. Shi, C.; Liu, M.; Ma, Q.; Zhao, T.; Liang, L.; Zhang, B. Impact of tetrapeptide-FSEY on oxidative and physical stability of hazelnut oil-in-water emulsion. *Foods* **2021**, *10*, 1400.
29. Vaisali, C.; Belur, P.D.; Regupathi, I. Comparison of antioxidant properties of phenolic compounds and their effectiveness in imparting oxidative stability to sardine oil during storage. *LWT-Food Sci. Technol.* **2016**, *69*, 153–160.

30. Di Mattia, C.D.; Sacchetti, G.; Mastrocola, D.; Sarker, D.K.; Pittia, P. Surface properties of phenolic compounds and their influence on the dispersion degree and oxidative stability of olive oil O/W emulsions. *Food Hydrocoll.* **2010**, *24*, 652–658.
31. Pei, Y.; Deng, Q.; McClements, D.; Li, J.; Li, B. Impact of Phytic Acid on the Physical and Oxidative Stability of Protein-Stabilized Oil-in-Water Emulsions. *Food Biophys.* **2020**, *15*, 433–441.
32. Zeng, Z. *Study on Structure-Activity Relationships of Antioxidant Peptide through Enzymatic by Quantum Chemistry*; Changsha University of Science & Technology: Changsha, China, 2013.
33. Huang, W.Y.; Majumder, K.; Wu, J. Oxygen radical absorbance capacity of peptides from egg white protein ovotransferrin and their interaction with phytochemicals. *Food Chem.* **2010**, *123*, 635–641.
34. Keil, B. *Specificity of Proteolysis*; Springer: Berlin/Heidelberg, Germany, 1992.
35. Tacias-Pascacio, V.G.; Morellon-Sterling, R.; Siar, E.-H.; Tavano, O.; Berenguer-Murcia, Á.; Fernandez-Lafuente, R. Use of alcalase in the production of bioactive peptides: A review. *Int. J. Biol. Macromol.* **2020**, *165*, 2143–2196.
36. Arise, A.K.; Alashi, A.M.; Nwachukwu, I.D.; Malomo, S.A.; Amonsou, E.O. Inhibitory properties of bambara groundnut protein hydrolysate and peptide fractions against angiotensin converting enzymes, renin and free radicals. *J. Sci. Food Agric.* **2016**, *97*, 2834–2841.
37. Ngamsuk, S.; Hsu, J.-L.; Huang, T.-C.; Suwannaporn, P. Ultrasonication of Milky Stage Rice Milk with Bioactive Peptides from Rice Bran: Its Bioactivities and Absorption. *Food Bioprocess Technol.* **2020**, *13*, 462–474.
38. Antonczak, S. Electronic description of four flavonoids revisited by DFT method. *J. Mol. Struc.-Theochem* **2008**, *856*, 38–45.
39. Wang, M.; Li, C.; Li, H.; Wu, Z.; Chen, B.; Lei, Y.; Shen, Y. In Vitro and In Silico Antioxidant Activity of Novel Peptides Prepared from Paeonia Ostii 'Feng Dan' Hydrolysate. *Antioxidants* **2019**, *8*, 433.
40. Srivastava, A.K.; Pandey, A.K.; Jain, S.; Misra, N. FT-IR spectroscopy, intra-molecular C−H···O interactions, HOMO, LUMO, MESP analysis and biological activity of two natural products, triclisine and rufescine: DFT and QTAIM approaches. *Spectrochim. Acta Part A Mol. Biomol. Spectrosc.* **2015**, *136*, 682–689.
41. Wu, R.; Huang, J.; Huan, R.; Chen, L.; Yi, C.; Liu, D.; Wang, M.; Liu, C.; He, H. New insights into the structure-activity relationships of antioxidative peptide PMRGGGGYHY. *Food Chem.* **2021**, *337*, 127678.
42. Hougland, J.L.; Darling, J.; Flynn, S. *Protein Posttranslational Modification*; John Wiley & Sons, Ltd.: Hoboken, NJ, USA, 2013.
43. Zheng, L.; Zhao, Y.; Dong, H.; Su, G.; Zhao, M. Structure activity relationship of antioxidant dipeptides: Dominant role of Tyr, Trp, Cys and Met residues. *J. Funct. Foods* **2016**, *21*, 485–496.
44. Uno, S.; Kodama, D.; Yukawa, H.; Shidara, H.; Akamatsu, M. Quantitative analysis of the relationship between structure and antioxidant activity of tripeptides. *J. Pept. Sci.* **2020**, *26*, e3238.
45. Wang, J.; Chen, C.; Xu, Y.; Jia, C.; Zhang, B.; Niu, M.; Zhao, S.; Xiong, S. Selection of Antioxidant Peptides from Gastrointestinal Hydrolysates of Fermented Rice Cake by Combining Peptidomics and Bioinformatics. *ACS Food Sci. Technol.* **2021**, *1*, 443–452.
46. Pihlanto, A. Antioxidative peptides derived from milk proteins. *Int. Dairy J.* **2006**, *16*, 1306–1314.
47. Wu, H. Structure-Activity Relationship of Antioxidative Peptides Derived from Whey Protien. Master's Thesis, Jiangnan University, Wuxi, China, 2011.
48. Berton-Carabin, C.C.; Ropers, M.H.; Genot, C. Lipid Oxidation in Oil-in-Water Emulsions: Involvement of the Interfacial Layer. *Compr. Rev. Food Sci. F* **2014**, *13*, 945–977.
49. Chen, H.; Muramoto, K.; Yamauchi, F. Structural analysis of antioxidative peptides from soybean beta-conglycinin. *J. Agric. Food. Chem.* **1995**, *43*, 574–578.
50. Suetsuna, K. Antioxidant Peptides from the Protease Digest of Prawn (*Penaeus japonicus*) Muscle. *Mar. Biotechnol.* **2000**, *2*, 5–10.
51. Raja, M.; Raj Muhamed, R.; Muthu, S.; Suresh, M.; Muthu, K. Synthesis, spectroscopic (FT-IR, FT-Raman, NMR, UV–Visible), Fukui function, antimicrobial and molecular docking study of (E)-1-(3-bromobenzylidene)semicarbazide by DFT method. *J. Mol. Struct.* **2017**, *1130*, 374–384.
52. Shen, S.; Chahal, B.; Majumder, K.; You, S.J.; Wu, J. Identification of Novel Antioxidative Peptides Derived from a Thermolytic Hydrolysate of Ovotransferrin by LC-MS/MS. *J. Agric. Food. Chem.* **2010**, *58*, 7664–7672.
53. Guo, H.; Kouzuma, Y.; Yonekura, M. Structures and properties of antioxidative peptides derived from royal jelly protein. *Food Chem.* **2009**, *113*, 238–245.
54. Torkova, A.; Koroleva, O.; Khrameeva, E.; Fedorova, T.; Tsentalovich, M. Structure-Functional Study of Tyrosine and Methionine Dipeptides: An Approach to Antioxidant Activity Prediction. *Int. J. Mol. Sci.* **2015**, *16*, 25353–25376.

Article

Improving Aqueous Solubility of Natural Antioxidant Mangiferin through Glycosylation by Maltogenic Amylase from *Parageobacillus galactosidasius* DSM 18751

Jiumn-Yih Wu [1,†], Hsiou-Yu Ding [2,†], Tzi-Yuan Wang [3,†], Yu-Li Tsai [4], Huei-Ju Ting [4] and Te-Sheng Chang [4,*]

[1] Department of Food Science, National Quemoy University, Kinmen County 892, Taiwan; wujy@nqu.edu.tw
[2] Department of Cosmetic Science, Chia Nan University of Pharmacy and Science, No. 60 Erh-Jen Rd., Sec. 1, Jen-Te District, Tainan 71710, Taiwan; ding8896@gmail.com
[3] Biodiversity Research Center, Academia Sinica, Taipei 11529, Taiwan; tziyuan@gmail.com
[4] Department of Biological Sciences and Technology, National University of Tainan, Tainan 70005, Taiwan; aa0920281529@gmail.com (Y.-L.T.); hting@mail.nutn.edu.tw (H.-J.T.)
* Correspondence: mozyme2001@gmail.com; Tel./Fax: +886-6-2602137
† Authors contributed equally to this manuscript.

Abstract: Mangiferin is a natural antioxidant *C*-glucosidic xanthone originally isolated from the *Mangifera indica* (mango) plant. Mangiferin exhibits a wide range of pharmaceutical activities. However, mangiferin's poor solubility limits its applications. To resolve this limitation of mangiferin, enzymatic glycosylation of mangiferin to produce more soluble mangiferin glucosides was evaluated. Herein, the recombinant maltogenic amylase (MA; E.C. 3.2.1.133) from a thermophile *Parageobacillus galactosidasius* DSM 18751^T (*Pg*MA) was cloned into *Escherichia coli* BL21 (DE3) via the expression plasmid pET-Duet-1. The recombinant *Pg*MA was purified via Ni^{2+} affinity chromatography. To evaluate its transglycosylation activity, 17 molecules, including mangiferin (as sugar acceptors), belonging to triterpenoids, saponins, flavonoids, and polyphenol glycosides, were assayed with β-CD (as the sugar donor). The results showed that puerarin and mangiferin are suitable sugar acceptors in the transglycosylation reaction. The glycosylation products from mangiferin by *Pg*MA were isolated using preparative high-performance liquid chromatography. Their chemical structures were glucosyl-α-(1→6)-mangiferin and maltosyl-α-(1→6)-mangiferin, determined by mass and nucleic magnetic resonance spectral analysis. The newly identified maltosyl-α-(1→6)-mangiferin showed 5500-fold higher aqueous solubility than that of mangiferin, and both mangiferin glucosides exhibited similar 1,1-diphenyl-2-picrylhydrazyl free radical scavenging activities compared to mangiferin. *Pg*MA is the first MA with glycosylation activity toward mangiferin, meaning mangiferin glucosides have potential future applications.

Keywords: mangiferin; maltogenic amylase; glycosylation; glucoside; *Parageobacillus galactosidasius*

Citation: Wu, J.-Y.; Ding, H.-Y.; Wang, T.-Y.; Tsai, Y.-L.; Ting, H.-J.; Chang, T.-S. Improving Aqueous Solubility of Natural Antioxidant Mangiferin through Glycosylation by Maltogenic Amylase from *Parageobacillus galactosidasius* DSM 18751. *Antioxidants* **2021**, *10*, 1817. https://doi.org/10.3390/antiox10111817

Academic Editors: Li Liang and Hao Cheng

Received: 23 October 2021
Accepted: 15 November 2021
Published: 16 November 2021

Publisher's Note: MDPI stays neutral with regard to jurisdictional claims in published maps and institutional affiliations.

Copyright: © 2021 by the authors. Licensee MDPI, Basel, Switzerland. This article is an open access article distributed under the terms and conditions of the Creative Commons Attribution (CC BY) license (https:// creativecommons.org/licenses/by/ 4.0/).

1. Introduction

Mangiferin is a natural *C*-glucosidic xanthone originally isolated from the *Mangifera indica* (mango) plant. Mangiferin has been reported to possess diverse health-promoting activities, such as antioxidant [1,2], anticancer [3,4], anti-inflammatory [5], and anti-osteoarthritis pain activities [6], allowing it to prevent memory impairment [7], neurodegeneration [8], and organ fibrosis [9]. Furthermore, it offers protection from the deleterious effects of heavy metals [10]. However, the pharmacological use of mangiferin is restricted owing to its poor solubility and low bioavailability [11,12]. As the glycosylates of small molecules have been proven to have better aqueous solubility and bioavailability than the original molecules [13,14], the glycosylation of mangiferin should be further improved for better usage.

Glycosylation of molecules can be achieved using chemical or enzymatic methods; however, enzymatic glycosylation using glycosyltransferases (GTs) and glycoside hydrolases (GHs) offers more advantages than chemical methods [15]. Moreover, GHs use cheaper sugars, such as starch, maltodextrin, maltose, and sucrose, as donors during glycosylation [16], whereas GTs use expensive uridine diphosphate-glucose (UDP-G). Therefore, GHs are preferred for the bioindustrial production of glycosylated molecules. According to the carbohydrate-activating enzyme (CAZy) database, a classification of GH in families based on amino acid sequence similarities and 117 GH families has been discovered to date [17].

Maltogenic amylase (MA; E.C. 3.2.1.133) belongs to the GH13 gene family and hydrolyzes starch to produce maltose [17]. Some specific features of MA were further identified. First, MAs were found to prefer cyclodextrin (CD) to starch as a substrate, whereas typical amylases do not catalyze CD. The sugar preference is due to 130 unique residues at the N-terminal of the MA protein, which would help the enzyme form a dimer and greatly increase its catalytic activities toward CD [18,19]. Second, MAs are intracellular proteins, whereas typical amylases are extracellular proteins. Third, MAs exhibit the bifunctions of hydrolysis and transglycosylation activities [18–28], whereas typical amylases are rarely reported to have transglycosylation activities [18].

Based on the dual functions of hydrolysis and transglycosylation and dual recognition sites on both the α-1,4 and α-1,6 glycosidic bonds, MAs have also been used in the fine chemical industry to produce novel and branched oligosaccharides from liquefied starch [19–28]. MAs could be further used for the glycosylation of bioactive molecules to develop new drugs in the clinical chemistry field.

In addition to sugars, MAs can glycosylate small and/or bioactive molecules. MAs exhibit transglycosylation reactions in the presence of various acceptor molecules, such as glucose, maltose, and acarbose, by forming α-1,3, α-1,4, and α-1,6 glycosidic linkages [18–28]. MAs have been proven to glycosylate some small molecules, such as hydroquinone [29], caffeic acid [30], ascorbic acid [31], puerarin [32–34], genistin [35], neohesperidin [36], and naringin [37]. For example, the glycosylation of puerarin by two MAs, *Tf*MA from archaeon *Thermofilum pendens* [32] and *Bs*MA from *Bacillus stearothermophilus* [34], has been studied.

In the present study, a maltogenic amylase gene from *Parageobacillus galactosidasius* (*Pg*MA) was cloned into *Escherichia coli* BL21 (DE3) via the expression plasmid pET-Duet-1, and the expressed *Pg*MA was purified. The purified *Pg*MA was characterized and found to glycosylate mangiferin. The novel mangiferin glucosides were isolated for the characterization of both the chemical structures and compounds.

2. Materials and Methods

2.1. Reagents and Chemicals

A Ni^{2+} affinity column (10 i.d. × 50 mm, Ni Sepharose 6 Fast Flow) used for the purification of the recombinant MA was purchased from GE Healthcare (Chicago, IL, USA). Isopropyl β-D-1-thiogalactopyranoside (IPTG), 1,1-diphenyl-2-picrylhydrazine (DPPH), dimethyl sulfoxide (DMSO), and maltodextrin (dextrose equivalent 4.0–7.0) were bought from Sigma (St. Louis, MO, USA). α-CD, β-CD, γ-CD, soluble starch, and pullulan were purchased from Tokyo Chemical Industry Co., Ltd. (Tokyo, Japan). Restriction enzymes and DNA-modified enzymes were obtained from New England Biolabs (Ipswich, MA, USA). All kits for molecular cloning, including the Geno Plus Genomic DNA Extraction Midiprep System, Mini Plus Plasmid DNA Extraction System, Gel Advanced Gel Extraction Miniprep System, and Midi Plus Ultrapure Plasmid Extraction System, were purchased from Viogene (Taipei, Taiwan). Other reagents and solvents used are commercially available.

2.2. Strains and Plasmids

P. galactosidasius DSM 18751T (BCRC 80657) was obtained from the Bioresources Collection and Research Center (BCRC; Food Industry Research and Development Institute, Hsinchu, Taiwan). *E. coli* BL21 (DE3) and the expression plasmid pET-Duet-1 were obtained from Novagen Inc. (Madison, WI, USA).

2.3. Aligned Amino Acid Sequences

In total, 588 amino acids of PgMA (OXB94089) and close-related BsMT (AAC46346) were aligned using Clustal W in MEGA X [38].

2.4. Construction of Expression Plasmids

P. galactosidasius DSM 18751T (BCRC 80657) was cultivated in accordance with the BCRC protocol. The genomic DNA of the bacterium was isolated using the Geno Plus Genomic DNA Extraction Midiprep System (Viogene) according to the manufacturer's protocol. The target gene (PgMA) was amplified from the genomic DNA with polymerase chain reaction (PCR). The primer set used in the PCR was as follows: forward 5′-gggggatccgttgaaagaagccatttatcatcg-3′ and reverse 5′-gggctcgagtcaattttctacttgatagaggag-3′, which contain BamHI and XhoI restriction sites (underlined mark) for cloning. The amplified DNA fragment (1.8 kb length) was cloned into the expression plasmid pET-Duet-1, named pETDuet-PgMA, which was then transformed into E. coli BL21 (DE3) for the recombinant PgMA.

2.5. Production and Purification of Recombinant PgMA in E. coli

The recombinant E. coli harboring the recombinant expression plasmid pETDuet-PgMA was cultivated in Luria–Bertani (LB) medium containing 1% (w/v) tryptone and sodium chloride and 0.5% (w/v) yeast extract to the optical density at 560 nm (OD_{560}) of 0.6 and then induced with 0.2 mM of IPTG. After further cultivation at 18 °C for 20 h, the cells were centrifuged at 4500× g and 4 °C for 20 min. The cell pellet was washed and spun down twice with 50 mM of phosphate buffer (PB) at pH 6.8 and then broken with sonication via a Branson S-450D Sonifier (Branson Ultrasonic Corp., Danbury, CT, USA). The sonication program was run for five cycles of 5 s on and 30 s off at 4 °C. The mixture was then centrifuged at 15,000× g and 4 °C for 20 min to remove the cell debris. A supernatant containing the recombinant PgMA fused with a His-tag in its N-terminal was applied in an Ni^{2+} affinity column. The His-tag-fused PgMA was washed with PB with 25 mM imidazole and eluted with PB containing 250 mM imidazole. The eluate was then concentrated and desalted through Macrosep 10 K centrifugal filters (Pall, Ann Arbor, MI, USA). The concentration of the purified PgMA was determined using the Bradford method [39] and analyzed with sodium dodecyl sulfate (SDS) polyacrylamide gel electrophoresis (PAGE). The purified PgMA was stored in a final concentration of 50% glycerol at −80 °C before use.

2.6. Assay of Hydrolysis Activity

The standard reaction was performed with 1% (w/v) β-CD, 5.6 µg/mL PgMA, and 50 mM of PB at pH 6 and 60 °C for 30 min. After the reaction was stopped by boiling, the amount of reducing sugars produced from each reaction was estimated using the dinitrosalicylic acid method [40]. One unit of MA activity was defined as the amount of the enzyme that released 1 µmol of reducing sugar as maltose per min under the assay condition described earlier. For optimal conditions, the reaction was further performed at different temperatures and pH values, including pH 5 (acetate buffer), pH 6–7 (PB), and pH 8–10 (glycine buffer). Accordingly, the substrate specificity was measured with 1% (w/v) of the studied sugars, including α-CD, β-CD, γ-CD, soluble starch, and pullulan, performed at pH 7 and 65 °C. To realize the effects of metal ions and DMSO on the hydrolysis activity of PgMA, 10 mM of tested metal ion or 5–20% (v/v) of DMSO was added into the reaction mixture.

2.7. Assay of Transglycosylation Activity

To determine the transglycosylation activity of the PgMA, β-CD was used as a sugar donor, and the molecules, which belonged to triterpenoids, saponins, flavonoids, or polyphenol glycosides, were tested as sugar acceptors. The reaction mixture containing 5% (w/v) β-CD, 5.6 µg/mL PgMA, and 1 mg/mL tested molecules, dissolved in DMSO with 50 mM of PB (pH 7), was incubated at 65 °C for 24 h. The reaction mixture was then mixed with an equal volume of methanol and analyzed with high-performance liquid chromatography (HPLC).

2.8. HPLC Analysis

HPLC was performed with the Agilent 1100 series HPLC system (Santa Clara, CA, USA) equipped with a gradient pump (Waters 600, Waters, Milford, MA, USA). The stationary phase was a C18 column (5 μm, 4.6 i.d. × 250 mm; Sharpsil H-C18, Sharpsil, Bei-jing, China), and the mobile phase was 1% acetic acid in water (A) and methanol (B). The elution condition was a linear gradient from 0 min with 40% B to 20 min with 70% B, an isocratic elution from 20 to 25 min with 70% B, a linear gradient from 25 min with 70% B to 28 min with 40% B, and an isocratic elution from 28 to 35 min with 40% B. All eluants were eluted at a flow rate of 1 mL/min. The sample volume was 10 μL. The detection condition was set at 254 nm.

2.9. Purification of Mangiferin Glycosides

The purification process was a previously described method [41]. A 100 mL reaction mixture containing 50% (w/v) maltodextrin, 1 mg/mL mangiferin, 5.6 μg/mL PgMA, and 50 mM PB (pH 7) was incubated at 65 °C for 24 h. After the large-scale reaction, an equal volume of methanol was added to stop the transglycosylation. The mixture was then filtrated through a 0.2 μm nylon membrane, and the filtrate was injected in a preparative YoungLin HPLC system (YL9100, YL Instrument, Gyeonggi-do, South Korea) equipped with a preparative C18 reversed-phase column (10 μm, 20.0 i.d. × 250 mm, ODS 3; Inertsil, GL Sciences, Eindhoven, The Netherlands) for the purification of biotransformation products. The operational conditions for the preparative HPLC analysis were the same as those in the HPLC analysis. The elution corresponding to the peak of the metabolite in the HPLC analysis was collected, condensed under a vacuum, and then crystallized by freeze drying. Finally, 20.1 mg of compound (**1**) and 9.3 mg of compound (**2**) were obtained, and the structures of the compounds were confirmed with nucleic magnetic resonance (NMR) and mass spectral analyses. The mass analysis was performed using the Finnigan LCQ Duo mass spectrometer (ThermoQuest Corp., San Jose, CA, USA) with electrospray ionization (ESI). ^{1}H- and ^{13}C-NMR, distortionless enhancement by polarization transfer (DEPT), heteronuclear single quantum coherence (HSQC), heteronuclear multiple bond connectivity (HMBC), correlation spectroscopy (COSY), and nuclear Overhauser effect spectroscopy (NOESY) spectra were recorded on a Bruker AV-700 NMR spectrometer at ambient temperature. Standard pulse sequences and parameters were used for the NMR experiments, and all chemical shifts were reported in parts per million (ppm, δ).

The composition of compound (**1**) was as follows: light yellow powder; mp 233–235 °C; ESI/MS *m/z*: 583.4 [M-H]$^-$, 565.3, 331.0, 300.9, 259.3; ^{1}H-NMR (DMSO-d_6, 700 MHz): Hδ 3.05 (1H, t, J = 5.6 Hz, H-4″), 3.15 (1H, d, J = 6.3 Hz, H-2″), 3.20 (1H, t, J = 9.1 Hz, H-3′), 3.29 (1H, t, J = 9.1 Hz, H-4′), 3.33 (1H, m, H-5′), 3.35 (1H, m, H-5″), 3.38 (1H, m, H-3″), 3.46 (1H, m, H-6″a), 3.52 (1H, d, J = 9.1 Hz, H-6″b), 3.62 (1H, d, J = 9.8 Hz, H-6′a), 3.70 (1H, dd, J = 11.2, 4.2 Hz, H-6′b), 4.02 (1H, br, H-2′), 4.58 (1H, d, J = 9.1 Hz, H-1′), 4.73 (1H, J = 4.2 Hz, H-1″), 6.36 (1H, s, H-4), 6.86 (1H, s, H-5), and 7.37 (1H, s, H-8). ^{13}C-NMR (DMSO-d_6, 175 MHz): Cδ 60.6 (C-6″), 66.9 (C-6′), 70.0 (C-4″), 70.2 (C-2′, 4′), 72.1 (C-2″), 72.5 (C-3″), 73.2 (C-1′), 73.3 (C-5″), 78.9 (C-3′), 79.7 (C-5′), 93.3 (C-4), 98.7 (C-1″), 101.3 (C-9a), 102.6 (C-5), 107.5 (C-2), 108.1 (C-8), 111.8 (C-8a), 143.7 (C-7), 150.8 (C-10a), 154.0 (C-6), 156.2 (C-4a), 161.7 (C-1), 163.8 (C-3), and 179.1 (C-9).

The composition of compound (**2**) was as follows: light yellow powder; mp 227–229 °C; ESI/MS *m/z*: 745.3 [M-H]$^-$, 727.3, 403.3, 385.0, 331.0, 313.2, 301.2; ^{1}H-NMR (DMSO-d_6, 700 MHz): Hδ 3.03 (1H, t, J = 9.1 Hz, H-4‴), 3.19 (1H, dd, J = 9.1, 3.5 Hz, H-2‴), 3.21 (1H, m, H-3′), 3.23 (1H, m, H-2″), 3.30 (1H, m, H-4′), 3.34 (1H, m, H-4″), 3.36 (1H, m, H-5′), 3.38 (1H, m, H-3‴), 3.41 (1H, br, H-2′), 3.42 (1H, m, H-5‴), 3.44 (2H, m, H-6‴), 3.46 (1H, m, H-5″), 3.55 (2H, m, H-6″), 3.60 (1H, m, H-3″), 3.64 (1H, m, H-6′a), 3.71 (1H, m, H-6′b), 4.58 (1H, d,

J = 9.8 Hz, H-1′), 4.75 (1H, d, J = 3.5 Hz, H-1″), 4.95 (1H, d, J = 3.5 Hz, H-1‴), 6.36 (1H, s, H-4), 6.86 (1H, s, H-5), and 7.37 (1H, s, H-8). ^{13}C-NMR (DMSO-d_6, 175 MHz): Cδ 60.0 (C-6″), 60.7 (C-6‴), 67.2 (C-6′), 69.8 (C-4‴), 70.2 (C-2′, 4′), 70.8 (C-5″), 71.6 (C-2″), 72.6 (C-2‴), 73.1 (C-3″), 73.2 (C-1′), 73.3 (C-3‴), 73.4 (C-5‴), 78.9 (C-3′), 79.7 (C-5′), 79.8 (C-4″), 93.3 (C-4), 98.6 (C-1″), 100.9 (C-1‴), 101.3 (C-9a), 102.6 (C-5), 107.4 (C-2), 108.1 (C-8), 111.7 (C-8a), 143.7 (C-7), 150.8 (C-10a), 154.0 (C-6), 156.2 (C-4a), 161.8 (C-1), 163.8 (C-3), and 179.1 (C-9).

2.10. Determination of Aqueous Solubility

The aqueous solubility of mangiferin and its glucoside derivative were examined as follows: each compound was vortexed in double-deionized H$_2$O for 1 h at 25° C. The mixture was centrifuged at 10,000× g for 30 min at 25 °C. The supernatant was filtrated with 0.2 µM of nylon membrane and analyzed with HPLC. Based on their peak areas, the concentrations of the tested compounds were determined by using calibration curves prepared with HPLC analyses of authentic samples.

2.11. Determination of Antiradical Activity Using a DPPH Assay

The assay was performed as previously described [42] with minor modifications. The tested sample (dissolved in DMSO) was added to the DPPH solution (1 mM in methanol) to a final volume of 0.1 mL. After 30 min of reaction, the absorbance of the reaction mixture was measured at 517 nm with a microplate reader (Sunrise, Tecan, Männedorf, Switzerland). Ascorbic acid (dissolved in DMSO) was used as a positive antioxidant standard. The DPPH free radical scavenging activity was calculated as follows: DPPH free radical scavenging activity = (OD$_{517}$ of the control reaction − OD$_{517}$ of the reaction)/(OD$_{517}$ of the control reaction).

3. Results and Discussion

3.1. Selection of Candidate MAs via Online Genome Sequences

Maltogenic amylase from *Bacillus stearothermophilus* (*Bs*MA, GenBank accession number: AAC46346) is a well-studied MA for the glycosylation of bioactive molecules [31,33,34,36,37]. Thus, the amino acid sequence of *Bs*MA was selected to search for new MAs from the NCBI GenBank. The highest but distinct MA sequences from bacterial whole genomes were further identified as our study's candidates. The available bacterial strains with known genome sequences in BCRC (Hsinchu, Taiwan) were also included for comparison. Accordingly, an α-glycosidase gene (GenBank accession number: OXB94089.1) from the genome data of *P. galactosidasius* DSM 18751T (GenBank accession number: PRJNA383662) showed the highest homology (79.1%) with *Bs*MA (Figure 1) and the top five candidates with the best-hit of *Bs*MA from NCBI GenBank in Table S1. Therefore, the α-glycosidase gene from the DSM 18,751 strain was identified as a suitable candidate in the present study. Figure 1 shows the α-glycosidase gene from *P. galactosidasius* DSM 18751, which was classified as a maltogenic amylase gene, and the gene product was named *Pg*MA in this study.

```
                Alpha-amylase_N
#OXB94089  MLKEAIYHRPKDNFAYAYNEKTLHIRLRTKKNDIETVYLLYADPYEWVNGVWQLNRAPMVKSGSDDLFDYWFIEVVPPYRRLRYGFELTAGSETIVYTEK
#AAC46346  .F..............D.Q.............V.H.R.I.G....E..H..VSYQS.Q...T.E......ALT.............SNN.Q......
                                                                                   Alpha-amylase (GH13)
#OXB94089  GFYNEAPTDDTAYYFCFPFLNKIDVFHAPEWVKDTIWYQIFPERFANGNPSLNPEGTLPWGSEEPTPTSFFGGDFEGIIQHLDYLVELGINGIYFTPIFY
#AAC46346  ...RT..M..........Q...................EA...A......AD.................K..H..D..V..........K

#OXB94089  APSNHKYDTIDYFEIDPHFGDKQTFKKLVDLCHQKGIRVMLDAVFNHCGYYFAPFQDVLKNGNNSKYKDWFHIHEFPLRTVPRPNYDTFAFVEQMPKLNT
#AAC46346  .S............Q......R..E.....A........S..E.P......Y.E....H....R....Q.........TPN......

#OXB94089  ENPEVKQYLLDVATYWIREFDIDGWRLDVANEVDHQFWREFRQAVKAIKPDVYILGEIWHDAMPWLRGDQFDAVMNYPFTNGVIRFFAKDEIRAEQFANI
#AAC46346  ..Q...N...............................T....A.....................TL....QH....S..VGM

#OXB94089  IMNVLHSYPANVNEVAFNLLGSHDTPRILTTCNNDVRKVKLLFLFQLSFTGSPCIYYGDEIGMTGGQDPGCRKCMIWDESKQNRDILEHVKKLISLRKTH
#AAC46346  MTH.....T...............L..L.KE.....A..S..........T............I..D..............HQ....ELFR..Q...A...AY
                Malt_amylase_C
#OXB94089  PALGSRGNITFIDANNETNHLIYTKTYKHETILVMINNNNKNIEVTVPLSLKGKRLTNLWTNEQFAVEAKTLRANIPPYGFLLYQVEN--
#AAC46346  K.F.N...LH.....D.......M..FEE.A.IFLV...EQE..I.L......L.........SA..D..KSTL.....FI.KI.DWL
```

Figure 1. *Pg*MA (OXB94089) protein sequence. The GH13 family gene sequence identity between *Pg*MA and BsMT (AAC46346) is 79.08% (465/588). The Pfam domains are as follows: (1) alpha-amylase_N (alpha amylase, N-terminal IG-like domain; pfam02903): 1–121 amino acids; (2) alpha-amylase (alpha amylase catalytic domain found in cyclomaltodextrinases and related proteins; cd11338): 173–469 amino acids; and (3) malt_amylase_C (maltogenic amylase, C-terminal domain; pfam16657), 507–583 amino acids. "." Denotes an identical amino acid; "-" denotes insertions and deletions.

3.2. Production of Recombinant PgMA in E. coli

To produce the recombinant *Pg*MA, an expression plasmid pETDuet-*Pg*MA was constructed (Figure 2a), and the recombinant *Pg*MA was produced in recombinant *E. coli* BL21 (DE3) and purified as a major band shown by SDS-PAGE with a Ni^{2+} affinity chromatography (Figure 2b). The purified *Pg*MA showed an estimated 68 kD molecular weight in the SDS-PAGE. The production yield was 18.73 mg/L, and the specific activity of the purified enzyme was determined to be 91.46 U/mg by using β-CD as a substrate at pH 7 and 60 °C.

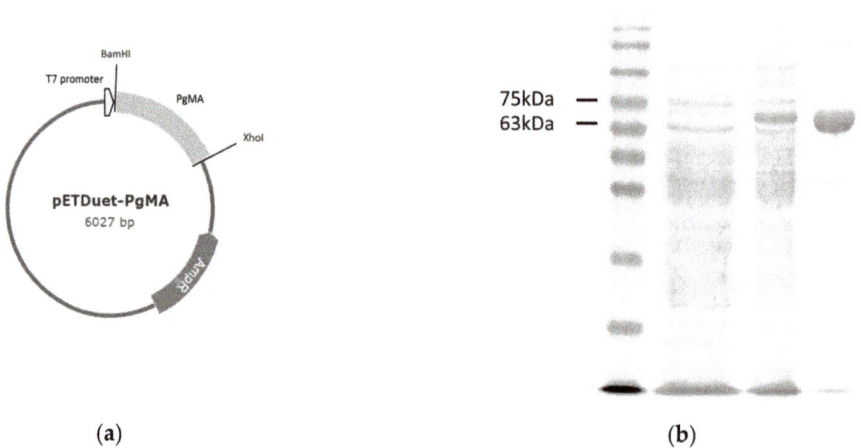

Figure 2. Expression of *Pg*MA from *Parageobacillus galactosidasius* DSM 18751T in *Escherichia coli* (DE3). (**a**) Recombinant repression plasmid pETDuet-*Pg*MA. (**b**) Sodium dodecyl sulfate-polyacrylamide gel electrophoresis (SDS-PAGE) analysis of expressed and purified proteins from recombinant *E. coli* harboring pETDuet-*Pg*MA. Lane 1: molecular marker; lane 2: total protein before induction; lane 3: total protein after 20 h of induction; and lane 4: purified protein.

3.3. Determination of Hydrolysis Activity by Recombinant PgMA

To determine the optimal pH and temperature, β-CD was used as a substrate, and the reaction was performed at different pH levels and temperatures. The results showed that the optimal pH and temperature for the PgMA catalytic reaction were pH 7 (Figure 3a) at 65 °C (Figure 3b). Moreover, the addition of Mg^{2+}, K^+, or ethylenediaminetetraacetic acid did not significantly affect the activity of PgMA; by contrast, DMSO decreased the activity of PgMA (Figure 3c).

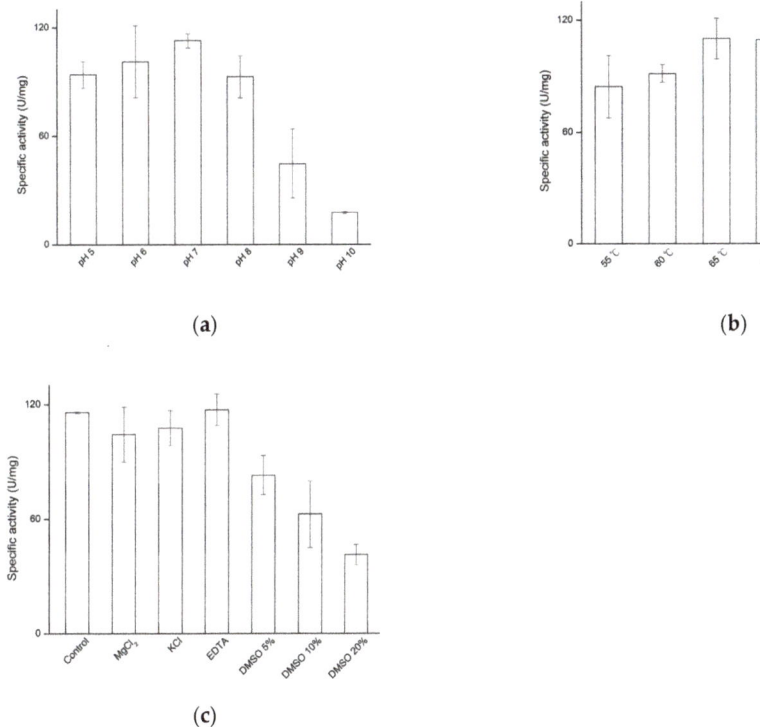

Figure 3. Effects of pH (**a**), temperature (**b**), and metal ion or dimethyl sulfoxide (DMSO) (**c**) on the hydrolytic activity of PgMA. The reaction was performed with 1% (w/v) β-cyclodextrin (CD), 5.6 µg/mL of PgMA, and 50 mM of a different buffer at the tested temperature in the absence or presence of the tested metal ion or DMSO for 30 min. After the reaction was stopped by boiling, the hydrolytic activity of PgMA was determined by measuring the reducing sugars produced from the reaction as described in Section 2.

One of the specific features of MA is that the enzyme prefers CD as its substrate over other polysaccharides, such as starch or pullulan. To characterize the PgMA hydrolysis activity, different polysaccharides were used as a substrate for the PgMA catalytic reaction. The results showed that PgMA exhibited almost equally high specific activities toward α-CD, γ-CD, and β-CD. The specific activities of PgMA toward CD were 65- and 650-fold higher than those toward pullulan and starch, respectively (Figure 4). This CD preference was consistent with other known MAs [20–27]. The 130 residues at the N-terminal of the MA are key for the enzymes to form dimers and largely increase hydrolysis activities toward CD [18,19]. In addition, some recombinant MAs have been purified using N-terminal His-tag fusion [18–27]. The N-terminal His-tag fusion did not seem to affect its dimerization; herein, the recombinant PgMA also remained as the CD preference.

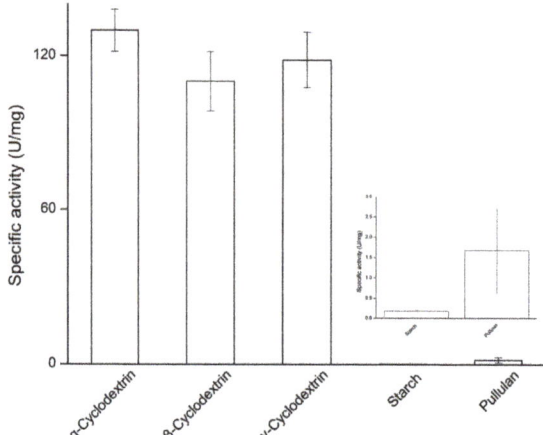

Figure 4. Substrate specificity of the hydrolytic activity of PgMA. The reaction was performed with 1% (w/v) of the tested sugar substrate, 5.6 µg/mL PgMA, and 50 mM of PB at pH 7 and 65 °C for 30 min. The hydrolytic activity of PgMA was determined as described in the legend of Figure 3.

3.4. Determination of Transglycosylation Activity by Recombinant PgMA

Transglycosylation activity is an important property of MAs for biotechnology applications. To clarify the transglycosylation activity of recombinant PgMA, 17 different molecules, including mangiferin (Table S2), belonging to triterpenoids, saponins, flavonoids, flavonoid glycosides, or xanthone glycoside, were used as sugar acceptors with 1% (w/v) β-CD (as the sugar donor) for activity. The reaction mixture was then analyzed with HPLC. The results showed that only puerarin (Figure 5a) and mangiferin (Figure 5b) could be glycosylated by PgMA.

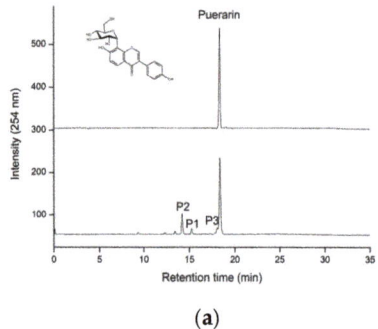

(a)

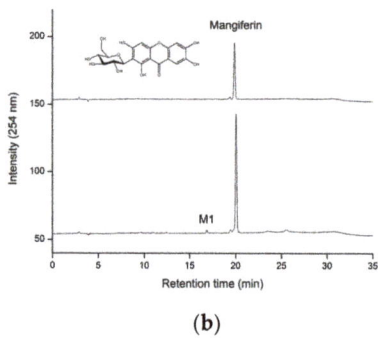

(b)

Figure 5. High-performance liquid chromatography (HPLC) analysis of the biotransformation products of puerarin (a) and mangiferin (b) by PgMA. The reaction was performed with 1% (w/v) β-CD, 5.6 µg/mL of PgMA, and 1 mg/mL of puerarin or mangiferin at 50 mM of PB (pH 7) and 65 °C for 24 h. After the reaction, the reaction mixture was analyzed with HPLC. The conditions for HPLC are described in Section 2.

Except mangiferin and puerarin, the other four tested triterpenoids, two triterpenoids saponins, nine flavonoid aglycones, and glycosides could not act as the sugar acceptors in the transglycosylation of PgMA. PgMA could transglycosylate puerarin, which has the isoflavone-8-C-glucosdie structure. However, PgMA could not transglycosylate isoflavone-7-O-glucoside (8-hydroxydaidzein-7-α-O-glucoside) or flavone-8-C-glucoside (vitexin). The results imply that PgMA has a narrow and/or specific substrate range. Nevertheless, the main finding is that *Pg*MA can glycosylate mangiferin, which will expand the biotechnological applications of MAs in the future. MAs have been proven to glycosylate some small molecules, such as hydroquinone [29], caffeic acid [30], ascorbic acid [31], puerarin [32–34], genistin [35], neohesperidin [36], and naringin [37]. Our results also showed that *Pg*MA glycosylated puerarin to three major products, P1, P2, and P3 (Figure 5a). These three major products were not identified in advance because the glycosylation of puerarin has been studied based only on known MAs [32,34]. Li et al. (2004) reported that *Bs*MA glycosylated puerarin to three products (T1, T2, and T3), two of which were identified as maltosyl-α-(1→6)-puerarin (T1) and glucosyl-α-(1→6)-puerarin (T2), while T3 was not identified [34]. Li et al. (2011) further reported that a maltogenic amylase (*Tf*MA) from the archaeon *T. pendens* glycosylated puerarin to a series of products containing glucosyl puerarin and maltosyl puerarin, although they did not identify the exact chemical structures of the products [32]. From the results of the two studies, the P1–P3 products might contain glucosyl and maltosyl puerarin.

The results revealed that *Pg*MA glycosylates mangiferin to produce low amounts of the M1 compound with a yield of 2.3% (Figure 5b). This is the first study to report that MA could glycosylate mangiferin, of which mangiferin glycoside may have better aqueous solubility for different applications. Therefore, we mainly focused on the unknown mangiferin glycoside by *Pg*MA in the following assays.

3.5. Optimization of Biotransformation of Mangiferin by PgMA

As the yield of the M1 compound from the biotransformation by *Pg*MA is too low to be easily purified, the glycosylation condition must be optimized. The glycosylation reaction was optimized with different sugar donors, concentrations of sugar donors, and reaction times. First, although the yield of the M1 using α-CD showed the best output (Figure 6a), maltodextrin was selected as the sugar donor for experiments due to its highly aqueous solubility at a much lower price.

Second, the GH enzymes contained both hydrolysis and transglycosylation activities. The transglycosylation activity of GH has been reported to increase under low water concentrations [43]. Thus, a solution with a higher sugar concentration and lower water concentration would increase its transglycosylation activity. The transglycosylation activity of *Pg*MA was indeed increased in 50% maltodextrin, the highest soluble concentration. When the maltodextrin concentration was increased from 1% to 50% (w/v), three compounds (M1, M2, and M3 in Figure 6b) were formed with higher yields (Figure 6c). The maximal yields of M1 and M2 reached 10% and 21%, respectively. The M3 compound was not completely separated with mangiferin, which is similar to the situation of the T3 compound with puerarin by *Bs*MA [34]. Therefore, only M1 and M2 were further studied.

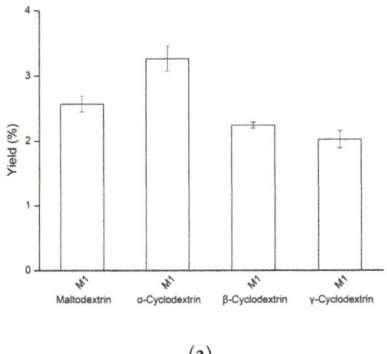

(a)

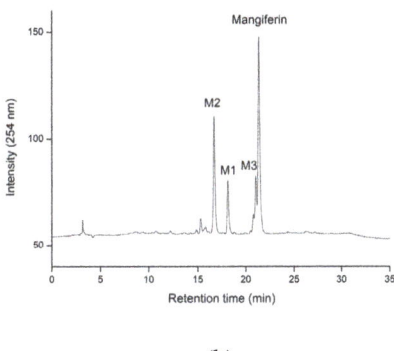

(b)

Figure 6. *Cont.*

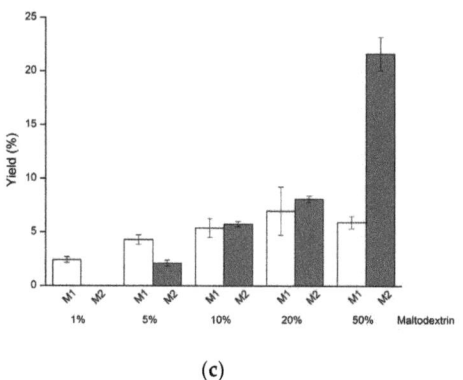

(c)

Figure 6. Effects of the sugar donor on the glycosylation of mangiferin by PgMA. (**a**) Different sugar donors were used in the glycosylation of mangiferin. (**b**) HPLC analysis of the glycosylation mixture of mangiferin by PgMA with 50% (w/v) maltodextrin. (**c**) Different maltodextrin concentrations were used in the glycosylation of mangiferin. The reaction was performed with 5.6 µg/mL PgMA, 1 mg/mL mangiferin, and the tested sugar donors at 50 mM of PB (pH 7) and 65 °C for 24 h. After the reaction, the reaction mixture was analyzed with HPLC. The conditions for HPLC are described in Section 2. The yield of the product was calculated by dividing the area of the product by that of mangiferin without an enzyme reaction (control reaction).

Third, the yields of M1 and M2 in 50% maltodextrin by PgMA were further determined under different time courses (Figure 7). The results showed that the yields plateau of M1 and M2 reached 13.2% at 168 h and 33.8% at 72 h, respectively.

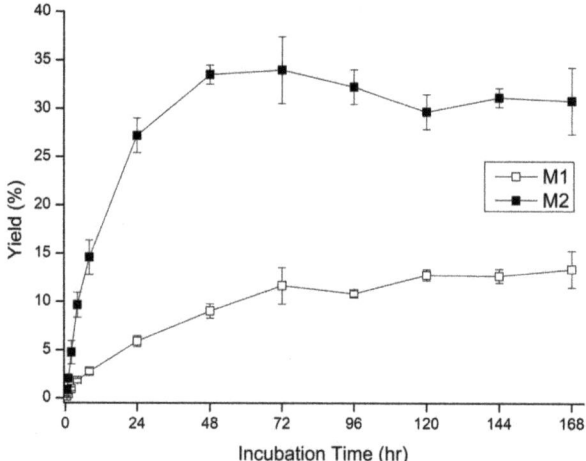

Figure 7. Time course of the glycosylation of mangiferin by PgMA. The reaction was conducted with 50% (w/v) maltodextrin, 5.6 µg/mL PgMA, and 1 mg/mL mangiferin at 50 mM of PB (pH 7) and 65 °C. At the interval time, the reaction mixture was analyzed with HPLC. The conditions for HPLC and the calculation of the yield were the same as those described in the legend to Figure 6.

3.6. Isolation and Identification of Mangiferin Glycosides by PgMA

The glycosylation of mangiferin by PgMA was scaled up to 100 mL. The products M1 and M2 were purified using preparative HPLC. From the 100 mL reaction, 20.1 mg of compound (**1**) (M1) and 9.3 mg of compound (**2**) (M2) were isolated. The molecular weights of the purified products were then determined with mass spectrometry. The mass spectrometry of compound (1) revealed an [M–H]$^-$ ion peak at *m/z*: 583.4 in the electrospray ionization mass spectrum (ESI-MS) corresponding to the molecular formula $C_{25}H_{28}O_{16}$ (Figure S1). The mass data imply that M1 contains one glucosyl moiety attached to the mangiferin structure. In the mass data of M2, an [M–H]$^-$ ion peak at *m/z*: 745.3 in the ESI-MS corresponded to the molecular formula $C_{31}H_{38}O_{21}$ (Figure S2), which implies that compound (2) contains two glucosyl moieties attached to the mangiferin structure. To identify the structures in advance, the structures of both compounds were determined using NMR spectroscopy. ^{1}H and ^{13}C NMR, including the DEPT, HSQC, HMBC, COSY, and NOESY spectra, were obtained.

The characteristic ^{1}H and ^{13}C NMR sugar signals in compound (1) were assigned to *C*-glucosyl and *O*-glucosyl moieties by one-dimensional (1-D) and 2-D NMR experiments. The ^{1}H spectrum of compound (1) in DMSO-d_6 showed three singlets at 6.36, 6.86, and 7.37 ppm and a complex 10-spin system between 3.0 and 5.0 ppm. Analysis of this second-order system revealed coupling constants typical of two glucose moieties. The compound (1) glucosidic linkage of the *C*-glucosyl moiety on the xanthone C-2 was revealed by the presence of HMBC correlations between C-2/H-1′ (107.5/4.58 ppm), and the anomeric proton H-1′ at 4.58 (d, *J* = 9.1 Hz) indicated a *C*-β- configuration of mangiferin that was confirmed by the data reported in the literature [44]. The mangiferin *O*-glucosyl moiety was a doublet signal at H-1″ (4.73 ppm, d, *J* = 4.2 Hz) with the corresponding carbon atom at C-1″ (98.7 ppm) assigned to the anomeric proton and indicating an *O*-α-configuration by HSQC, which is in the *O*-α-configuration. The H-1″ (δ = 4.73 ppm) of mangiferin and the HMBC cross signaled H-1″/C-6′ (4.73/66.9 ppm) and H-6′a, 6′b/C-1″ (3.46, 3.52/98.7 ppm). The significant downfield shift of the ^{13}C signal of C-6′ indicated the connection of the second glucosyl moiety, which confirmed the α-(1→6) linkage of the second glucosyl moiety. The NMR signals were identified as shown in Table S3. The compound (1) was thus confirmed as glucosyl-α-(1→6)-mangiferin (Figures S3–S9).

The ^{1}H spectrum of compound (2) in the same compound (1) solvent also showed three singlets at 6.36, 6.86, and 7.37 ppm and a complex 11-spin system between 3.0 and 5.0 ppm. Analysis of this second-order system revealed coupling constants typical of three glycose moieties, which included the chemical shifts listed in Table S3. The glucosyl moiety chemical shifts of C-2 at 107.4 ppm and H-1′ at 4.58 ppm (d, *J* = 9.8 Hz) according to the corresponding HMBC indicated a C-C bond between the sugar and the aglycone of mangiferin (*C*-glucosyl-xanthone) and were confirmed by the data reported in the literature [44]. The *O*-maltosyl moiety connected to mangiferin was confirmed by HMBC from the anomeric carbon C-6′ (66.9 ppm), and the corresponding anomeric proton H-1″ at 4.75 (d, *J* = 3.5 Hz) indicated an *O*-α-configuration. The maltose doublet signal at $δ_H$ H-1″ (d, *J* = 3.5 Hz) and H-1‴ 4.95 (d, *J* = 3.5 Hz) with the corresponding carbon atom at C-1″ (98.6 ppm) and C-1‴ (100.9 ppm) was assigned to the anomeric proton and indicated two *O*-α-configurations by HSQC. The HMBC cross peaks of C-1″/H-6′ (98.6/3.64, 3.71 ppm) and C-1‴/H-4″ (100.9/3.34 ppm) confirmed the α-(1→4) between the two-glucosyl moiety. Our experimental ^{1}H and ^{13}C chemical shifts listed in Table S2 confirmed compound (2) as maltosyl-α-(1→6)-mangiferin (Figures S10–S16). Figure 8 summarizes the biotransformation process of mangiferin by *Pg*MA.

3.7. Characterizations of Mangiferin Glucosides

The low aqueous solubility of mangiferin restricts its usage as a pharmaceutical agent. However, the glycosylation of mangiferin may mitigate such a restriction. Only a few studies have reported different glycosylation agents for mangiferin. Wu et al. (2013) used β-fructofuranosidase (E.C. 3.2.1.26; GH 32 family) to glycosylate mangiferin into fructosyl-β-(2→6)-mangiferin and found that its DPPH radical scavenging activity was similar to that of mangiferin [45]. Nguyen et al. (2020) used dextransucrase (E.C. 2.4.1.5; GH 70 family) from *Leuconostoc mesenteroides* to glycosylate mangiferin into glucosyl-α-(1→6)-mangiferin (1) [46]. They found that the aqueous solubility of glucosyl-α-(1→6)-mangiferin (1) was 2300-fold higher than that of mangiferin. In this study, the amount of purified glucosyl-α-

(1→6)-mangiferin (**1**) was too low to repeat the solubility experiment. The solubility of the newly identified maltosyl-α-(1→6)-mangiferin (**2**) was determined. The results showed that the solubility of maltosyl-α-(1→6)-mangiferin (**2**) was 5500-fold higher than that of mangiferin (Table 1). Thus, maltosyl-α-(1→6)-mangiferin (**2**) possesses higher solubility than glucosyl-α-(1→6)-mangiferin (**1**). It has been reported that the more sugar moieties in the glycosylated compounds, the higher the aqueous solubility of the glycosylated compounds [29–37].

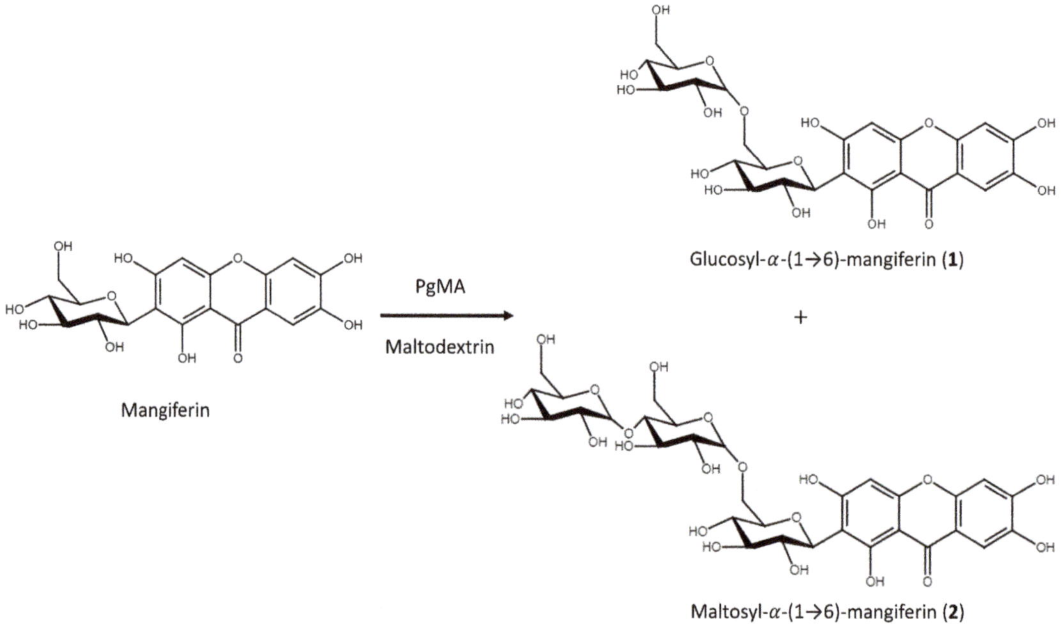

Figure 8. Biotransformation process of mangiferin by *Pg*MA.

Table 1. Aqueous solubility of mangiferin and maltosyl-α-(1→6)-mangiferin (**2**).

Compound	Aqueous Solubility (mg/L) [1]	Fold [2]
Mangiferin	92.2 ± 4.60	1.0
Maltosyl-α-(1→6)-mangiferin (**2**)	$5.11 \times 10^5 \pm 2.64 \times 10^3$	5.5×10^3

[1] The mean (n = 2) is shown, and the standard deviations are represented by error bars. [2] The folds of the aqueous solubilities of the mangiferin glucoside derivatives are expressed as relative to that of mangiferin normalized to 1.

Mangiferin exhibits a wide pharmacological profile, and its antioxidant property is well known from previous studies. Furthermore, it has been associated with the redox aromatic system of the xanthone nucleus [3–10,12]. Thus, the antioxidative activities of the two mangiferin glucosides were determined using DPPH free radical scavenging assay. The assay showed that the antioxidant activity levels of mangiferin and its two glucosides were all higher than those of ascorbic acid (Figure 9). In other words, the antioxidant activities of the two mangiferin glucosides are comparable with those of mangiferin. The *ortho*-dihydroxyl groups on the benzene ring of the mangiferin structure have been reported to play a key role in exerting its antioxidant activity [1,2]. Both mangiferin glucosides remained the key functional groups after glycosylation; therefore, most of the antioxidant activity remained in the mangiferin derivatives. These glycoside derivatives (glucoside and fructoside) might possess different pharmacological properties. A futher study will focus on the bioactivities and bioavailability of these mangiferin derivatives

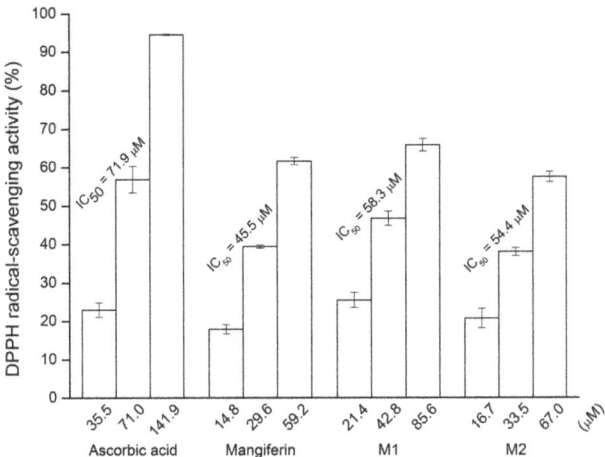

Figure 9. The 1,1-diphenyl-2-picrylhydrazyl (DPPH) free radical scavenging activity of mangiferin, mangiferin glucosides, and ascorbic acid. The DPPH scavenging activity was determined as described in Section 2. The IC$_{50}$ values represent the concentrations required for 50% DPPH free radical scavenging activity. The mean (n = 3) is shown, and the standard deviations are represented by error bars. M1 and M2 are glucosyl-α-(1→6)-mangiferin and maltosyl-α-(1→6)-mangiferin, respectively.

4. Conclusions

The recombinant *Pg*MA from *P. galactosidasius* DSM 18751^T was confirmed to exhibit the bifunctions of hydrolysis and transglycosylation activities. The recombinant *Pg*MA can glycosylate mangiferin and produce glucosyl-α-(1→6)-mangiferin and maltosyl-α-(1→6)-mangiferin with a high maltodextrin concentration. The novel maltosyl-α-(1→6)-mangiferin showed much higher aqueous solubility than that of mangiferin. The two mangiferin glucosides exhibited similar DPPH antioxidative activity compared to mangiferin. To our knowledge, *Pg*MA is the first MA identified with glycosylation activity toward mangiferin. With higher water solubility and compatible antioxidant activity, the two mangiferin glucoside derivatives have better pharmaceutical applicability.

Supplementary Materials: The following materials are available online at https://www.mdpi.com/article/10.3390/antiox10111817/s1. Table S1: Top five candidates with best-hit of *Bs*MA from NCBI GenBank. Table S2: The list of the tested molecules in the transglycosylation reaction of *Pg*MA. Table S3: ^{1}H and ^{13}C NMR assignments in DMSO-d_6 at 700 and 175 MHz for compounds (**1**) and (**2**). Figure S1: Mass–mass analysis of glucosyl-α-(1→6)-mangiferin (**1**) in the negative mode. Figure S2: Mass–mass analysis of maltosyl-α-(1→6)-mangiferin (**2**) in the negative mode. Figure S3: One-dimensional NMR spectrum (^{1}H-NMR, 700 MHz, DMSO-d_6) of the glucosyl-α-(1→6)-mangiferin (**1**). Figure S4: One-dimensional NMR spectrum (^{13}C-NMR, 175 MHz, DMSO-d_6) of the glucosyl-α-(1→6)-mangiferin (**1**). Figure S5: One-dimensional NMR spectrum (DEPT-135, 175 MHz, DMSO-d_6) of the glucosyl-α-(1→6)-mangiferin (**1**). Figure S6: Two-dimensional NMR spectrum (^{1}H-^{13}C HSQC, 700 MHz, DMSO-d_6) of the glucosyl-α-(1→6)-mangiferin (**1**). Figure S7: Two-dimensional NMR spectrum (^{1}H-^{13}C HMBC, 700 MHz, DMSO-d_6) of the glucosyl-α-(1→6)-mangiferin (**1**). Figure S8: Two-dimensional NMR spectrum (^{1}H-^{1}H COSY, 700 MHz, DMSO-d_6) of the glucosyl-α-(1→6)-mangiferin (**1**). Figure S9: Two-dimensional NMR spectrum (^{1}H-^{1}H NOESY, 700 MHz, DMSO-d_6) of the glucosyl-α-(1→6)-mangiferin (**1**). Figure S10: One-dimensional NMR spectrum (^{1}H-NMR, 700 MHz, DMSO-d_6) of the maltosyl-α-(1→6)-mangiferin (**2**). Figure S11: One-dimensional NMR spectrum (^{13}C-NMR, 175 MHz, DMSO-d_6) of the maltosyl-α-(1→6)-mangiferin (**2**). Figure S12: One-dimensional NMR spectrum (DEPT-135, 175 MHz, DMSO-d_6) of the maltosyl-α-(1→6)-mangiferin (**2**). Figure S13: Two-dimensional NMR spectrum (^{1}H-^{13}C HSQC, 700 MHz, DMSO-d_6) of the maltosyl-α-(1→6)-mangiferin (**2**). Figure S14: Two-dimensional NMR spectrum (^{1}H-^{13}C HMBC, 700 MHz, DMSO-d_6) of the maltosyl-α-(1→6)-mangiferin (**2**). Figure S15: Two-dimensional NMR spectrum (^{1}H-

^{1}H COSY, 700 MHz, DMSO-d_6) of the maltosyl-α-(1→6)-mangiferin (**2**). Figure S16: Two-dimensional NMR spectrum (^{1}H-^{1}H NOESY, 700 MHz, DMSO-d_6) of the maltosyl-α-(1→6)-mangiferin (**2**).

Author Contributions: Conceptualization: T.-S.C.; data curation and methodology: Y.-L.T., T.-S.C. and H.-Y.D.; project administration: T.-S.C. and J.-Y.W.; writing—original draft, review, and editing: T.-S.C., T.-Y.W., J.-Y.W., H.-J.T. and H.-Y.D. All authors have read and agreed to the published version of the manuscript.

Funding: This research was funded by the Ministry of Science and Technology of Taiwan under grant number MOST 110-2221-E-024-002 to T.-S.C. and grant number MOST 110-2221-E-507-002 to J.-Y.W.

Institutional Review Board Statement: Not applicable.

Informed Consent Statement: Not applicable.

Data Availability Statement: The data presented in this study are available in the article or supplementary material.

Conflicts of Interest: The authors declare no conflict of interest.

References

1. Imran, M.; Arshad, M.S.; Butt, M.S.; Kwon, J.H.; Arshad, M.U.; Sultan, M.T. Mangiferin: Aa natural miracle bioactive compound against lifestyle related disorders. *Lipids Health Dis.* **2017**, *16*, 84. [CrossRef] [PubMed]
2. Du, S.; Liu, H.; Lei, T.; Xie, X.; Wang, H.; He, X.; Tong, R.; Wang, Y. Mangiferin: Anan effective therapeutic agent against several disorders (Review). *Mol. Med. Rep.* **2018**, *18*, 4775–4786. [CrossRef] [PubMed]
3. Gold-Smith, F.; Fernandez, A.; Bishop, K. Mangiferin and cancer: Mechanismsmechanisms of action. *Nutrients* **2016**, *8*, 396. [CrossRef]
4. Morozkina, S.N.; Nhung Vu, T.H.; Generalova, Y.E.; Snetkov, P.P.; Uspenskaya, M.V. Mangiferin as new potential anti-cancer agent and mangiferin-integrated polymer systems—a novel research direction. *Biomolecules* **2021**, *11*, 79. [CrossRef] [PubMed]
5. Saha, S.; Sadhukhan, P.; Sil, P.C. Mangiferin: Aa xanthonoid with multipotent anti-inflammatory potential. *Biofactors* **2016**, *42*, 459–474. [CrossRef] [PubMed]
6. Garrido-Suarez, B.B.; Garrido, G.; Pineros, O.; Delgado-Hernandez, R. Mangiferin: Possiblepossible uses in the prevention and treatment of mixed osteoarthritic pain. *Phytother. Res.* **2020**, *34*, 505–525. [CrossRef]
7. Lum, P.T.; Sekar, M.; Gan, S.H.; Pandy, V.; Bonam, S.R. Protective effect of mangiferin on memory impairment: Aa systematic review. *Saudi. J. Biol. Sci.* **2021**, *28*, 917–927. [CrossRef] [PubMed]
8. Feng, S.T.; Wang, Z.Z.; Yuan, Y.H.; Sun, H.M.; Chen, N.H.; Zhang, Y. Mangiferin: Aa multipotent natural product preventing neurodegeneration in Alzheimer's and Parkinson's disease models. *Pharmacol. Res.* **2019**, *146*, 104336. [CrossRef]
9. Zhang, L.; Huang, C.; Fan, S. Mangiferin and organ fibrosis: Aa mini review. *Biofactors* **2021**, *47*, 59–68. [CrossRef] [PubMed]
10. Naraki, K.; Rezaee, R.; Mashayekhi-Sardoo, H.; Hayes, A.W.; Karimi, G. Mangiferin offers protection against deleterious effects of pharmaceuticals, heavy metals, and environmental chemicals. *Phytother. Res.* **2021**, *35*, 810–822. [CrossRef]
11. Ma, H.; Chen, H.; Sun, L.; Tong, L.; Zhang, T. Improving permeability and oral absorption of mangiferin by phospholipid complexation. *Fitoterapia* **2014**, *93*, 54–61. [CrossRef]
12. Tian, X.; Gao, Y.; Xu, Z.; Lian, S.; Ma, Y.; Guo, X.; Hu, P.; Li, Z.; Huang, C. Pharmacokinetics of mangiferin and its metabolite—Norathyriol, Part 1: Systemic evaluation of hepatic first-pass effect in vitro and in vivo. *Biofactors* **2016**, *42*, 533–544. [CrossRef] [PubMed]
13. Zhao, J.; Yang, J.; Xie, Y. Improvement strategies for the oral bioavailability of poorly water-soluble flavonoids: Anan overview. *Int. J. Pharm.* **2019**, *570*, 118642. [CrossRef] [PubMed]
14. Fu, J.; Wu, Z.; Zhang, L. Clinical applications of the naturally occurring or synthetic glycosylated low molecular weight. *Prog. Mol. Biol. Transl. Sci.* **2019**, *163*, 487–522. [PubMed]
15. Hofer, B. Recent developments in the enzymatic O-glycosylation of flavonoids. *Appl. Microbiol. Biotechnol.* **2016**, *100*, 4269–4281. [CrossRef] [PubMed]
16. Moulis, C.; Guieysse, D.; Morel, S.; Severac, E.; Remaud-Simeon, M. Natural and engineered transglycosylases: Green tools for the enzyme-based synthesis of glycoproducts. *Curr. Opin. Chem. Biol.* **2021**, *61*, 96–106. [CrossRef] [PubMed]
17. Cantarel, B.; Coutinho, P.M.; Rancurel, C.; Bernard, T.; Lombard, V.; Henrissat, B. The carbohydrate-active enzymes database (CAZy): An expert resource for glycogenomics. *Nucleic Acids Res.* **2009**, *37*, D233–D238. [CrossRef] [PubMed]
18. Park, K.H.; Kim, T.J.; Cheong, T.K.; Kim, J.W.; Oh, B.H.; Svensson, B. Structure, specificity and function of cyclomaltodetrinase, a multispecific enzyme of the α-amylase family. *Biochim. Biophy. Acta.* **2000**, *1478*, 165–185. [CrossRef]
19. Mehta, D.; Satyanarayana, T. Dimerization mediates thermo-adaptation, substrate affinity and transglycosylation in a highly thermostable maltogenic amylase of *Geobacillus thermoleovorans*. *PLoS ONE* **2013**, *8*, e73612.
20. Cha, H.J.; Yoon, H.G.; Kim, Y.W.; Lee, H.S.; Kim, J.W.; Kweon, K.S.; Oh, B.H.; Park, K.H. Molecular and enzymatic characterization of a maltogenic amylase that hydrolyzes and transglycosylates acarbose. *Eur. J. Biochem.* **1998**, *253*, 251–262. [CrossRef] [PubMed]

21. Li, X.; Li, D.; Yin, Y.; Park, K.H. Characterization of a recombinant amylolytic enzyme of hyperthermophilic archaeon *Thermofilum pendens* with extremely thermostable maltogenic amylase activity. *Appl. Microbiol. Biotechnol.* **2010**, *85*, 1821–1830. [CrossRef] [PubMed]
22. Li, D.; Park, J.T.; Li, X.; Kim, S.; Lee, S.; Shim, J.H.; Park, S.H.; Cha, J.; Lee, B.H.; Kim, J.W.; et al. Overexpression and characterization of an extremely thermostable maltogenic amylase, with an optimal temperature of 100°C, from the hyperthermophilic archaeon *Staphylothermus marinus*. *New Biotechnol.* **2010**, *27*, 300–307. [CrossRef] [PubMed]
23. Zhoui, J.; Li, Z.; Zhang, H.; Wu, J.; Ye, X.; Dong, W.; Jiang, M.; Huang, Y.; Cui, Z. Novel maltogenic amylase CoMA from *Corallococcus* sp. strain EGB catalyzes the conversion of maltooligosaccharides and soluble starch to maltose. *Appl. Envir. Microbiol.* **2018**, *84*, e00152-18. [CrossRef] [PubMed]
24. Manas, N.H.B.; Pachelles, S.; Mahadi, N.M.; Illias, R.M. The characterization of an alkali-stable maltogenic amylase from *Bacillus lehensis* G1 and improved malto-oligosaccharide production by hydrolysis suppression. *PLoS ONE* **2014**, *9*, e106481.
25. Cheong, K.A.; Tang, S.Y.; Cheong, T.K.; Cha, H.; Kim, J.W.; Park, K.H. Thermostable and alkalophilic maltogenic amylase of *Bacillus thermoalkalophilus* ET2 in monomer-dimer equilibrium. *Biocat. Biotrans.* **2005**, *23*, 79–87. [CrossRef]
26. Kim, T.J.; Kim, M.J.; Kim, B.C.; Kim, J.C.; Cheong, T.K.; Kim, J.W.; Park, K.H. Modes of action of acaarbose hydrolysis and transglycosylation catalyzed by a thermostable maltogenic amylase, the gene for which was cloned from a Thermus strain. *Appl. Environ. Microbiol.* **1999**, *65*, 1644–1651. [CrossRef] [PubMed]
27. Kim, I.C.; Cha, J.H.; Kim, J.R.; Jang, S.Y.; Seo, B.C.; Cheong, T.K.; Lee, D.S.; Cho, Y.D.; Park, K.H. Catalytic properties of the cloned amylase from *Bacillus licheniformis*. *J. Biol. Chem.* **1992**, *267*, 22108–22114. [CrossRef]
28. Ruan, Y.; Xu, Y.; Zhang, W.; Zhang, R. A new maltogenic amylase from *Bacillus licheniformis* R-53 significantly improves bread quality and extends shelf life. *Food Chem.* **2021**, *344*, 128599. [CrossRef]
29. Nishimura, T.; Kometani, T.; Takii, H.; Terada, Y.; Okada, S. Purification and some properties of a-amylase from *Bacillus subtilis* X-23 that glucosylates phenolic compounds such as hydroquinone. *J. Fer. Bioeng.* **1994**, *78*, 31–36. [CrossRef]
30. Nishimura, T.; Kometani, T.; Takii, H.; Terada, Y.; Okada, S. Glucosylation of caffeic acid with *Bacillus subtilis* X-23 a-amylase and a description of the glucosides. *J. Fer. Bioeng.* **1995**, *80*, 18–23. [CrossRef]
31. Lee, S.B.; Nam, K.C.; Lee, S.J.; Lee, J.H.; Inouye, K.; Park, K.H. Antioxidative effects of glycosyl-ascorbic acids synthesized by maltogenic amylase to reduce lipid oxidation and volatiles production in cooked chicken meat. *Biosci. Biotechnol. Biochem.* **2004**, *68*, 36–43. [CrossRef] [PubMed]
32. Li, X.; Li, D.; Park, S.-H.; Gao, C.; Park, K.-H.; Gu, L. Identification and antioxidative properties of transglycosylated puerarins synthesised by an archaeal maltogenic amylase. *Food Chem.* **2011**, *124*, 603–608. [CrossRef]
33. Choi, C.H.; Kim, S.H.; Jang, J.H.; Park, J.T.; Shim, J.H.; Kim, Y.W.; Park, K.H. Enzymatic synthesis of glycosylated puerarin using maltogenic amylase from *Bacillus stearothermophilus* expressed in *Bacillus subtilis*. *J. Sci. Food. Agric.* **2010**, *90*, 1179–1184. [CrossRef]
34. Li, D.; Park, S.H.; Shim, J.H.; Lee, H.S.; Tang, S.Y.; Park, C.S.; Park, K.H. In vitro enzymatic modification of puerarin to puerarin glycosides by maltogenic amylase. *Carbohydr Res.* **2004**, *339*, 2789–2797. [CrossRef]
35. Li, X.; Wang, Y.; Park, J.T.; Gu, L.; Li, D. An extremely thermostable maltogenic amylase from *Staphylothermus marinus*: Bacillus expression of the gene and its application in genistin glycosylation. *Int. J. Biol. Macromol.* **2018**, *107*, 413–417. [CrossRef] [PubMed]
36. Cho, J.S.; Yoo, S.S.; Cheong, T.K.; Kim, M.J.; Kim, Y.; Park, K.H. Transglycosylation of neohesperidin dihydrochalcone by *Bacillus stearothermophilus* maltogenic amylase. *J. Agric. Food Chem.* **2000**, *48*, 152–154. [CrossRef] [PubMed]
37. Lee, S.J.; Kim, J.C.; Kim, M.J.; Kitaoka, M.; Park, C.S.; Lee, S.Y.; Ra, M.J.; Moon, T.W.; Robyt, J.F.; Park, K.H. Transglycosylation of naringin by *Bacillus stearothermophilus* maltogenic amylase to give glycosylated naringin. *J. Agric. Food Chem.* **1999**, *47*, 3669–3674. [CrossRef] [PubMed]
38. Kumar, S.; Stecher, G.; Li, M.; Knyaz, C.; Tamura, K. MEGA X: Molecular evolutionary genetics analysis across computing platforms. *Mol. Biol. Evol.* **2018**, *35*, 1547–1549. [CrossRef] [PubMed]
39. Bradford, M.M. A rapid and sensitive method for the quantitation of microgram quantities of protein utilizing the principle of protein-dye binding. *Anal. Biochem.* **1976**, *72*, 248–254. [CrossRef]
40. Bernfeld, P. Amylases alpha and beta. *Methods Enzymol.* **1995**, *1*, 140–146.
41. Chang, T.-S.; Chiang, C.-M.; Wang, T.-Y.; Tsai, Y.-L.; Wu, Y.-W.; Ting, H.-J.; Wu, J.-Y. One-pot bi-enzymatic cascade synthesis of novel *Ganoderma* triterpenoid saponins. *Catalysts* **2021**, *11*, 580. [CrossRef]
42. Chang, T.S.; Wang, T.Y.; Chiang, C.M.; Lin, Y.J.; Chen, H.L.; Wu, Y.W.; Ting, H.J.; Wu, J.Y. Biotransformation of celastrol to a novel, well-soluble, low-toxic and anti-oxidative celastrol-29-O-beta-glucoside by *Bacillus* glycosyltransferases. *J. Biosci. Bioeng.* **2021**, *131*, 176–182. [CrossRef]
43. Abdul Manas, N.H.; Md Illias, R.; Mahadi, N.M. Strategy in manipulating transglycosylation activity of glycosyl hydrolase for oligosaccharide production. *Crit. Rev. Biotechnol.* **2018**, *38*, 272–293. [CrossRef] [PubMed]
44. Talamond, P.; Mondolot, L.; Gargadennec, A.; Kochko, A.D.; Hamon, S.; Fruchier, A.; Campa, C. First report on mangiferin (C-glucosyl-xanthone) isolated from leaves of a wild coffee plant, *Coffea pseudozanguebariae* (Rubiaceae). *Acta Bot. Gallica.* **2008**, *155*, 513–519. [CrossRef]

45. Wu, X.; Chu, J.; Liang, J.; He, B. Efficient enzymatic synthesis of mangiferin glycosides in hydrophilic organic solvents. *RSC Advances* **2013**, *3*, 19027–19032. [CrossRef]
46. Nguyen, H.; Lim, S.; Lee, S.; Park, B.; Kwak, S.; Park, S.; Kim, S.B.; Kim, D. Enzymatic synthesis and biological characterization of a novel mangiferin glucoside. *Enz. Microbial Technol.* **2020**, *134*, 109479.

Article

The Fine-Tuned Release of Antioxidant from Superparamagnetic Nanocarriers under the Combination of Stationary and Alternating Magnetic Fields

Lucija Mandić, Anja Sadžak, Ina Erceg, Goran Baranović and Suzana Šegota *

Ruđer Bošković Institute, 10000 Zagreb, Croatia; Lucija.Mandic@irb.hr (L.M.); Anja.Sadzak@irb.hr (A.S.); Ina.Erceg@irb.hr (I.E.); Goran.Baranovic@irb.hr (G.B.)
* Correspondence: ssegota@irb.hr

Abstract: Superparamagnetic magnetite nanoparticles (MNPs) with excellent biocompatibility and negligible toxicity were prepared by solvothermal method and stabilized by widely used and biocompatible polymer poly(ethylene glycol) PEG-4000 Da. The unique properties of the synthesized MNPs enable them to host the unstable and water-insoluble quercetin as well as deliver and localize quercetin directly to the desired site. The chemical and physical properties were validated by X-ray powder diffraction (XRPD), field emission scanning electron microscopy (FE–SEM), atomic force microscopy (AFM), superconducting quantum interference device (SQUID) magnetometer, FTIR spectroscopy and dynamic light scattering (DLS). The kinetics of in vitro quercetin release from MNPs followed by UV/VIS spectroscopy was controlled by employing combined stationary and alternating magnetic fields. The obtained results have shown an increased response of quercetin from superparamagnetic MNPs under a lower stationary magnetic field and s higher frequency of alternating magnetic field. The achieved findings suggested that we designed promising targeted quercetin delivery with fine-tuning drug release from magnetic MNPs.

Keywords: release kinetics; superparamagnetism; stationary magnetic field; alternating magnetic fields; magnetic nanoparticles; quercetin

1. Introduction

Quercetin ($C_{15}H_{10}O_7$, 3,3′,4′,5,7-pentahydroxyflavone) is a major member of the flavonols, a subclass of flavonoids, natural polyphenols [1]. It is an important component of the human's daily diet and widely distributed in vegetables and fruits such as onions, tomatoes, berries, grapes, nuts, as well as in many flowers and leaves [2,3]. In addition, quercetin exhibits a wide range of biological and pharmacological activities, including antioxidant, anti-inflammatory, antibacterial, anti-anaemic, and anticarcinogenic activities [1,3–5]. Extensive studies have reported that quercetin can inhibit the proliferation of several types of cancers such as lung, prostate, breast cancer, and pancreatic tumour cells [1]. In addition, the main feature of quercetin is its antioxidant potential of OH groups in the structure that can bind to reactive oxygen species (ROS) and maintain cell viability. Quercetin has been shown to decrease the activity of antioxidant and apoptotic proteins and increase the levels of antiapoptotic proteins [6]. However, its therapeutic and clinical properties are limited due to its hydrophobic nature and low stability in the physiological medium. The problem with instability, low solubility and poor bioavailability can be successfully overcome by their loading in drug delivery systems including nanoparticles (NPs) [3,5,7,8]. The rapid growth of nanotechnology is the key to a revolutionary platform for chemical, physical, biological and mechanical properties of various materials [9,10]. There is tremendous interest in nanomaterials or NPs in the biomedical field [11]. For example, a variety of NPs is envisioned to be used in medical applications such as cancer detection, magnetic resonance imaging, cardiovascular and neurological treatment diseases, targeted drug delivery, hyperthermia, bioseparation, and gene transfer [3,9,10]. In recent years, magnetic nanoparticles (MNPs) have been highlighted among the many types of NPs [12–14]. Essentially, researchers were attracted by their excellent unique physical, chemical and magnetic properties [15]. In 1957, Gilchrist and coworkers showed the first use of magnetic particles for inductive heating of lymph nodes in dogs [16,17].

In 1983, Widder and coworkers reported the first use of MNPs containing doxorubicin to treat Yoshida rat sarcoma with an external magnetic field. These results represented a compelling advance in chemotherapy treatment, with complete cancer remission demonstrated in 77% of animals in the magnetically localized doxorubicin-magnetite microparticles [17,18]. MNPs have received much attention as target drugs that can replace traditional chemotherapy without the side effects [19]. Several inorganic magnetic nanoparticles (MNPs) have the potential to be used for drug delivery, but MNPs are the only magnetic materials approved by the Food and Drug Administration (FDA) for human use [3,19–21]. Numerous physical, biological and chemical preparation methods have been accepted for the MNP's synthesis [22]. Magnetite NPs are most commonly used in biological applications due to their unique physicochemical properties such as particle size, size distribution, shape and high surface area [23,24]. They exhibit interesting properties such as superparamagnetism, high field irreversibility and high saturation field [24–26]. Due to these properties, the superparamagnetic NPs can become magnetized when the external magnetic field is used and they do not remain magnetized when the field is turned off. The localization of drugs with MNPs in combination with an external magnetic field and their retention until the completion of therapy [26] represent a promising strategy of drug delivery with the controlled release [27]. The effectiveness of magnetic delivery systems includes the field strength, gradient, magnetic and physicochemical properties of the NPs [28]. Moreover, the main challenge of bare MNPs is to avoid their agglomeration due to their van der Waals and magnetic dipole–dipole attraction forces [13,29]. Considering their hydrophobic surface and rapid clearance from the blood through the reticuloendothelial system (RES), they are not suitable for drug delivery systems [22]. Therefore, to overcome this inconvenience, it was necessary to coat the magnetite NPs to reduce the aggregation tendency, protect their surface from oxidation, and make the particles biocompatible and stable [30].

Various polymers have been used for drug delivery, with polyethylene glycol (PEG) being the gold standard and the most commonly used polymer [31]. PEG is approved by the Food and Drug Administration (FDA) for internal use in humans and its products have been on the market for 20 years [23,31]. Since 1994, Gref and co-workers have reported on the PEG coating and demonstrated that the naked particles were removed from the liver only 5 min after injection [31]. PEG coating is extensively used in the preparation of NPs for biomedical applications due to its many advantages, such as stability in physiological media, prolonged half-life in the body, biocompatibility, and water solubility. Moreover, PEG coating prevents or reduces aggregation and confers better physical stability to drugs through steric and hydric repulsion [19,31].

In this study, we synthesized MNPs, which are known to have excellent biocompatibility and negligible toxicity [26], allowing their application in therapy. Prepared by the solvothermal method and stabilized by the widely used PEG, MNPs possess unique properties, such as colloidal stability, dispersibility, high porosity, high loading capacity, and specific morphological, thermal, and magnetic properties, especially superparamagnetism, which enable them to host the unstable or water-insoluble drugs and to direct and localize the drugs to the specific site in the tissue. The synthesized MNPs were fully characterized in terms of structural, morphological and magnetic properties. In addition, the kinetics of quercetin from the MNPs were controlled in vitro by simply varying stationary and alternating magnetic fields, resulting in fine-tuned manipulation of the released quercetin as a model drug. It should be noted that in this study cytotoxicity has been considered, but it was not a priority at this stage of our research. Regarding the measurement and result in the cytotoxicity literature of MNPs, we quite rightly expected at least the same cytotoxicity as the results obtained by Barreto et al. [3], Hua et al. [32] and Luo et al. [33]. It was shown that the system developed provides prolonged quercetin release, which is an important characteristic of targeted drug delivery systems. The enhanced quercetin release at the lower stationary magnetic field and higher frequency alternating magnetic field, together with the synergism of chemical and physical, i.e., superparamagnetic properties of MNPs, demonstrate the great potential of MNPs as a promising targeted drug delivery system with high potential for their, both therapeutic and diagnostic activity

2. Materials and Methods

2.1. Chemicals

Iron (III) chloride hexahydrate (97%) was purchased from Alfa Easar (Ottawa, ON, Canada). Ammonium acetate and polyethylene glycol (PEG, Mw = 4000 Da) were obtained from Sigma Aldrich (St. Louis, MO, USA). Ethylene glycol and ethanol (96%) were purchased from Lach-ner (Neratovice, Czech Republic). Compressed nitrogen was purchased from Messer (Bad Soden am Taunos, Germany). Silicon oil was received from Acros organics (Waltham, MA, USA). Phosphate buffered saline (PBS) buffer (PBS tablets, pH 7.4, $I_c = 150 \times 10^{-3}$ mol dm^{-3}) were purchased from Sigma Aldrich (St. Louis, MO, USA). Deionized water Millipore mili Q-H2O was used to prepare the PBS medium. Quercetin ($\geq$99%) was supplied by Lach-ner (Neratovice, Czech Republic). A molecular weight cut-off dialysis bag (MWCO, 8Kd) was purchased from Thermo Fisher Scientific (Waltham, MA, USA).

2.2. Synthesis of Bare and PEG Coated Magnetite MNPs

The modified solvothermal reaction was used for the preparation of mesoporous magnetic nanoparticles [33,34]. Briefly, 1.35 g FeCl$_3$ × 6H$_2$O, 3.85 g of CH$_3$COONH$_4$, 1.0 g PEG (M = 4 kD) and 70 mL of ethylene glycol were added to a 250 mL two-necked flask equipped with a magnetic rod. The mixture was stirred vigorously for 1 h at 160 °C with a Heidolph MR Hei-Standard mixer (Schwabach, Germany). The chemical reaction was under the protection of an inert nitrogen atmosphere to form a homogeneous brownish solution (see Figure 1). After an hour, the system was stopped and cooled to room temperature. The mixture was transferred into a Teflon-coated stainless-steel autoclave (BHL 800 Berghof, Eningen, Germany) which is connected to a temperature controller. The autoclave was heated to 200 °C and maintained for 19 h and, afterwards, it was cooled to 50 °C. To remove the solvent, MNPs were centrifugated (Universal 320 Hettich Zentrifugen, Tuttlingen, Germany) at 8000 rpm for 10 min. After separation, MNPs were washed several times with ethanol and dispersed with a shaker (IKA Shaker Vortex 1, Staufen, Germany) between each wash. Finally, MNPs were left to dry in a desiccator for further characterization. For comparison purpose and characterization of NPs, bare magnetite MNPs were resynthesized each time using the same procedure.

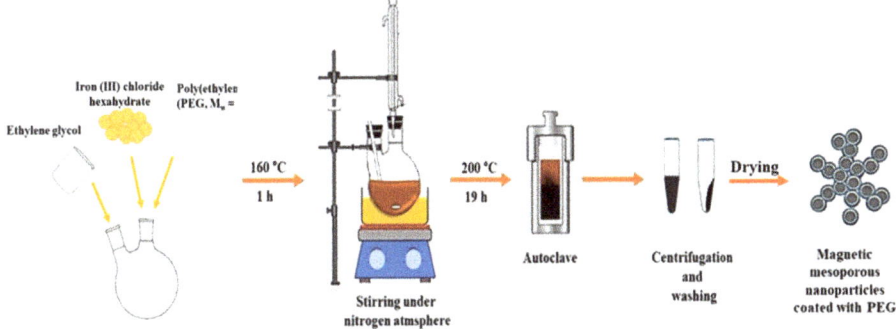

Figure 1. Schematic illustration of mesoporous MNPs preparation using solvothermal method.

2.3. Characterization of Synthesized Magnetite MNPs

The structural characteristics were determined by X-ray powder diffraction using Philips MPD 1880 diffractometer (Brooklyn, NY, USA) with monochromatic CuKα radiation (λ = 0.154 nm) at room temperature. The structural features of prepared samples were recorded at 2θ angles in the range of 10°–70° with a step of 0.02° and fixed time of 10 s per step. X-ray diffraction in polycrystalline is used to confirm crystalline size and structure of bare and coated magnetite MNPs obtained by solvothermal method. Field emission scanning electron microscope, JEOL JSM-7000F (Tokyo, Japan), was used to determine the morphology, particle size distribution and surface texture of bare and PEG-coated magnetite MNPs. FE-SEM was linked to the EDS/INCA 350 (energy dispersive X-ray analyzer) manufactured by Oxford Instruments Ltd., London, UK. The morphology of the MNPs has been further investigated using atomic force microscopy (AFM). The samples for AFM imaging were prepared by deposition of a magnetite MNPs suspension on the mica substrate. The MNPs are rinsed three times with 50 µL of MiliQ water to remove all residual impurities. AFM images were obtained by scanning the magnetite MNPs on the mica surface in the air using MultiMode Scanning Probe Microscope with Nanoscope IIIa controller (Bruker, Billerica, MA, USA) with SJV-JV-130 V ("J" scanner with vertical engagement); Vertical engagement (JV) 125 µm scanner (Bruker Instruments, Inc., Bruker, Billerica, MA, USA). Tapping mode was performed for imaging using silicon tip (R-TESPA, Bruker, Nom. Freq. 300 kHz, Nom. spring constant of 40 N/m) at 25.0 °C, allowing thermal equilibration before each imaging. AFM images were collected using random spot surface sampling (at least two areas per sample) for each analysed sample. All the images were processed by first-order flattening only and analysed using the NanoScope Analysis software (Version 5.31r1). Morphology analysis was investigated by Transmission Electron Microscope (TEM), Zeiss EM10 (Oberkochen, Germany), operated at 100 kV. In this propose, MNPs coated with PEG were dispersed in Milli-Q H_2O and bare MNPs were dispersed in ethanol, ultrasonicated and placed on carbon coated copper grids. After air drying, samples were photographed by transmission electron microscopy. Samples were analysed with ImageJ and 50 particles were counted for each image.

The MNPs mesoporosity determination has been performed using nitrogen adsorption–desorption measurements on an ASAP2020 (Micromeritics, Norcross, GA, USA) accelerated surface area analyzer at 77 K. Before measuring, the samples were degassed at reduced pressure andat 120 °C for at least 6 h. All measurements have been made in duplicate. The specific surface area, the pore volume and the pore size are determined using Brunauer–Emmett–Teller (BET) analysis. To confirm the superparamagnetic property of the synthesized MNPs, magnetization measurements were performed. Magnetization of the powder samples of MNPs was measured using a commercial Quantum Design MPMS-5 SQUID magnetometer (San Diego, CA, USA). The powder samples were placed into a small ampoule whose diamagnetic contribution was properly subtracted. In addition, the field dependence of the magnetization (M(H)), including magnetic hysteresis loops, was measured at 300 K under fields up to 10 kOe.

2.4. The Determination of Loaded Quercetin into Magnetic Mesoporous MNPs

The quercetin loading efficiency into mesoporous MNPs has been confirmed using Brunauer–Emmett–Teller (BET) analysis. The difference in the specific surface area, the pore volume and the pore size before and after immersing the MNPs into the quercetin solution confirmed the loading of quercetin into MNPs. FTIR spectra were obtained by Alpha-T FTIR Spectrometer (Bruker, Billerica, MA, USA). All spectra were recorded between 4000 and 350 cm^{-1} at a nominal resolution of 4 cm^{-1} at 25 °C, and the total number of recordings was 16. Dried samples were mixed with KBr powder and then they were pressed to produce pellets. TG analysis data were carried out on a TG/DTA simultaneous analyser DTG-60H with a 10 °C/min heating rate under a nitrogen atmosphere. The measurements were recorded in a range of room temperature up to 1200 °C.

UV/VIS spectrophotometer was used to study the quercetin loading efficiency of the synthesized MNPs. The loading efficiency was calculated by measuring the absorbance of the supernatant with the WTW photoLab® 7600 UV-VIS Spectrophotometer (Xylem, Rye Brook, NY, USA) at 374 nm. Measurements were performed in a rectangular quartz cuvette with a 10 mm optical path length and covers at 25 °C.

A total of 500 mg of quercetin was dissolved in 100 mL of ethanol and suspended in an ultrasonic bath (Bandelin Sonorex Super RK 100 H, Berlin, Germany) for 1 h at room temperature. The 25 mL aliquot of quercetin solution and 100 mg of coated MNPs were transferred into a 50 mL Falcon conical centrifuge tube. The mixture was mechanically stirred at a thermocontrol shaker (Barnstead Lab-line 4450 e-class) for 24 h at 200 rpm and 25 °C. Afterwards, the quercetin-loaded MNPs were separated from unloaded quercetin by centrifugation (Universal 320 Hettich Zentrifugen, Tuttlingen, Germany) at 8000 rpm for 15 min. Compared with quercetin concentration supernatant before adding the synthesized MNPs, the concentration loss was determined using a calibration curve in pure EtOH. The coefficient of determination was 0.9978, and the determined molar absorption coefficient of quercetin at temperature 298 K and 374 nm is 19,131 mol^{-1} cm^{-1} dm^3. The loading efficiency (LE) was calculated by measuring the absorbance of the supernatant with the Photolab 7600 UV-VIS spectrophotometer (Xylem, New York, NY, USA) at λ = 374 nm. The absorbance (λ = 374 nm) was collected and converted to concentration by using the equation from the calibration curve. Therefore, the drug loading efficiency was calculated as the following equation:

$$LE = \frac{m_{\text{embedded}}}{m_{\text{NP}}} \times 100\% \qquad (1)$$

where m_{embedded} represent the mass of quercetin incorporated in nanoparticles, and m_{NP} is the total mass of MNPs which is used for loading.

Size Distribution of Magnetic MNPs Using Dynamic and Electrophoretic Light Scattering

Hydrodynamic diameter (d_H) and zeta potential (ζ) of suspended MNPs were measured by photon correlation spectrophotometer, Zetasizer Nano ZS (Malvern, UK) with green laser (λ = 532 nm) using the M3-PALS technique. All measurements were conducted at 25 °C in PBS (pH = 7.4) buffer. The hydrodynamic diameter was determined from the peak maximum of the volume size function. The zeta potential (ζ) was calculated from the electrophoretic mobility using a Smoluchowski approximation ($f(\kappa a)$ = 1.5). The d_H values were obtained as an average value of 10 measurements, while the zeta potential values were reported as an average of 3 independent measurements. The results were collected by the Zetasizer software 6.32 (Malvern Instruments, Malvern, UK).

2.5. Release Study of Quercetin under Stationary and Alternating Magnetic Fields

The apparatus setup scheme shown in Figure 2 enables the fine-tuning of the quercetin-release kinetic profile under applied combined stationary and alternating magnetic fields. The magnetic field enforced in the experiment is a combination of stationary and alternating magnetic fields. The position of the permanent magnet fixed and stable towards the fixed magnetic coil placed within the reactor ensures the permanent and alternating magnetic fields perpendicular to each other [35]. Without going into more detail, we showed in previous work [35] that, by applying external magnetic fields, MNPs behave as Brownian particles with quasi-periodic movement that enables loaded drug molecules to became enhanced released from the MNPs. A magnetic field is an effective stimulus with deep penetration capacity. The high-frequency alternating magnetic field (HF-AMF), from 50 to 400 kHz, and low-frequency alternating magnetic field (LF-AMF), from 0.1 to 5 kHz, have been employed to trigger the release of drugs from nanocarriers [36]. In our previous work, we employed the LF-AMF to induce the flavonoid release from the magnetic aggregates [35]. To get more insights into the effect of the combination of applied permanent and AMF to drug release, we designed sophisticated HF-AMF instrument to induce increased drug release at HF-AMF having in mind Brezovitch criterion [37]. He proposed a safety limit where the product of magnetic field and amplitude frequency ($H_0 \times f$) should not exceed 4.85×10^8 A m^{-1} s^{-1} to avoid any harmful effect on the organism. In our study, the product amounts 4 mA m^{-1} s^{-1}, 1 mA m^{-1} s^{-1} and 0.3 mA m^{-1} for 100 kHz, 50 kHz and 10 kHz, respectively, indicating that we chose good frequencies and applied magnetic fields for any possible safe application to patients. In consideration of the design of our experiment, where there were relatively high frequencies of alternating magnetic fields (from 10 to 100 kHz), it was expected that the release of quercetin would be significantly enhanced by the influence of higher frequency.

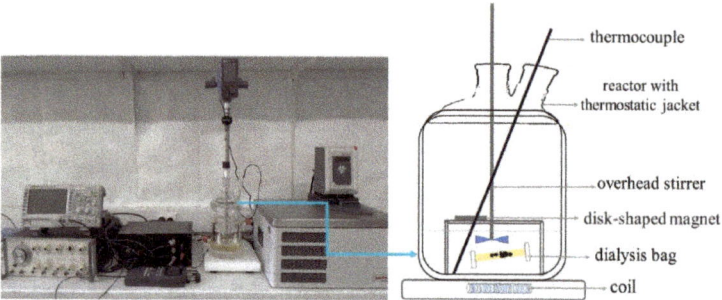

Figure 2. Experimental setup of the release kinetics and schematic illustration of a reactor.

The drug release was calculated by measuring the absorbance of quercetin released in the supernatant at λ = 330 nm by the UV-Vis spectrophotometer. The releasing kinetics of quercetin from magnetite MNPs was investigated at three temperatures (25 °C, 30 °C and 37 °C) in the mixture PBS/EtOH (vol. 50/50) in which the solubility of quercetin totals to 5.66 mg/mL, 5.83 mg/mL and 6.02 mg/mL at 25 °C, 30 °C and 37 °C, respectively [38]. For the use of NPs as a drug delivery system, a physiological temperature of 37 ° C is of crucial importance. However, before the application of MNPs, they must be adequately stored. There are studies on the storage of MNPs at low temperatures, and the influence of low temperatures on the magnetic properties of MNPs [39]. Although the temperature effects on the magnetic properties of magnetite are very weak, MNPs contain a coating of organic material, in our case PEG 4000. Since PEG 4000 has crystallization temperatures at around 32 °C [40], which means that by changing the temperature of the MNPs suspension from the storage temperature to the ambient temperature, and further to the physiological temperature of 37 °C, the surface properties of MNPs and the relaxation kinetics could be changed. For this reason, measurements were made not just at 37 °C, but also at 25 and 30 °C to see how much the temperature change affects the drug-release kinetics. In addition, similar studies of the mechanism of release kinetics from NPs at different temperatures (10, 22 and 37 °C) have been conducted, for example, in study by Gronczewska [41], in that drug release and matrix degradation of polymer microspheres with different glass transition temperatures were investigated at various temperatures in order to clarify the effect of temperature on mechanisms of drug release. Then, 100 mg of quercetin-loaded MNPs were transferred in a dialysis bag (MWCO 8 kD, Thermo Fisher Scientific (Waltham, MA, USA)), after which 1 mL of PBS/EtOH (Vol. % = 50:50) medium was added and closed with dialysis bag clip holders. The dialysis bag was placed in a glass cylindrical reactor with a thermostatic jacket and flange size LF 100 containing 150 mL of PBS/EtOH mixture. The overhead stirrer (Ministar 20 control, IKA-Werke GmbH&Co (Staufen, Germany)) is going through the centre neck flat flange lid. The stirrer was set to 200 rpm. At selected time intervals, 1 mL of supernatant was replaced with fresh PBS/EtOH mixture through one angled side neck flat flange lid. The reactor is connected to a refrigerated–heating circulator (Corio CD-201F Julabo GmbH (Seelbach, Germany)) to control the appropriate temperature. The controlled release of quercetin was tested using appropriate external alternating (10 kHz, 50 kHz and 100 kHz) and stationary magnetic fields (B = 7.9 mT and 11.0 mT) at controlled electric current (I = 100 mA). An external alternating magnetic field was achieved with a function generator (Wavetek 164 30 MhZ (San Diego, CA, USA) connected to the coil (N = 270, l = 4 cm). Experiments were performed using the magnetic field system set up from permanent disk-shaped magnets (rare earth) and solenoid with permalloy core connected to signal generator alternating 100 mA current. A reactor with the sample was placed among the two magnetic fields. Defining the O_{xy} plane as the surface of the liquid, the permanent field was along the O_z axis and the alternating field along the O_x axis. Weak fields were applied in all experiments: the strength of the static permanent magnetic fields at the appropriate distance between the membrane dialysis bag and the permanent magnet was B = 7.9 mT and 11.0 mT. They were placed on a stand inside the reactor and used as sources of the permanent magnetic field. The function generator is connected to the oscilloscope (DS1000Z, Rigol Technologies (Beijing, China), which allowed the observation of the sinus waveform signal. The release kinetics of quercetin from magnetite MNPs was quantified by UV/VIS spectrophotometry (Photolab 7600 UV/VIS spectrophotometer Xylem (New York, NY, USA)). The linearity of the calibration was found to be valid from 1×10^{-6} mol dm^{-3} to 1×10^{-4} mol dm^{-3} with correlation coefficients for quercetin all approaching 1.00. All release kinetics experiments have been performed in duplicate.

3. Results and Discussion

3.1. Characterization of Synthesized Magnetite MNPs

The X-ray powder diffraction patterns of the synthesized bare and PEG-coated MNPs are presented in Figure 3A,B, respectively. Characteristic peaks exhibited in the XRPD pattern are well-matched with the magnetite diffraction peaks and confirm the cubic inverse spinel structure of MNPs. The sharp diffraction of three characteristic peaks (220), (311) and (400) also indicate the spinel structure of magnetite [42]. The formation of pure cubic magnetite is confirmed by the value of the calculated lattice parameter "a" which has been determined to be 8.389 Å [43]. Using Scherrer's equation, the average crystallite size of bare and PEG-coated MNPs were calculated to be 25 nm and 19 nm, respectively. In addition, the decreased average crystallite size, another piece of evidence suggesting that PEG decreases the crystallinity of MNPs at a lower intensity and with broader diffraction peaks than PEG-coated MNPs.

The size and morphology of bare and PEG-coated mesoporous MNPs were observed by field-emission scanning electron microscope (FE-SEM), atomic force microscopy (AFM) and transmission electron microscopy (TEM), as shown in Figure 3A,B,G–J,K–L). Both bare and PEG-coated MNPs maintain a uniform spherical shape, some of them agglomerated due to magneto–dipole interactions between MNPs. While bare MNPs showed cluster structure with a very rough surface, the process of PEG coating revealed the flattened surface of mesoporous MNPs (Figure 3I,J). These findings have been confirmed by AFM imaging where the roughness of the MNPs surface has been increased by PEG coating, for almost 100%, from 5.58 ± 1.06 nm to 10.9 ± 2.1 nm, confirming the effective coverage of the bare MNPs by PEG [43]. The less agglomerated texture of the PEG-coated MNPs can be related to the effect of the PEG layer during the synthesis of MNPs. In addition, a size histogram of bare mesoporous MNPs obtained using SEM micrographs shows a broader size distribution than PEG-coated MNPs (d_{ave} = 103.4 ± 0.7 nm and 101.0 ± 0.9 nm for bare and PEG-coated MNPs, respectively), indicating that polymer decreased the magnetic interaction among the particles and prevent their agglomeration. The cluster structure of mesoporous MNPs has been confirmed using AFM. The average size of bare MNPs and PEG-coated MNPs corresponds to results obtained by SEM, it was even 10% larger due to the convolution effect. However, in both 2D height images, it is shown that the cluster structure MNPs consists of smaller substructures, in the range of 15 to 35 nm, which roughly correspond to the dimension of the crystallite size obtained by X-ray powder diffraction. Hence, AFM revealed that only several nanometers increase in roughness is observed when PEG 4 kD is used, suggesting that the PEG layer is largely twisted the magnetite surface, rather than stretch out linearly [44]. Furthermore, this PEG layer should decrease the magnetic interactions among the MNPs and prevents their agglomeration. The mean diameter of bare MNPs obtained by TEM is larger (143 ± 30 nm) than obtained by SEM (103.4 ± 0.7 nm), indicating a higher degree of polydispersity. However, the mean diameter of PEG-coated MNPs (96 ± 10 nm) corresponds the value obtained by SEM (101.0 ± 0.9 nm). Increasing the size of the PEG-coated MNPs can be attributed to the successful coating of PEG. As shown in Figure 3K,L, both MNPs maintain a typical spherical shape.

Nitrogen adsorption–desorption isotherms of bare MNPs confirmed their mesoporosity (see Figure 4). The surface area, pore size and total pore volume were calculated to be 20.5 $m^2 g^{-1}$, 17.7 nm and 0.09 $cm^3 g^{-1}$, respectively, strongly supporting the fact that the bare MNPs have mesoporous structure. The PEG coating of bare MNPs led to a slight decrease in surface area (19.3 $m^2 g^{-1}$) but an increase in pore size and total pore volume (24.6 nm^1 and 0.11 $cm^3 g^{-1}$, respectively).

FTIR spectrum (Figure 5A) of bare magnetite MNPs shows bands at 585 and 395 cm^{-1} corresponding to the symmetric stretching vibration mode of the Fe-O bond. The absorption maxima at 3452 cm^{-1} and 1644 cm^{-1} suggest the presence of O-H linkages. In the pure PEG, the bands at 1341 and 1100 cm^{-1} belong to the C-O-C ether bond asymmetric and symmetric stretching vibrations. The band at 2890 cm^{-1} is attributed to $-CH_2-$ stretching vibration in PEG. In addition, absorption bands at 1283 and 1465 cm^{-1} attributed to the vibration of $-CH_2$ and at 964 cm^{-1} corresponds to the CH out-of-plane vibration [18]. The hydroxyl groups were also confirmed at 3444 and 1631 cm^{-1}. The presence of characteristic FTIR bands of PEG in PEG-coated MNPs spectrum confirmed the successful coating of PEG on the surface of magnetite MNPs. The PEG-coated MNPs spectrum shows a strong C-O-C ether stretch at 1110 and 1381 cm^{-1} [44]. In addition, the O-H linkages at the 1642 and 3445 cm^{-1} bands exhibited enhanced intensity which also indicates that PEG modified the surface of MNPs. The transmittance bands at 589 and 399 cm^{-1} confirm the symmetric stretching mode of the Fe-O bond. The results indicate that PEG is successfully functionalizing the surface of MNPs.

In order to confirm the superparamagnetic properties of synthesized bare and PEG-coated MNPs, measurements of the magnetization curve have been performed. The S-shaped hysteresis loops are a typical feature of superparamagnetic MNPs and obtained result is very similar to our previous work (77 emu g^{-1}) on Fe_3O_4 MNPs [35]. The magnetization curve clearly shows that magnetization depends on the applied magnetic field, but not on the sign of the applied field. Magnetic characterization at 300 K indicates that the bare MNPs have saturation magnetization at the maximum field of 5 kOe value of 76.71 emu g^{-1}, which is lower than obtained value for the bulk Fe_3O_4 (M_s = 92 emu g^{-1}) [42] or Ms = 96.43 emu g^{-1} [19].

The observed decrease in the saturation magnetization could be explained either by the use of PEG for the surface modification process [43] that causes the softening of the magnetization or by the difference in particle size [45]. In addition, the saturation magnetization of PEG-coated MNPs at the maximum field of 5 kOe is 74.75 emu g^{-1}, somewhat lower than bare MNPs (Figure 5B). However, the magnetization measurement of synthesized MNPs confirmed their superparamagnetic properties, thus confirming their further application in release studies under applied magnetic fields.

3.2. Loading of Flavonoid into Magnetite MNPs

The FTIR spectrum (Figure 6A,B) of quercetin detected OH group stretching at 3403 cm^{-1} and 3328 cm^{-1}. The band at 1664 cm^{-1} corresponds to the C = O aryl ketonic stretching and the bands at 1617 cm^{-1}, 1558 cm^{-1} and 1520 cm^{-1} correspond to C = C aromatic ring stretching. OH bending of the phenol functional group at 1375 cm^{-1} and 1314 cm^{-1} belongs to the in-plane bending of C–H in aromatic hydrocarbon. Bands at 933 cm^{-1}, 820 cm^{-1}, 639 cm^{-1} and 602 cm^{-1} correspond to aromatic C–H out-of-plane bendings. The C–O stretching in the aryl ether ring and the C–O stretching in phenol corresponds to 1244 cm^{-1} and 1210 cm^{-1} transmittance maxima. The band at 1167 cm^{-1} is attributed to the C–CO–C stretch and bending mode in ketone, respectively. The FTIR spectra of quercetin-loaded MNPs show the broadening of the OH band at 3446 cm^{-1}, which confirms the entrapment of quercetin in MNPs [46]. The interval from 1560 to 816 cm^{-1} matches very well with pure quercetin peaks and indicates successful loading [46].

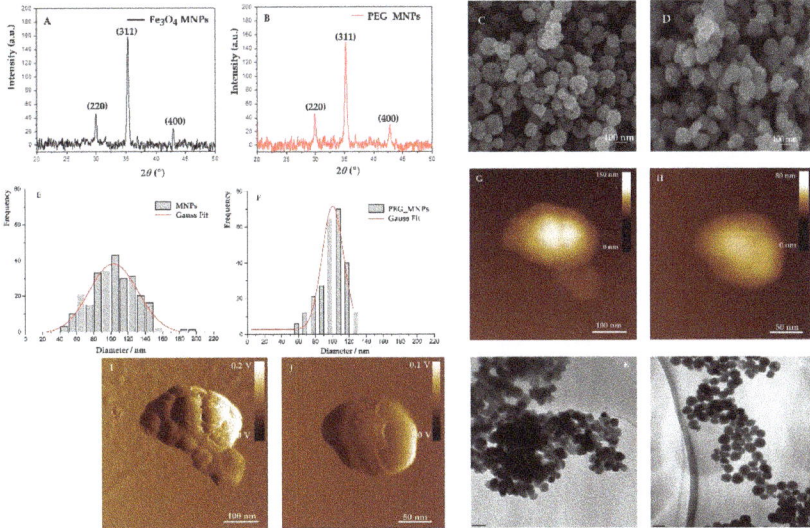

Figure 3. XRD pattern of (**A**) mesoporous bare MNPs Fe3O4; (**B**) PEG-coated MNPs; SEM images of (**C**) bare Fe3O4 MNPs showing the rough rounded cluster with extensive open porosity ranging from 30 to 200 nm in diameter; (**D**) SEM image of the surface of PEG-coated Fe3O4 MNPs ranging in di-ameter from 50–130 nm; Histogram of bare MNPs Fe3O4 (**E**) and PEG-coated Fe3O4 MNPs (**F**); (2D height topographic AFM image of single (**E**) bare magnetite Fe3O4 MNPs (**G**) and (**F**) PEG-coated Fe3O4 MNPs (**H**) showing cluster structure details; 2D-amplitude AFM image of single (**G**) bare magnetite (**I**) and (**H**) PEG-coated MNPs (**J**) showing subcluster structures in size range from 10 to 30 nm; TEM images of (**K**) bare Fe3O4 MNPSs; (**L**) TEM image of the surface of PEG-coated Fe3O4 MNPs.

The loading of quercetin has been also confirmed using nitrogen adsorption–desorption isotherms of quercetin-loaded PEG MNPs. In comparison to PEG covered MNPs, the surface area has been decreased by almost 19% to 15.7 m^2 g^{-1}, pore size decreased to 14.2 nm for 42.4%, while total pore volume amounting 0.06 cm^3 g^{-1}, decreased for 49% to 0.06 cm^3 g^{-1}, strongly supporting the fact that the quercetin has been effectively loaded into MNPs.

The thermogravimetric study was performed to confirm the quercetin loading in MNPs. Figure 6C shows comparative weight loss for quercetin and quercetin-loaded MNPs. A thermogravimetric study of quercetin reveals that the compound undergoes a three-stage thermal decomposition. The first stage of mass loss begins at 30 °C and continues up to 133 °C. A mass loss of 3.52% is attributed to dehydration or loss of water molecules on the surface of quercetin. In the temperature range 133 °C to 385 °C, the compound experiences a weight loss of 28.5% due to the melting of quercetin. The final thermal decomposition is observed in the temperature range of 385 °C to 1110 °C, and the weight loss of quercetin is 67% [47]. In the case of quercetin-loaded MNPs, the weight loss in the temperature range from 30 °C to 1200 °C is about 12.2%, which is attributed to the decomposition of organic compounds from MNPs. This data results in the great thermal stability of quercetin when it has been encapsulated in MNPs. In the case of

PEG-coated MNPs, there is an increase in the weight gain resulting from the burning of the PEG and oxygenation of Fe_3O_4 starting at 270 °C under the continuous flow of oxygen at high temperatures and finishing at 450 °C [48]. However, the weight loss of PEG from the PEG-coated MNPs amounts to 4%. It is assumed that the thermal decomposition of PEG occurs at both C-O and C-C bonds of the backbone chain [49]. The influence of the quercetin loading into magnetite MNPs on the size and morphology of the MNPs has been further investigated. 2D height AFM image (Figure 6D) and 2D-amplitude AFM image (Figure 6E) revealed that distinct subcluster structure containing MNPs has been retained irrespective of quercetin loading, with size grains from 20 to 50 nm in diameter. The roughness value after quercetin loading decreased from (10.9 ± 2.1) nm to (4.86 ± 1.1) nm additionally confirming the successful loading of quercetin to PEG loaded MNPs.

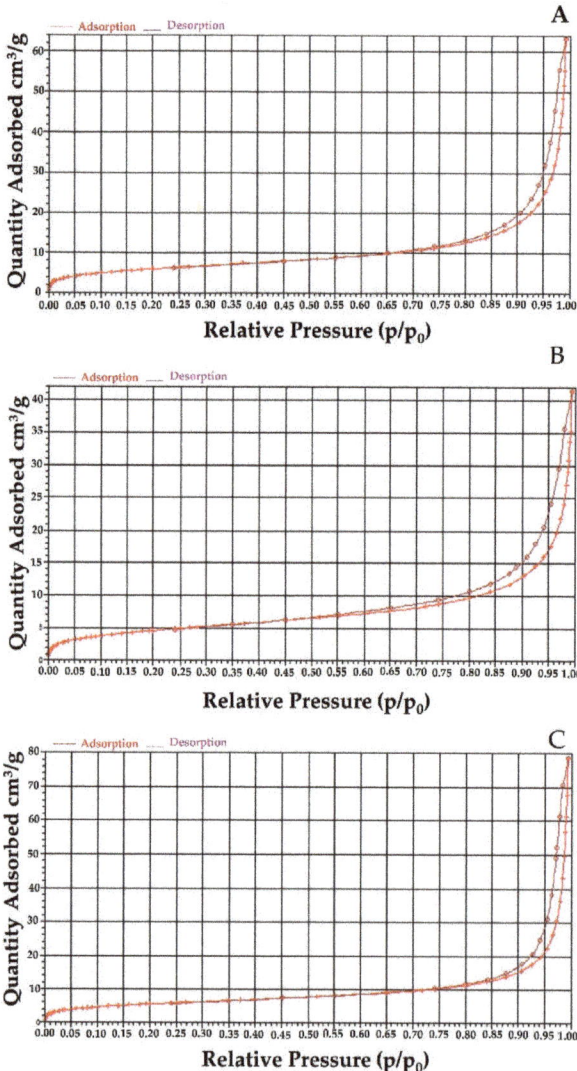

Figure 4. N_2 absorption-desorption isotherm for (**A**) bare MNPs, (**B**) PEG-coated MNPs and (**C**) quercetin-loaded PEG_MNPs.

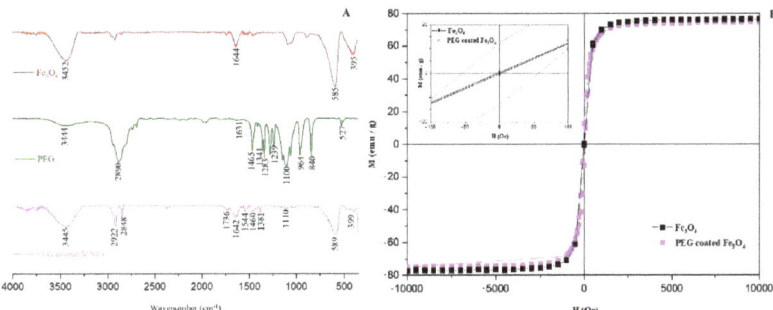

Figure 5. (**A**) FTIR spectra of MNPs, PEG and PEG-coated MNPs (**B**) Magnetic behavior of MNPs. The hysteresis loop shows a slight decrease in the magnetization behavior after a thin PEG layer coating. The magnetic hysteresis loops of mesoporous MNPs at 300 K. Inset within Figure 5 (**B**) shows the magnetic coercivity H_c = 53.3 Oe.

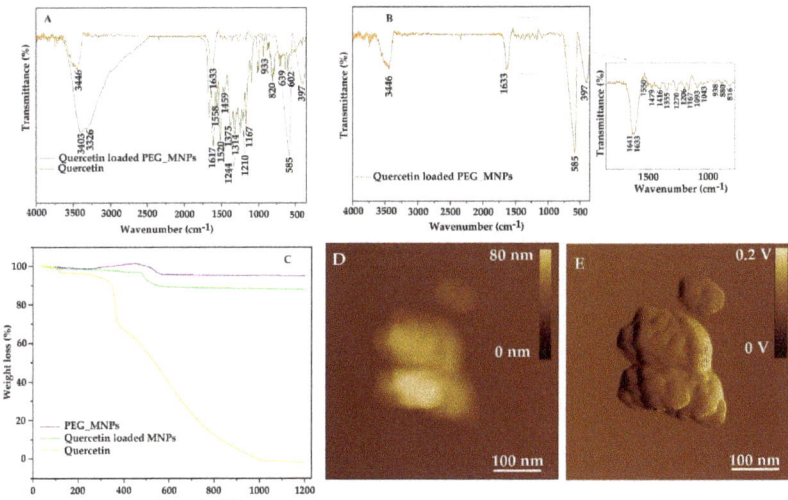

Figure 6. FTIR spectra of (**A**) quercetin and quercetin-loaded MNPs; (**B**) zoomed spectrum of quercetin-loaded MNPs in the region of quercetin reach bands; (**C**) TGA curves of PEG-coated MNPs, quercetin and quercetin-loaded MNPs. Morphology of quercetin-loaded MNPs on the 2D height AFM image (**D**) and 2D-amplitude AFM image (**E**).

In addition, UV/VIS spectroscopy is used to study quercetin loading efficiency. Compared with quercetin concentration before adding the synthesized MNPs, the concentration loss was determined using a calibration curve in pure ethanol (EtOH). The coefficient of determination was 0.9978, and the determined molar absorption coefficient of quercetin at 298 K and 374 nm is 19,131 mol^{-1} dm^3 cm^{-1}. The loading efficiency (LE) was determined to be (20.2 ± 1.3%) calculated from 17 independent experiments. Our results suggest significant improvement of the loading efficiency of the quercetin compared with a loading efficiency of solid lipid NPs (13.20 ± 0.18%) [1]. However, the limited LE is due to the reduced specific loading site of quercetin induced by PEG coatings [50]. Despite this, the PEG –MNPs provide the capability to load antioxidant quercetin with low aqueous solubility which reflects the potential of the synthesized MNPs as drug delivery carriers.

3.3. Homogeneity and Stability of Synthesized Mesoporous MNPs

One of the essential features for successful drug delivery is stability and homogeneous dispersion of NPs in buffer. Zeta potential was used to determine the stability of the colloidal suspension of bare MNPs, PEG-coated MNPs and quercetin-loaded MNPs. It has been shown that PEG-stabilized NPs exhibit longer bloodstream circulation time and higher resistance to protein binding [44]. Due to its unique properties and its biocompatibility, PEG is selected as the stabilizing agent in this study. The surface charge of NPs has an important effect on the blood circulation time, the pharmacokinetics of NPs and the zeta potential above ±30 mV indicated to be relevant for stability of NPs in aqueous suspensions [12,51,52]. The zeta potential of MNPs (Table 1) was determined in phosphate-buffered solution (PBS). Bare MNPs exhibited a zeta potential of (-30.6 ± 0.7) mV, while PEG coating increased the absolute zeta potential value to (-35.1 ± 1.5) mV. Indeed, the higher negative zeta potential value indicated that PEG-coated MNPs possessed higher stability after PEG coating. The zeta potential of quercetin-loaded MNPs (-31.3 ± 0.8) mV decreased slightly in comparison to PEG-coated MNPs indicating quercetin adsorption on the PEG layer of MNPs. However, the observed change in zeta potential value did not decrease the stability of MNPs. Moreover, the successful loading of quercetin into PEG-coated MNPs has been confirmed also by electrophoretic measurements.

Furthermore, the hydrodynamic diameter is lower which could be attributed that PEG coating while quercetin loading provides better colloidal stability and reduces aggregation. The volume size distributions of all samples, i.e., bare MNPs, PEG-coated MNPs and quercetin-loaded MNPs were unimodal. In Table 1, it can be seen that the highest polydispersity index was observed to be 0.54 ± 0.1 for the bare synthesized MNPs suspension. This result is consistent with the results obtained from the size distribution of bare MNPs using SEM where the size distribution of bare MNPs was broader than the size distribution of the PEG-coated MNPs. The observed discrepancy between SEM and DLS data, particularly in the polydispersity index (PDI), can be explained by the fact that the SEM micrographs were taken in a dried state, whereas the DLS experiment was carried out in suspension. The PDI results reported in Table 1 also support the conclusion that PEG coatings, even presented in suspension, decreased the process of aggregation of bare MNPs and break down the cluster. The average hydrodynamic diameter of both, bare and PEG-coated MNPs also supports the above findings. In Figure 7, it is shown the volume size distributions of bare MNPs, PEG-coated MNPs and quercetin-loaded PEG_MNPs. The PDI of PEG-coated MNPs is lower than PDI of loaded MNPs (PDI = 0.47 ± 0.1 and 0.52 ± 0.07, respectively) indicating narrower size distribution of the PEG-coated MNPs than quercetin-loaded MNPs (see Figure 7). However, the cluster sizes of quercetin-loaded MNPs were within the range of (681 ± 73) nm, which is the smallest range of all analysed MNPs. Thus, the PEG coating on the surface of Fe_3O_4 decreased the size compared to the bare MNPs and indicates that the PEG coating prevents or reduces aggregation to some extent.

Table 1. Zeta potential and hydrodynamic diameter of bare MNPs, PEG-coated MNPs and quercetin-loaded MNPs.

	Bare MNPs	PEG Coated MNPs	Quercetin Loaded PEG_MNPs
ζ/mV	-30.6 ± 0.7	-35.1 ± 1.5	-31.3 ± 0.8
d_h/nm	877 ± 105	762 ± 92	681 ± 73
PDI	0.54 ± 0.10	0.47 ± 0.10	0.52 ± 0.07

3.4. Release Study

Before we start to evaluate the quercetin-release kinetics from MNPs into PBS/EtOH (Vol. % 50:50), we want to emphasize that the results were obtained by applying external stationary and alternating magnetic fields. The underlying idea was to enable fine-tuned and controlled quercetin-release kinetics, which is certainly very important in a widespread, high-demand application of high-demand drug delivery via nanocarriers.

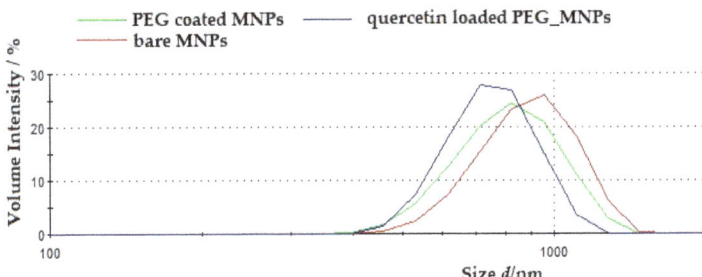

Figure 7. Volume size distributions of bare MNPs, PEG-coated MNPs and quercetin-loaded PEG_MNPs.

The in vitro quercetin release profile from synthesized and carefully designed MNPs was studied in duration up to 8 h with a dialysis membrane and in the presence of external magnetic fields. The results are shown as the cumulative released mass of quercetin in Figure 8. Within 8 h, only up to 5% of quercetin has been released depending on the experimental conditions.

A similar sustained-release profile of quercetin from alginate NPs at pH 7.4 has been reported earlier [53], where only 10% of quercetin release was observed after 12 h, while after 9 days, only 50% of the quercetin got released. In another study, the quercetin release from the functionalized magnetite NPs was conducted in acidic and basic pH and cumulative release reached 3.7% after 6 h [14]. A similar prolonged release has also been found for quercetin release from polylactide NPs, which showed almost 60% quercetin released after 4 days [54]. The observed slow quercetin-release kinetics offers prolonged exposure to the drug and improves its efficiency compared with free drugs [55]. This is considered as an advantage because the burst release of drugs leads to a significant premature quantity of the drug that can result in toxicity [21]. For example, the quercetin at a concentration between 0 and 200×10^{-6} mol dm^{-3} could decrease antioxidant activity while quercetin at a concentration of $(0.2–1) \times 10^{-6}$ mol dm^{-3} possesses pro-oxidant activity [56]. In addition, quercetin release was affected by oxidative degradation process in PBS solution after continuously stirring for 6 h; this is yet another reason to employ nanocarriers for quercetin delivery [14,57,58]. Therefore, we prepared MNPs onto which quercetin easily adsorbs and has a prolonged stability and duration. The net result is quercetin release for a prolonged time. A first-order release profile of quercetin from Fe_3O_4-quercetin-copolymer NPs was revealed by Barreto et al. [3]. On the other hand, the release rate of quercetin from superparamagnetic magnetite NPs coated with chitosan, PEG and dextran was found to be of zero-order kinetics (linear with time) [41].

The assumptions in release kinetics experiments were as follows:

(i) The total amount of quercetin remained practically constant during the whole release experiment; (ii) both the quercetin solution within the membrane interior and in the membrane exterior were homogeneous due to the constant stirring and the fact that quercetin solution was never saturated; (iii) the thickness of the membrane provides the equal rate constants from the membrane interior to the membrane exterior, and vice versa; (iv) the volume within the membrane interior V_i and the volume exterior to the membrane V_o were constant during all performed release experiments. This is because V_i = 1 mL and V_o = 150 mL, $V_i < V_o$.

As a preliminary approach in working out the data, a simple model of single exponential decay was meant to be used in which the fraction of the released drug is $\Phi = 1 - exp(-kt)$, where Φ is a fraction of drug present in the outer volume V_0 (other fractions are in the inner volume V_i, Φ_i, and in the membrane, Φ_m) [59]. The coefficient k is related to the apparent kinetics and, as such, cannot provide information about the actual release rate from the nanocarriers into the interior volume V_i. However, the problem with this simple formula is the equilibrium value $\Phi_o = 1$ which is achieved when the elapsed time is sufficiently large ($t \to +\infty$). In our release kinetics experiments, it is invariably between 0.04 and 0.10. Thus, the dialysis bag with its content has to be considered as a source of drug molecules. It is not possible to know the total mass of the drug that is amenable to be released, but it can be estimated from the value of m_0, which occurs in another simple model:

$$m(t) = m_0\left(1 - e^{-kt}\right), \quad \frac{m}{m_0} = 1 - e^{-kt} \quad (2)$$

where $m(t)$ is the released mass, not a fraction of, at time t, m_0 is the total released quercetin mass from the dialysis bag after infinite time, and k is the rate coefficient of actual release kinetics of the dialysis bag membrane. The problem is that m_0 is not experimentally well defined, i.e., it is only approximately constant across the series of experiments. Fitting the release experimental data obtained at various magnetic fields and at three different temperatures using this simple model (Table 2 and Figures 8 and 9) resulted as expected in a fairly narrow interval of m_0 values with average $m_0 = (1.48 \pm 0.34)$ mg.

Each experiment was done in duplicate. This was fully justified to avoid averaging the measurement results and use the mean values of the two fitting procedures because the product, kt, was always rather small (Table 2). This was the case because, if $m_1 = m_{0,1}\left(1 - e^{-k_1 t}\right)$, $m_2 = m_{0,2}\left(1 - e^{-k_2 t}\right)$ and $m = m_0\left(1 - e^{-kt}\right)$ where $m = 1/2(m_1 + m_2)$ and $m_0 = 1/2(m_{0.1} + m_{0.2})$, the following is obtained:

$$e^{-kt} = \frac{m_{0.1}}{m_{0.1} + m_{0.2}} e^{-k_1 t} + \frac{m_{0.2}}{m_{0.1} + m_{0.2}} e^{-k_2 t} \quad (3)$$

Since the product kt is always very small, it turns out that a simple formula,

$$k = \frac{m_{0.1}}{m_{0.1} + m_{0.2}} k_1 + \frac{m_{0.2}}{m_{0.1} + m_{0.2}} k_2 \quad (4)$$

can be used. Furthermore, it is very often $m_{0,1} \approx m_{0,2}$ which gives $k \approx \frac{1}{2}(k_1 + k_2)$.

It is important to emphasize that the intent of this study was not only to estimate the time required for the complete quercetin release from nanocarriers, but also to investigate and demonstrate how the stationary and alternating field affect the rate of quercetin release. In our previous work [35], we have shown how the simultaneous application of stationary and alternating field can accelerate the release of drug from aggregates of MNPs. Being relatively unstable and dysfunctional, aggregates vigorously moved under the influence of the alternating field which resulted in drug release enhancement. MNPs that were additionally functionalized and stabilized by the PEG layer were used for this purpose in the present study. Figure 8 shows the release kinetics of quercetin at the temperature of 30 °C in the absence of the magnetic field and under an alternating field of 10 kHz, 50 kHz and 100 kHz at a constant stationary magnetic field $B = 11$ mT.

Since we used a dialysis membrane bag, the first step was to perform calibration experiments with free quercetin following the same protocol as in experiments with MNPs to get information about membrane permeation kinetics during the first several hours ($k = 6.617 \times 10^{-3}$ min^{-1}; $k = 9.637 \times 10^{-3}$ min^{-1} and $k = 14.592 \times 10^{-3}$ min^{-1} at 25 °C, 30 °C and 37 °C, respectively). We selected the MWCO cellulose membrane (8 kD) membrane based on the porosity of the dialysis membrane as well as to avoid possible adverse interactions between quercetin and the membrane materials. The cellulose membrane has recently been used in the measurement of both, drug diffusion and drug release rates from varied formulations, such as creams and hydrogels [7].

The rate constant of the membrane when there was non-saturated quercetin solution in the membrane bag indicated the barrier effects of dialysis membrane and showed faster membrane permeation kinetics at higher temperatures ($k = 0.0066$ min^{-1}, 0.0096 min^{-1} and 0.0146 min^{-1} at 25 °C, 30 °C and 37 °C, respectively), as expected. The rate constants obtained in our experiments are larger than those obtained in release kinetics of doxorubicin by Yu et al., where $k = 0.019 \pm 0.003\ h^{-1} = 0.0003\ min^{-1}$ [60].

With no magnetic field and at 30 °C, the rate constant is $k = 0.0019 \pm 0.0001\ min^{-1}$. At 10 kHz, 50 kHz and 100 kHz and under $B = 11.0$ mT the rate constant values were $k = 0.0032 \pm 0.0009\ min^{-1}$, $k = 0.0019 \pm 0.0001\ min^{-1}$ and $k = 0.0034 \pm 0.0013\ min^{-1}$, respectively. Thus, the release of the quercetin is the faster at the highest field frequency. It is evident that the alternating magnetic field can accelerate the quercetin release at a given stationary magnetic field.

Table 2. The experimental release kinetics under the permanent magnetic fields of 7.9 mT and 11.0 mT and three frequencies (10 kHz, 50 kHz and 100 kHz) at temperatures 25 °C, 30 °C and 37 °C.

B/mT	$f_{alt.mag.f.}$/kHz	θ/°C	k/min^{-1}	m_0/mg
11.0	100	25	0.0029 ± 0.0004	0.82 ± 0.01
		30	0.0034 ± 0.0013	1.38 ± 0.30
		37	0.0038 ± 0.0013	1.88 ± 0.25
	50	25	0.0022 ± 0.0004	1.03 ± 0.30
		30	0.0019 ± 0.0001	1.41 ± 0.01
		37	0.0025 ± 0.0015	3.3 ± 1.4
	10	25	0.0024 ± 0.0012	1.07 ± 0.34
		30	0.0032 ± 0.0009	0.97 ± 0.20
		37	0.0033 ± 0.0003	1.73 ± 0.22
7.9	100	25	0.0015 ± 0.0002	1.79 ± 0.14
		30	0.0032 ± 0.0003	1.15 ± 0.44
		37	0.0052 ± 0.0011	1.50 ± 0.39
	50	25	0.0022 ± 0.0007	1.20 ± 0.26
		30	0.0022 ± 0.0008	1.86 ± 0.35
		37	0.0025 ± 0.0002	1.21 ± 0.01
	10	25	0.0043 ± 0.0003	1.13 ± 0.08
		30	0.0021 ± 0.0006	1.30 ± 0.24
		37	0.0016 ± 0.0005	1.64 ± 0.52
0	0	25	0.0022 ± 0.0004	1.03 ± 0.30
		30	0.0019 ± 0.0001	1.41 ± 0.01
		37	0.0025 ± 0.0015	2.25 ± 1.44

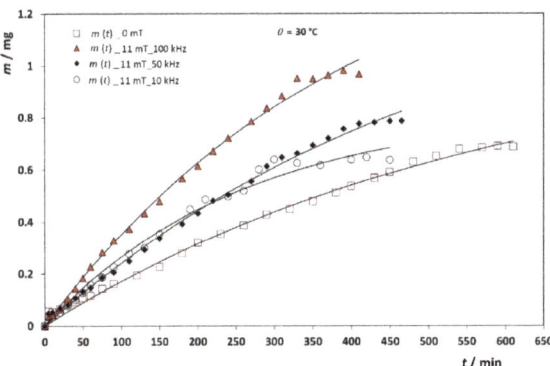

Figure 8. The representative cumulative release profiles for the quercetin from MNPs through dialysis membrane under the stationary magnetic field $B = 11$ mT and alternating field with frequencies 10 kHz, 50 kHz and 100 kHz at 30 °C.

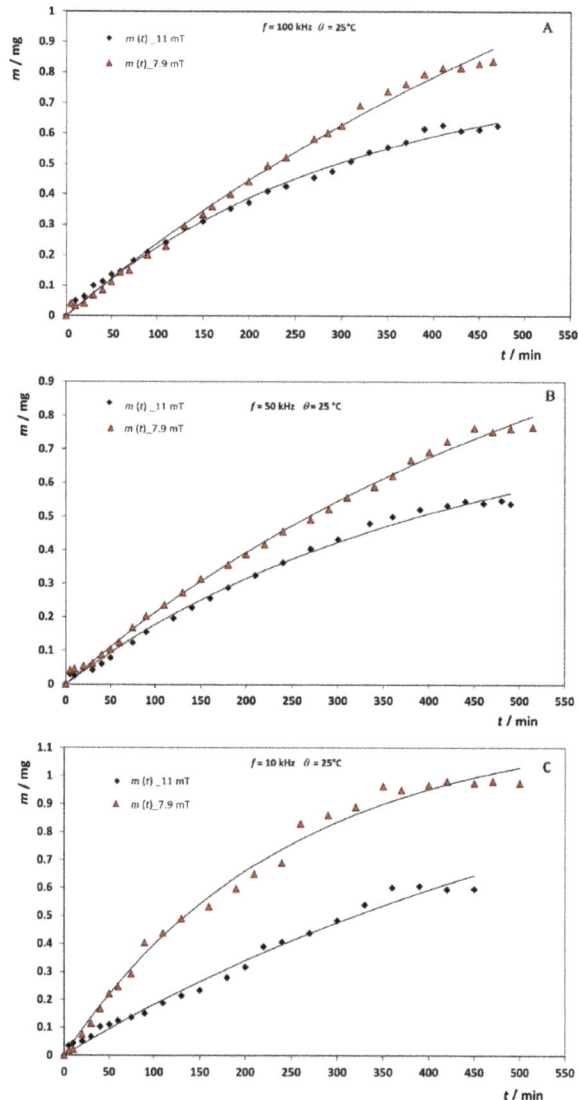

Figure 9. The representative cumulative release profiles for the quercetin from MNPs through dialysis membrane at 25 °C using two stationary magnetic fields and three different frequencies of alternating magnetic field (**A**) 100 kHz, (**B**) 50 kHz and (**C**) 10 kHz.

Our next task was to see the effect a stationary magnetic field on release kinetics. Figure 9 shows the dependence of quercetin-release kinetics at 25 °C under the alternating field frequency of 10 kHz and stationary magnetic field of B = 7.9 mT and 11 mT (Figure 9C). Under the stationary magnetic field of 7.9 mT, the membrane bag has released quercetin with a rate constant k = 0.0043 ± 0.0003 min^{-1}. When a stronger stationary field of 11 mT is applied, the release kinetics becomes slower, k = 0.0024 ± 0.0012 min^{-1}. The opposite effect on the release kinetics was observed at the highest 100 kHz frequency (Figure 9A), where an increase in the rate constant with increasing stationary field by almost 95% was observed (at B = 7.9 mT and B = 11 mT, k = 0.0015 ± 0.0002 min^{-1}; k = 0.0029 ± 0.0004 min^{-1}, respectively).

Since the opposite effects of the influence of a stationary magnetic field on the rate constant are obtained at different frequencies of the alternating field, our next analysis focuses on the measurements of the rate constants at the same stationary field and frequency, but at different temperatures. Comparing the constants of the apparent release rate of quercetin at a stationary field of 7.9 mT with an increase in temperature from 25 to 37 °C ($k = 0.0022 \pm 0.0004$ min^{-1}, $k = 0.0022 \pm 0.0008$ min^{-1} and $k = 0.0025 \pm 0.0002$ min^{-1} for 25 °C, 30 °C and 37 °C, respectively), it can be seen that the rate constant slightly increases with increasing temperature only above 30 °C. The same effect was observed under the stronger stationary field ($B = 11$ mT, see Table 2). At higher temperatures kinetic energy of quercetin molecules is larger or, putting differently, their diffusivity is larger and more quercetine is released and detected within the same time interval. If we compare the rate constants at the same temperature (e.g., 30 °C), the rate constant obtained at a stronger stationary field ($B = 11$ mT) and 50 KHz, is smaller ($k = 0.0019 \pm 0.0001$ min^{-1}) than at a weaker field of 7.9 mT ($k = 0.0022 \pm 0.0008$ min^{-1}). It is obvious that when overcoming stationary field, the thermal energy of the MNPs is large enough to increase the movement of the MNPs and increase quercetin release. The influence of the temperature under the constant stationary field and the frequency of the alternating field clearly shows that the amount of quercetin released increases with increasing temperature at almost all applied frequencies of the alternating fields except at $B = 11$ mT and frequencies $f = 100$ and 50 kHz and at $B = 7.9$ mT and $f = 50$ kHz, which is also reflected in the magnitudes of the quercetin release rate constants. For example, at $B = 11$ mT and $f = 100$ kHz, $k = 0.0029 \pm 0.0004$ min^{-1}; $k = 0.0034 \pm 0.0013$ min^{-1} and $k = 0.0038 \pm 0.0013$ min^{-1} for 25 °C, 30 °C and 37 °C, respectively.

Thus, it was shown here that by choosing the temperature, the quercetin release rates can cover a wide range of values. We have shown that the synthesized MNPs are suitable nanocarriers for quercetin, especially because the required drug dose can be delivered in a prolonged time. Since the average half-life of quercetin absorbed in the human body is 3.5 h [61], this study represents a significant improvement for flavonoid delivery, which, when loaded into MNPs, remains stable in a prolonged period of time.

4. Conclusions

In summary, superparamagnetic magnetite nanoparticles (MNPs) were prepared by solvothermal method and stabilized by biocompatible poly (ethylene glycol) PEG-4000 Da. The X-ray powder diffraction patterns of the synthesized bare and PEG-coated MNPs confirmed the cubic inverse spinel structure of MNPs. The size and morphology of bare and PEG-coated MNPs have been obtained by SEM, TEM and AFM analysis. By AFM is showed that bare MNPs have cluster structure with a very rough surface and when bare MNPs is coated with PEG, the roughness of the MNPs surface has been increased by PEG coating, for almost 100%, from 5.58 ± 1.06 nm to 10.9 ± 2.1 nm, confirming the effective coverage of the bare MNPs by PEG. A size histogram of bare mesoporous MNPs obtained using SEM micrographs shows a broader size distribution than PEG-coated MNPs ($d_{ave} = 103.4 \pm 0.7$ nm and 101.0 ± 0.9 nm for bare and PEG-coated MNPs, respectively), indicating that PEG decreased the magnetic interaction among the particles and prevent their agglomeration. Nitrogen adsorption–desorption of bare MNPs and PEG-coated MNPs confirmed their mesoporosity. The PEG molecules were successfully coated on the surface of MNPs, as revealed by FTIR spectroscopy. The PEG-coated MNPs spectrum showed a strong C-O-C ether stretch at 1110 and 1381 cm^{-1}. The results of the saturation magnetization confirmed the superparamagnetic properties of synthesized bare and PEG-coated MNPs. The stability of MNPs improved after PEG modification, indicated by the increase in zeta potential from (-30.6 ± 0.7) mV to (-35.1 ± 1.5) mV. The loading of quercetin into MNPs was confirmed by FTIR spectroscopy and thermogravimetric analysis. The UV/VIS spectra of the supernatant revealed a loading efficiency of (20.2 ± 1.3%). The quercetin release studies in vitro followed by UV/VIS spectroscopy have shown the prolonged quercetin-release kinetics from MNPs that could be controlled by using combined stationary and alternating magnetic fields. The prolonged quercetin release, as an important characteristic for targeted drug delivery, the study of the kinetic parameters of the quercetin release process and the increased response of quercetin release under both the lower stationary magnetic field (7.9 mT) and the higher frequency of alternating magnetic field (100 kHz) suggest that the fine tuning of the release as desired along with synergism of physicochemical and superparamagnetic properties enables the great potential of MNPs as a promising targeted flavonoid delivery system.

Author Contributions: S.Š. designed research; L.M., I.E. and A.S. performed research; L.M., G.B. and S.Š. analysed data and contributed to the discussion; L.M., S.Š. and G.B. wrote the paper; all authors approved the final version of the paper. All authors have read and agreed to the published version of the manuscript.

Funding: This work was supported by the Croatian Science Foundation under the project IP-2016-06-8415. The funders had no role in the design and conduct of this study, collection and interpretation of the data, or preparation and approval of the manuscript.

Institutional Review Board Statement: Not applicable.

Informed Consent Statement: Not applicable.

Data Availability Statement: Data available on request due to restrictions e.g., privacy or ethical. The data presented in this study are available on request from the corresponding author. The data are not publicly available due to extreme quality of data.

Acknowledgments: The authors thank T. Mrla for experimental setup and instrument development for controlling the electric current (I = 100 mA).

Conflicts of Interest: The authors declare no conflict of interest.

References

1. Li, H.; Zhao, X.; Ma, Y.; Zhai, G.; Li, L.; Lou, H. Enhancement of gastrointestinal absorption of quercetin by solid lipid nanoparticles. *J. Control. Release* **2009**, *133*, 238–244. [CrossRef]
2. Pawlikowska-Pawlęga, B.; Dziubinska, H.; Król, E.; Trebacz, K.; Jarosz-Wilkolazka, A.; Paduch, R.; Gawron, A.; Gruszecki, W. Characteristics of quercetin interactions with liposomal and vacuolar membranes. *Biochim. Biophys. Acta* **2014**, *1838*, 254–265. [CrossRef]
3. Barreto, A.C.H.; Santiago, V.R.; Mazzetto, S.E.; DeNardin, J.C.; Lavín, R.; Mele, G.; Ribeiro, M.E.N.P.; Vieira, I.G.P.; Gonçalves, T.; Ricardo, N.M.P.S.; et al. Magnetic nanoparticles for a new drug delivery system to control quercetin releasing for cancer chemotherapy. *J. Nanoparticle Res.* **2011**, *13*, 6545–6553. [CrossRef]
4. Mandić, L.; Sadžak, A.; Strasser, V.; Baranović, G.; Jurašin, D.D.; Sikirić, M.D.; Šegota, S. Enhanced Protection of Biological Membranes during Lipid Peroxidation: Study of the Interactions between Flavonoid Loaded Mesoporous Silica Nanoparticles and Model Cell Membranes. *Int. J. Mol. Sci.* **2019**, *20*, 2709. [CrossRef] [PubMed]
5. Wang, Y.; Tao, B.; Wan, Y.; Sun, Y.; Wang, L.; Sun, J.; Li, C. Drug delivery based pharmacological enhancement and current insights of quercetin with therapeutic potential against oral diseases. *Biomed. Pharmacother.* **2020**, *128*, 110372. [CrossRef]
6. Amanzadeh, E.; Esmaeili, A.; Rahgozar, S.; Nourbakhshnia, M. Application of quercetin in neurological disorders: From nutrition to nanomedicine. *Rev. Neurosci.* **2019**, *30*, 555–572. [CrossRef] [PubMed]
7. Berlier, G.; Gastaldi, L.; Ugazio, E.; Miletto, I.; Iliade, P.; Sapino, S. Stabilization of quercetin flavonoid in MCM-41 mesoporous silica: Positive effect of surface functionalization. *J. Colloid Interface Sci.* **2013**, *393*, 109–118. [CrossRef] [PubMed]
8. Elmasry, S.A.; Elgawish, M.A.; El-Shawy, O.E.; Askar, M.A.; Helmy, E.A.; Rashed, L.A. Biologically Synthesized Quercetin Loaded Magnetite Nanoparticles Enhanced Cytotoxicity and Radiosensitivity of Cancer Cells in Vitro. *J. Chem. Pharm. Res.* **2016**, *8*, 758–771.
9. Hu, J.; Obayemi, J.; Malatesta, K.; Kosmrlj, A.; Soboyejo, W. Enhanced cellular uptake of LHRH-conjugated PEG-coated magnetite nanoparticles for specific targeting of triple negative breast cancer cells. *Mater. Sci. Eng. C* **2018**, *88*, 32–45. [CrossRef]
10. Krukemeyer, M.G.; Krenn, V.; Huebner, F.; Wagner, W.; Resch, R. History and Possible Uses of Nanomedicine Based on Nanoparticles and Nanotechnological Progress. *J. Nanomed. Nanotechnol.* **2015**, *6*, 336. [CrossRef]
11. Stevanović, M.; Lukić, M.J.; Stanković, A.; Filipović, N.; Kuzmanović, M.; Janićijević, Ž. Biomedical inorganic nanoparticles: Preparation, properties, and perspectives. *Biomed. Mater. Eng.* **2019**, 1–46. [CrossRef]
12. Favela-Camacho, S.E.; Samaniego, J.E.; Godínez-García, A.; Avilés-Arellano, L.M.; Pérez-Robles, J.F. How to decrease the agglomeration of magnetite nanoparticles and increase their stability using surface properties. *Colloids Surf. A Physicochem. Eng. Asp.* **2019**, *574*, 29–35. [CrossRef]

13. Unsoy, G.; Gunduz, U.; Oprea, O.; Ficai, D.; Sonmez, M.; Radulescu, M.; Alexie, M.; Ficai, A. Magnetite: From Synthesis to Applications. *Curr. Top. Med. Chem.* **2015**, *15*, 1622–1640. [CrossRef] [PubMed]
14. Lee, X.J.; Lim, H.N.; Gowthaman, N.; Rahman, M.B.A.; Abdullah, C.A.C.; Muthoosamy, K. In-situ surface functionalization of superparamagnetic reduced graphene oxide—Fe$_3$O$_4$ nanocomposite via Ganoderma lucidum extract for targeted cancer therapy application. *Appl. Surf. Sci.* **2020**, *512*, 145738. [CrossRef]
15. Thiesen, B.; Jordan, A. Clinical applications of magnetic nanoparticles for hyperthermia. *Int. J. Hyperth.* **2008**, *24*, 467–474. [CrossRef]
16. Jesus, P.D.C.C.D.; Pellosi, D.; Tedesco, A. Magnetic nanoparticles: Applications in biomedical processes as synergic drug-delivery systems. *Mater. Biomed. Eng.* **2019**, 371–396. [CrossRef]
17. Widder, K.J.; Morris, R.M.; Poore, G.A.; Howard, D.P.; Senyei, A.E. Selective targeting of magnetic albumin microspheres containing low-dose doxorubicin: Total remission in Yoshida sarcoma-bearing rats. *Eur. J. Cancer Clin. Oncol.* **1983**, *19*, 135–139. [CrossRef]
18. Mohanta, S.C.; Saha, A.; Devi, P.S. PEGylated Iron Oxide Nanoparticles for pH Responsive Drug Delivery Application. *Mater. Today Proc.* **2018**, *5*, 9715–9725. [CrossRef]
19. García-Jimeno, S.; Estelrich, J. Ferrofluid based on polyethylene glycol-coated iron oxide nanoparticles: Characterization and properties. *Colloids Surfaces A Physicochem. Eng. Asp.* **2013**, *420*, 74–81. [CrossRef]
20. Revia, R.; Zhang, M. Magnetite nanoparticles for cancer diagnosis, treatment, and treatment monitoring: Recent advances. *Mater. Today* **2016**, *19*, 157–168. [CrossRef]
21. Wahajuddin, S.A. Superparamagnetic iron oxide nanoparticles: Magnetic nanoplatforms as drug carriers. *Int. J. Nanomed.* **2012**, *7*, 3445–3471. [CrossRef] [PubMed]
22. Assa, F.; Jafarizadeh-Malmiri, H.; Ajamein, H.; Anarjan, N.; Vaghari, H.; Sayyar, Z.; Berenjian, A. A biotechnological perspective on the application of iron oxide nanoparticles. *Nano Res.* **2016**, *9*, 2203–2225. [CrossRef]
23. Kurczewska, J.; Cegłowski, M.; Schroeder, G. Preparation of multifunctional cascade iron oxide nanoparticles for drug delivery. *Mater. Chem. Phys.* **2018**, *211*, 34–41. [CrossRef]
24. Tietze, R.; Zaloga, J.; Unterweger, H.; Lyer, S.; Friedrich, R.P.; Janko, C.; Pöttler, M.; Dürr, S.; Alexiou, C. Magnetic Nanoparticle-based drug delivery for cancer therapy, Magnetic Nanoparticle-based Drug Delivery for Cancer Therapy. *Biochem. Biophys. Res. Commun.* **2015**, *468*, 463–470. [CrossRef]
25. Akbarzadeh, A.; Samiei, M.; Davaran, S. Magnetic nanoparticles: Preparation, physical properties, and applications in biomedicine. *Nanoscale Res. Lett.* **2012**, *7*, 144. [CrossRef]
26. Tansık, G.; Yakar, A.; Gündüz, U. Tailoring magnetic PLGA nanoparticles suitable for doxorubicin delivery. *J. Nanoparticle Res.* **2013**, *16*, 2171. [CrossRef]
27. Podaru, G.; Chikan, V. Magnetism in Nanomaterials: Heat and Force from Colloidal Magnetic Particles. In *Magnetic Nanomaterials: Applications in Catalysis and Life Sciences*; Bossmann, H.S., Wang, H., Eds.; The Royal Society of Chemistry: Croydon, UK, 2017; pp. 1–24. [CrossRef]
28. Kharisov, B.I.; Dias, H.V.R.; Kharissova, O.V.; Vázquez, A.; Peña, Y.; Gómez, I. Solubilization, dispersion and stabilization of magnetic nanoparticles in water and non-aqueous solvents: Recent trends. *RSC Adv.* **2014**, *4*, 45354–45381. [CrossRef]
29. Ali, A.; Zafar, H.; Zia, M.; ul Haq, I.; Phull, A.R.; Ali, J.S.; Hussain, A. Synthesis, characterization, applications, and challenges of iron oxide nanoparticles. *Nanotechnol. Sci. Appl.* **2016**, *9*, 49–67. [CrossRef]
30. Knop, K.; Hoogenboom, R.; Fischer, D.; Schubert, U.S. Poly(ethylene glycol) in Drug Delivery: Pros and Cons as Well as Potential Alternatives. *Angew. Chem. Int. Ed.* **2010**, *49*, 6288–6308. [CrossRef]
31. Rabanel, J.M.; Hildgen, P.; Banquy, X. Assessment of PEG on polymeric particles surface, a key step in drug carrier translation. *J. Control. Release* **2014**, *185*, 71–87. [CrossRef] [PubMed]
32. Hua, X.; Yang, Q.; Dong, Z.; Zhang, J.; Zhang, W.; Wang, Q.; Tan, S.; Smyth, H.D.C. Magnetically triggered drug release from nanoparticles and its applications in anti-tumor treatment. *Drug Deliv.* **2017**, *24*, 511–518. [CrossRef]
33. Luo, B.; Xu, S.; Luo, A.; Wang, W.-R.; Wang, S.-L.; Guo, J.; Lin, Y.; Zhao, D.; Wang, C.-C. Mesoporous Biocompatible and Acid-Degradable Magnetic Colloidal Nanocrystal Clusters with Sustainable Stability and High Hydrophobic Drug Loading Capacity. *ACS Nano* **2011**, *5*, 1428–1435. [CrossRef]
34. Luo, B.; Xu, S.; Ma, W.-F.; Wang, W.-R.; Wang, S.-L.; Guo, J.; Yang, W.-L.; Hu, J.-H.; Wang, C.-C. Fabrication of magnetite hollow porous nanocrystal shells as a drug carrier for paclitaxel. *J. Mater. Chem.* **2010**, *20*, 7107–7113. [CrossRef]
35. Šegota, S.; Baranović, G.; Mustapić, M.; Strasser, V.; Domazet Jurašin, D.; Crnolatac, I.; Al Hossain, M.S.; Dutour Sikirić, M. The role of spin-phonon coupling in enhanced desorption kinetics of antioxidant flavonols from magnetic nanoparticles aggregates. *J. Magn. Magn. Mater.* **2019**, *490*, 165530. [CrossRef]
36. Obaidat, I.M.; Issa, B.; Haik, Y. Magnetic Properties of Magnetic Nanoparticles for Efficient Hyperthermia. *Nanomaterials* **2015**, *5*, 63. [CrossRef] [PubMed]
37. Mertz, D.; Sandre, O.; Bégin-Colin, S. Drug releasing nanoplatforms activated by alternating magnetic fields. *Biochim. Biophys. Acta Gen. Subj.* **2017**, *1861*, 1617–1641. [CrossRef] [PubMed]
38. Razmara, R.S.; Daneshfar, A.; Sahraei, R. Solubility of Quercetin in Water + Methanol and Water + Ethanol from (292.8 to 333.8) K. *J. Chem. Eng. Data* **2010**, *55*, 3934–3936. [CrossRef]

39. Widdrat, M.; Kumari, M.; Tompa, Éva; Pósfai, M.; Hirt, A.; Faivre, D. Keeping Nanoparticles Fully Functional: Long-Term Storage and Alteration of Magnetite. *ChemPlusChem* **2014**, *79*, 1225–1233. [CrossRef] [PubMed]
40. Kou, Y.; Wang, S.; Luo, J.; Sun, K.; Zhang, J.; Tan, Z.; Shi, Q. Thermal analysis and heat capacity study of polyethylene glycol (PEG) phase change materials for thermal energy storage applications. *J. Chem. Thermodyn.* **2019**, *128*, 259–274. [CrossRef]
41. Gronczewska, E.; Defort, A.; Kozioł, J. Kinetics of Ibuprofen Release from Magnetic Nanoparticles Coated with Chitosan, Peg and Dextran. *Pharm. Chem. J.* **2016**, *50*, 491–499. [CrossRef]
42. Hariani, P.L.; Faizal, M.; Ridwan, R.; Marsi, M.; Setiabudidaya, D. Synthesis and Properties of Fe_3O_4 Nanoparticles by Co-precipitation Method to Removal Procion Dye. *Int. J. Environ. Sci. Dev.* **2013**, *4*, 336–340. [CrossRef]
43. Yang, J.; Zou, P.; Yang, L.; Cao, J.; Sun, Y.; Han, D.; Yang, S.; Wang, Z.; Chen, G.; Wang, B.; et al. A comprehensive study on the synthesis and paramagnetic properties of PEG-coated Fe_3O_4 nanoparticles. *Appl. Surf. Sci.* **2014**, *303*, 425–432. [CrossRef]
44. Masoudi, A.; Hosseini, H.R.M.; Shokrgozar, M.A.; Ahmadi, R.; Oghabian, M.A. The effect of poly(ethylene glycol) coating on colloidal stability of superparamagnetic iron oxide nanoparticles as potential MRI contrast agent. *Int. J. Pharm.* **2012**, *433*, 129–141. [CrossRef] [PubMed]
45. Wu, W.; He, Q.; Jiang, C. Magnetic Iron Oxide Nanoparticles: Synthesis and Surface Functionalization Strategies. *Nanoscale Res. Lett.* **2008**, *3*, 397–415. [CrossRef]
46. Catauro, M.; Papale, F.; Bollino, F.; Piccolella, S.; Marciano, S.; Nocera, P.; Pacifico, S. Silica/quercetin sol–gel hybrids as antioxidant dental implant materials. *Sci. Technol. Adv. Mater.* **2015**, *16*, 035001. [CrossRef] [PubMed]
47. Sathishkumar, P.; Li, Z.; Govindan, R.; Jayakumar, R.; Wang, C.; Gu, F.L. Zinc oxide-quercetin nanocomposite as a smart nano-drug delivery system: Molecular-level interaction studies. *Appl. Surf. Sci.* **2021**, *536*, 147741. [CrossRef]
48. Dong, Y.; Hu, M.; Ma, R.; Cheng, H.; Yang, S.; Li, Y.Y.; Zapien, J.A. Evaporation-induced synthesis of carbon-supported Fe_3O_4 nanocomposites as anode material for lithium-ion batteries. *CrystEngComm* **2013**, *15*, 1324–1331. [CrossRef]
49. Park, J.S.; Kim, T.S.; Son, K.S.; Maeng, W.-J.; Kim, H.-S.; Ryu, M.; Lee, S.Y. The effect of UV-assisted cleaning on the performance and stability of amorphous oxide semiconductor thin-film transistors under illumination. *Appl. Phys. Lett.* **2011**, *98*, 12107. [CrossRef]
50. Swain, A.K.; Pradhan, L.; Bahadur, D. Polymer Stabilized Fe_3O_4-Graphene as an Amphiphilic Drug Carrier for Thermo-Chemotherapy of Cancer. *ACS Appl. Mater. Interfaces* **2015**, *7*, 8013–8022. [CrossRef] [PubMed]
51. Singh, R.; Lillard, J.W., Jr. Nanoparticle-based targeted drug delivery. *Exp. Mol. Pathol.* **2009**, *86*, 215–223. [CrossRef] [PubMed]
52. Avedian, N.; Zaaeri, F.; Daryasari, M.P.; Javar, H.A.; Khoobi, M. pH-sensitive biocompatible mesoporous magnetic nanoparticles labeled with folic acid as an efficient carrier for controlled anticancer drug delivery. *J. Drug Deliv. Sci. Technol.* **2018**, *44*, 323–332. [CrossRef]
53. Selvaraj, S.; Shanmugasundaram, S.; Maruthamuthu, M.; Venkidasamy, B.; Shanmugasundaram, S. Facile Synthesis and Characterization of Quercetin-Loaded Alginate Nanoparticles for Enhanced In Vitro Anticancer Effect Against Human Leukemic Cancer U937 Cells. *J. Clust. Sci.* **2020**, 1–12. [CrossRef]
54. Kumari, A.; Yadav, S.K.; Pakade, Y.B.; Singh, B.; Yadav, S.C. Development of biodegradable nanoparticles for delivery of quercetin. *Colloids Surf. B Biointerfaces* **2010**, *80*, 184–192. [CrossRef] [PubMed]
55. Usacheva, M.; Layek, B.; Nirzhor, S.S.R.; Prabha, S. Nanoparticle-Mediated Photodynamic Therapy for Mixed Biofilms. *J. Nanomater.* **2016**, *2016*, 1–11. [CrossRef]
56. Engen, A.; Maeda, J.; Wozniak, D.E.; Brents, C.A.; Bell, J.J.; Uesaka, M.; Aizawa, Y.; Kato, T.A. Induction of cytotoxic and genotoxic responses by natural and novel quercetin glycosides. *Mutat. Res. Toxicol. Environ. Mutagen.* **2015**, *784–785*, 15–22. [CrossRef]
57. Jurasekova, Z.; Domingo, C.; Garcia-Ramos, J.V.; Sanchez-Cortes, S. Effect of pH on the chemical modification of quercetin and structurally related flavonoids characterized by optical (UV-visible and Raman) spectroscopy. *Phys. Chem. Chem. Phys.* **2014**, *16*, 12802–12811. [CrossRef] [PubMed]
58. Sokolová, R.; Degano, I.; Ramešová, Š.; Bulíčková, J.; Hromadová, M.; Gál, M.; Fiedler, J.; Valášek, M. The oxidation mechanism of the antioxidant quercetin in nonaqueous media. *Electrochim. Acta* **2011**, *56*, 7421–7427. [CrossRef]
59. Schwarzl, R.; Du, F.; Haag, R.; Netz, R.R. General method for the quantification of drug loading and release kinetics of nanocarriers. *Eur. J. Pharm. Biopharm.* **2017**, *116*, 131–137. [CrossRef]
60. Yu, M.; Yuan, W.; Li, D.; Schwendeman, A.; Schwendeman, S.P. Predicting drug release kinetics from nanocarriers inside dialysis bags. *J. Control. Release* **2019**, *315*, 23–30. [CrossRef] [PubMed]
61. Li, Y.; Yao, J.; Han, C.; Yang, J.; Chaudhry, M.T.; Wang, S.; Liu, H.; Yin, Y. Quercetin, Inflammation and Immunity. *Nutrients* **2016**, *8*, 167. [CrossRef]

Article

Construction of *Polygonatum sibiricum* Polysaccharide Functionalized Selenium Nanoparticles for the Enhancement of Stability and Antioxidant Activity

Wanwen Chen [1,2], Hao Cheng [1,*] and Wenshui Xia [1,2]

1 State Key Laboratory of Food Science and Technology, School of Food Science and Technology, Jiangnan University, Wuxi 214122, China; wwchen@jiangnan.edu.cn (W.C.); xiaws@jiangnan.edu.cn (W.X.)
2 Collaborative Innovation Center of Food Safety and Quality Control in Jiangsu Province, Jiangnan University, Wuxi 214122, China
* Correspondence: haocheng@jiangnan.edu.cn

Abstract: Although selenium nanoparticles (SeNPs) have attracted great attention due to their potential antioxidant activity and low toxicity, the application of SeNPs is still restricted by their poor stability. A combination of polysaccharides and SeNPs is an effective strategy to overcome the limitations. In this study, *Polygonatum sibiricum* polysaccharide (PSP) was used as a stabilizer to fabricate SeNPs under a simple redox system. Dynamic light scattering, transmission electron microscopy, energy dispersive X-ray, ultraviolet-visible spectroscopy, Fourier transform infrared, and X-ray photoelectron spectrometer were applied to characterize the synthesized PSP-SeNPs. The stability and the antioxidant activity of PSP-SeNPs were also investigated. The results revealed that the zero-valent and well-dispersed spherical PSP-SeNPs with an average size of 105 nm and a negative ζ-potential of −34.9 mV were successfully synthesized using 0.1 mg/mL PSP as a stabilizer. The prepared PSP-SeNPs were stable for 30 days at 4 °C. The decoration of the nanoparticle surface with PSP significantly improved the free radical scavenging ability of SeNPs. Compared to the H_2O_2-induced oxidative stress model group, the viability of PC-12 cells pretreated with 20 µg/mL PSP-SeNPs increased from 56% to 98%. Moreover, PSP-SeNPs exhibited a higher protective effect on the H_2O_2-induced oxidative damage on PC-12 cells and lower cytotoxicity than sodium selenite and SeNPs. In summary, these results suggest the great potential of PSP-SeNPs as a novel antioxidant agent in the food or nutraceuticals area.

Keywords: selenium nanoparticles; *Polygonatum sibiricum* polysaccharide; stability; antioxidant

1. Introduction

Selenium is an essential micronutrient for humans and animals [1]. It is an integral component of more than 30 kinds of selenoproteins and selenium-containing enzymes, such as selenoprotein P (SelP), selenoprotein S (SelS), selenoprotein M (SelM), subfamilies of thioredoxin reductases (TrxR), glutathione peroxidases (GPx), and iodothyronine deiodinases (ID), that play a key role in regulating redox balance and preventing cellular damage from radicals [2,3]. However, at least one billion people in the world are at risk of selenium deficiency at present because the intake of selenium is insufficient to meet the daily requirement [4]. Epidemiological studies established that selenium deficiency is associated with many diseases, including premature aging, a decline in sperm motility, myocardial failure, neurological diseases, endemic osteoarthropathy (Keshan disease), and ischemic heart disease [5]. Although high-dose sodium selenite, methyl selenium, and selenocysteine exhibit excellent bioactivities, they can also result in serious toxicity problems, leading to many diseases [6]. Thus, it is of great importance to seek novel selenium species as food supplements or additives.

Selenium nanoparticles (SeNPs) have gained much attention owing to their unique physical, chemical, and antioxidant activities [7]. Moreover, SeNPs have higher bioavailability and lower toxicity in comparison to other chemical forms of selenium, making them the promising alternative selenium source in food dietary [8]. However, SeNPs alone with valence state zero are highly unstable in an aqueous solution and easily transform to aggregate, resulting in lower bioactivity and further limiting their practical application [9]. Many efforts have been made to develop a simple, efficient, and green strategy for the dispersion and stabilization of SeNPs using bioactive templates [10]. Natural polysaccharides not only have complex structures, large specific surface areas, and ionizable functional groups but also possess excellent biocompatibility and biodegradability [11]. These features could decrease the surface energy of SeNPs, further preventing aggregation through electrostatic interaction or hydrogen bonds. Thus, polysaccharides applied as carriers to fabricate SeNPs with desired characteristics, such as stability and functionality, using the green chemical method is drawing much attention recently. For example, numerous studies reported that chitosan (CS) could be used as templates to prepare uniform SeNPs and the ligated SeNPs remain stable for over 1 month [12]. However, the superior properties of CS are limited due to its water insolubility and our previous research also found that CS-SeNPs aggregated under alkaline conditions (pH $\geq$ 9) [13]. Several polysaccharides derived from fungi [14], fruit [15], and medicinal plants [16] have been demonstrated to enhance the antioxidant activity of SeNPs. Recently, medicinal plant polysaccharides have attracted increasing attention due to their significant bioactivities with no side effects [17]. Therefore, it can be expected that the combination of medicinal plant polysaccharides with SeNPs will reduce the inherent limitations and enhance the benefits of selenium and polysaccharides.

Polygonatum sibiricum is a traditional Chinese herbal medicine, belonging to the *Liliaceae* family, which has been introduced in the 2015 edition of pharmacopeia [18]. China has abundant resources of *Polygonatum sibiricum*, especially in the south of the Yangtze River [19]. The constituents of *P. sibiricum* include polysaccharides, saponins, flavonoids, alkaloids, lignin, vitamins, and a variety of trace elements, of which polysaccharides are the major pharmacologically active ingredients [20]. In the last three years, *Polygonatum sibiricum* polysaccharides (PSP) are demonstrated to exhibit a wide range of pharmacological activity [21], such as osteogenic activity [22], anti-diabetes [23], immunological activity [24], and especially antioxidant activity, which makes them suitable for application in functional foods and therapeutic agents. PSP demonstrated strong antioxidant properties, which could attenuate D-gal-induced heart aging [25] and protect the mice livers against ethanol-induced oxidative damage via inhibiting oxidative stress [26]. However, no study has been reported using PSP as a decorator to functionalize SeNPs.

In this study, considering the antioxidant activity of PSP as well as the drawbacks of SeNPs, a combined strategy was conducted to fabricate SeNPs using PSP as a stabilizer in the redox system of sodium selenite (Na_2SeO_3) and ascorbic acid (Vc) through a simple chemistry approach. The synthesized PSP functionalized SeNPs (PSP-SeNPs) were characterized by dynamic light scattering (DLS), transmission electron microscopy (TEM), energy dispersive X-ray (EDX), ultraviolet-visible spectroscopy (UV-vis), Fourier transform infrared (FTIR), and X-ray photoelectron spectrometer (XPS). The physicochemical stabilities of synthesized nanoparticles under varying conditions, including ionic strength, pH, and temperature, were analyzed. In addition, the antioxidant activity of PSP and PSP-SeNPs was quantified by ABTS and DDPH free radical scavenging assays. Moreover, the protective effect on the H_2O_2-induced cell death was also investigated by MTT assay.

2. Materials and Methods

2.1. Reagents

Commercial *Polygonatum sibiricum* polysaccharide (PSP) with a purity of 95% and a molecular weight of 14 kDa was obtained from Qiannuo Biotechnology Co. Ltd. (Xi'an, China), sodium selenite (Na_2SeO_3), hydrogen peroxide (H_2O_2), ascorbic acid (Vc), potassium persulfate ($K_2S_2O_8$), 1, 1-diphenyl-2-picrylhydrazyl (DPPH), 2, 2-azinobis (3-ethylbenzothiazoline-6-

sulfonic acid) and diammonium salt (ABTS) were purchased from Sinopharm Chemical Reagent Co., Ltd. (Shanghai, China). All chemicals used were of analytical grade, and the water used in all experiments was obtained from the Milli-Q system.

2.2. Preparation of SeNPs and PSP Stabilized SeNPs

PSP-SeNPs were prepared according to the procedure described by Ye et al. with minor modification [8]. PSP stock solution (5 mg/mL) was freshly prepared. Where 1 mL of sodium selenite solution (50 mM) was mixed with various volumes of PSP solution under stirring for 5 min. Then 1 mL of ascorbic acid solution (200 mM) was added dropwise into the mixture, and it was reconstituted to a final volume of 10 mL with Milli-Q water. The reaction was carried out at room temperature for 30 min. Finally, the solution was dialyzed using regenerated cellulose tubes (Mw cutoff 3500 Da) against ultrapure water for 48 h at 4 °C. The final concentrations of PSP were 0.01, 0.05, 0.075, 0.1, 0.125, 0.15, 0.25 mg/mL. SeNPs were synthesized in the absence of PSP through the same procedure as above. The resultant products were lyophilized to obtain the freeze-dried nanocomposites. The concentration of selenium was determined by the Optima 8300 inductively coupled plasma optical emission spectrometer (ICP-OES, PerkinElmer, Billerica, MA, USA).

2.3. Characterization

The mean diameter, size distribution, and ζ-potential of nanocomposites were determined using a Zetasizer Nano ZS analyzer (Malvern Instruments Corporation, Worcestershire, UK). The morphology was observed using transmission electron microscopy (TEM) (JEOL, JEM-2100, Tokyo, Japan). Samples for TEM observation were prepared by placing one drop of SeNPs and PSP-SeNPs aqueous solution on a carbon-coated copper grid and dried at room temperature. The micrographs were acquired at the accelerating voltage of 200 kV. The elemental composition and distribution were determined by the energy dispersive X-ray (EDX) analysis performed on a high-resolution transmission electron microscopy (HRTEM) (JEOL, JEM-2100, Tokyo, Japan). The ultraviolet-visible (UV-vis) spectrophotometer (UV-1800, Shimadzu Corporation, Tokyo, Japan) was used to measure the UV-vis absorption spectra of SeNPs and PSP-SeNPs solutions in the wavelength range of 190–800 nm with an interval of 1.0 nm. The Fourier transform infrared (FTIR) spectra were recorded on a Nicolet iS 10 instrument (Thermo Fisher Scientific, Waltham, MA, USA). Each sample was grounded with KBr, pressed into uniform pellets, and scanned in the wavenumber range of 4000–400 cm^{-1} with a resolution of 4.0 cm^{-1} using pure KBr as the background. The X-ray photoelectron spectrometer (XPS) was used to analyze the valence states of the elements. The XPS patterns were operated on a Thermo Scientific ESCALab 250Xi+ (Thermo Fisher Scientific, Waltham, MA, USA) using 150 W monochromated Al Kα radiation.

2.4. Stability of PSP-SeNPs

The stability of PSP-SeNPs under various conditions was investigated according to the methods described previously [27]. To determine the effect of ionic concentration on stability, 10 mL of PSP-SeNPs were mixed with different concentrations of NaCl solution (10, 50, and 100 mM). The effect of pH on the stability of NPs was analyzed by adjusting the pH of PSP-SeNPs to 2, 3, 4, 5, 6, 7, 8, 9, and 10 using 0.1 M HCl or NaOH. Where 10 mL of PSP-SeNPs were incubated in a water bath at different temperatures (25, 50, 70, and 90 °C) to investigate the effect of temperature on the stability of PSP-SeNPs. After being stabilized for 1 h, their mean diameter and ζ-potential were determined using a Zetasizer Nano ZS analyzer. In addition, PSP-SeNPs solutions were stored at 4 °C for 30 days to investigate the short-term storage stability by determining the mean diameter and ζ-potential.

2.5. Antioxidant Assays

2.5.1. DPPH Radical Scavenging Ability

The DPPH radical scavenging activity was determined referring to the methods reported previously with minor modifications [14]. Various concentrations of PSP, SeNPs,

PSP-SeNPs, and Vc at 0.01, 0.05, 0.1, 0.25, 0.5, 0.75, 1.0 mg/mL were prepared. Further, 2 mL of the sample solutions were mixed with an equal volume of freshly prepared DPPH solution (50 mg/L) in ethanol. The mixture was shaken vigorously and incubated in darkness at 33 °C for 30 min. The absorbance was measured at 517 nm using a UV-vis spectrophotometer. Vc was used as a positive control. The scavenging activity was calculated as follows:

$$\text{DPPH radical scavenging ability } (\%) = \left(1 - \frac{A_a - A_b}{A_c}\right) \times 100 \quad (1)$$

where A_a is the absorbance of the sample mixed with DPPH solution, A_b is the absorbance of the sample in the absence of the DPPH solution, A_c is the absorbance of the DPPH solution without the sample as a blank control.

2.5.2. ABTS Radical Cation Decolonization Assay

The assay of ABTS radical cation scavenging ability was performed as described previously with some modification [28]. ABTS and potassium persulfate ($K_2S_2O_8$) were dissolved in distilled water. A stock solution of ABTS$^{•+}$ was prepared by mixing 7.4 mM ABTS solution with 2.6 mM $K_2S_2O_8$ solution. The mixture was incubated for 12 h in the dark to reach equilibrium. The ABTS$^{•+}$ stock solution was diluted with sodium phosphate buffer (pH 7.4) to obtain an optical density of 0.70 ± 0.02 at 734 nm. Then 1 mL of different concentrations of PSP, SeNPs, PSP-SeNPs, and Vc (0.01, 0.05, 0.1, 0.25, 0.5, 0.75, 1.0 mg/mL) was added to 4 mL of diluted ABTS$^{•+}$ solution. The mixture was vigorously blended and incubated at room temperature for 6 min in darkness. The absorbance was measured at 734 nm using a UV-vis spectrophotometer. The ability to scavenge ABTS$^{•+}$ was calculated by Equation (2).

$$\text{ABTS}^{•+}\text{ radical scavenging ability } (\%) = \left(1 - \frac{A_d - A_e}{A_f}\right) \times 100 \quad (2)$$

where A_d is the absorbance of the sample mixed with the ABTS$^{•+}$ solution, A_e is the absorbance of the sample in the absence of the ABTS$^{•+}$ solution, A_f is the absorbance of the ABTS$^{•+}$ solution without the sample.

2.6. Cells Culture and MTT Assays

PC-12 cells were cultured in Dulbecco's modified Eagle's medium (DMEM) supplemented with 10% fetal bovine serum (FBS) and 1% antibiotic mixture (100 U/mL penicillin and 100 µg/mL streptomycin). The cytotoxic effects of different selenium concentrations of PSP-SeNPs, SeNPs, and Na_2SeO_3 on cells were tested using MTT assays [15]. Cells were seeded in a 96-well plate at a density of 1×10^4 cells/well and incubated at 37 °C in a CO_2 incubator (5% CO_2 and 95% relative humidity) for 24 h. Then the medium was removed and cells were treated with different concentrations of samples prepared in DMEM with 10% FBS for an additional 24 h. After incubation, 20 µL of MTT (5 mg/mL) was added to each well and incubated at 37 °C for 3 h. Then the supernatant was removed and 150 µL of DMSO was added. The absorbance was measured by a microplate reader at 570 nm. The cell viability was calculated by Equation (3).

$$\text{Cell viability } (\%) = OD_{sample}/OD_{control} \times 100 \quad (3)$$

where OD_{sample} is the absorbance of the treated cells and $OD_{control}$ is the absorbance of the control cells.

To determine the protective effect of PSP-SeNPs, SeNPs, and Na_2SeO_3 on H_2O_2-induced cell cytotoxicity, cells were pre-incubated with different selenium concentrations of samples prepared in DMEM with 10% FBS for 24 h. After incubation, the medium was removed and cells were washed with PBS. Then cells were treated with a medium containing 500 μM H_2O_2 for 12 h. The medium was removed and the cell viability was determined by MTT assay as described above.

2.7. Statistical Analysis

All the experiments were performed at least in triplicate. The results were expressed as mean ± standard deviation (SD). Statistical analysis was carried out using paired t-tests for comparing means of two samples by the SPSS 20.0 statistical software (IBM, Armonk, NY, USA). Statistical differences between samples were performed with a level of significance of $p < 0.05$.

3. Results

3.1. The Synthesis of SeNPs and PSP-SeNPs

In the present study, SeNPs and PSP-SeNPs were prepared using a simple redox system of ascorbic acid and sodium selenite in the absence and presence of PSP as the stabilizer and capping agent. The visual color of the reaction solution is an indicator to preliminary infer the formation of selenium nanoparticles [29]. As shown in Figure 1, the red color of the solution indicated the SeO_3^{2-} was successfully reduced to either monoclinic or amorphous SeNPs [16]. In addition, the SeNPs in the presence of PSP showed a uniform red color and were stable in the aqueous solution. However, SeNPs without the decoration of PSP aggregated into precipitation after 1 day of storage, whereas no significant changes were observed in the solution of PSP-SeNPs. This might be attributed to the high surface energy, leading to the aggregation of SeNPs [9]. Hence, PSP plays a key role in the formation and stabilization of SeNPs.

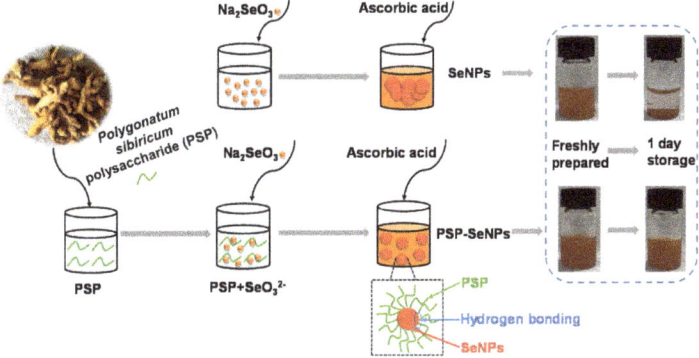

Figure 1. Synthetic scheme for the preparation of selenium nanoparticles (SeNPs) and Polygonatum sibiricum polysaccharide stabilized selenium nanoparticles (PSP-SeNPs) and images of the dispersions before and after storage for 1 day.

3.2. The Size and ζ-Potential Analysis of SeNPs and PSP-SeNPs

The concentration of the polysaccharides is an important factor that influences the size of SeNPs, further affecting their functionality in food or medical application [30]. Thus, the effect of PSP concentrations on the hydrated particle size and the corresponding polydispersity index (PDI), as well as the ζ-potential of nanoparticles in the aqueous solution was investigated first. The particle size of barely SeNPs was up to 157 nm (Figure 2A). The addition of PSP at different concentrations could decrease the average size of SeNPs. The average diameter of PSP-SeNPs significantly decreased from 151 to 132 nm as the concentration of PSP increased from 0.01 to 0.075 mg/mL. PSP-SeNPs showed the smallest average size of 114 nm at the PSP concentration of 0.1 mg/mL, whereas further increases in PSP concentration from 0.125 to 0.25 mg/mL resulted in an increase in the size from 123 to 152 nm. It might be due to PSP at a low concentration was not enough to control the formation of SeNPs and prevent them from aggregation [31]. On the other hand, too high

PSP concentration represented more PSP chains coated on the surface of SeNPs, resulting in a larger hydration particle size [32]. As shown in Figure 2B, SeNPs in the absence of PSP exhibited a negative ζ-potential at −20.3 mV. The ζ-potential values of PSP-SeNPs were determined to be approximately −24.7, −26.6, −29.6, −30.4, −32.8, −34.9 mV at the PSP concentration of 0.01, 0.05, 0.075, 0.1, 0.125, 0.25 mg/mL. The absolute ζ-potential values of PSP-SeNPs increased with the PSP concentration increasing, further demonstrating that negatively charged PSP was exposed on the surface of SeNPs. Moreover, the higher magnitude of ζ-potential represents greater stability of nanoparticles [13], suggesting that the SeNPs decorated with PSP possess higher stability than barely SeNPs. PSP-SeNPs prepared by 0.1 mg/mL PSP were used in the following experiments.

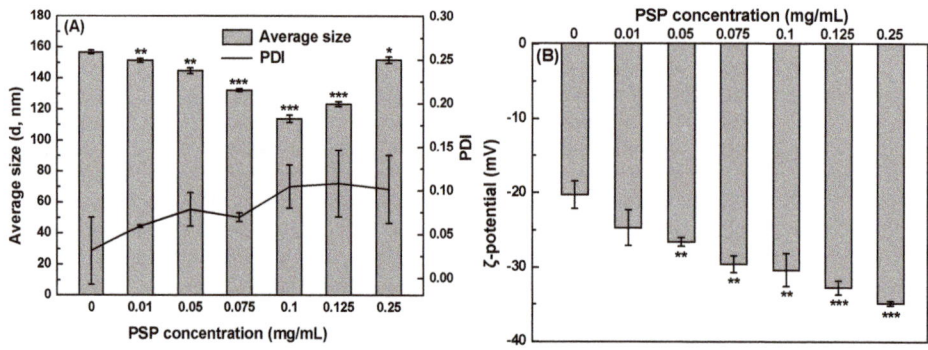

Figure 2. Size distribution (**A**) and ζ-potential (**B**) of SeNPs and PSP-SeNPs prepared with different concentrations of PSP (0.01–0.25 mg/mL). Values marked with *: $p < 0.05$, **: $p < 0.01$, and ***: $p < 0.001$ indicated significant differences when compared to SeNPs.

3.3. Morphological and Structural Characterizations of SeNPs and PSP-SeNPs

The morphology and size of SeNPs and PSP-SeNPs were further characterized by TEM. Figure 3A,B exhibited the TEM images of SeNPs in the absence of PSP. The results showed that adjacent SeNPs agglomerated together and presented a dendritic structure. The large-sized cluster and aggregates can also be easily visualized. However, the SeNPs in the presence of 0.1 mg/mL PSP (Figure 3C,D) exhibited a homogeneous and monodisperse spherical structure with an average size of about 105 nm, confirming the important role of PSP in regulating and stabilizing SeNPs. It should be pointed out that the hydrodynamic radius of the nanoparticles provided in the DLS analysis was larger than the size observed in the TEM image, which was sensitive to the electron-rich nanoparticles. The HRTEM image (Figure 3E) of an individual PSP-SeNPs showed a distinct lattice fringe with an interplanar spacing of 0.43 nm, revealing the excellent crystallinity of PSP-SeNPs. The elemental composition and distribution of the PSP-SeNPs were further determined by EDX. As shown in Figure 3F, the strong C, O, and Se element peaks were observed in EDX spectra. The PSP-SeNPs had a 63.10% weight percentage of C atom, together with 10.95% O atom and 25.94% Se atom. Furthermore, no other peaks for other elements were detected, indicating that PSP was successfully coated on the surface of SeNPs and confirming the purity of PSP-SeNPs [33].

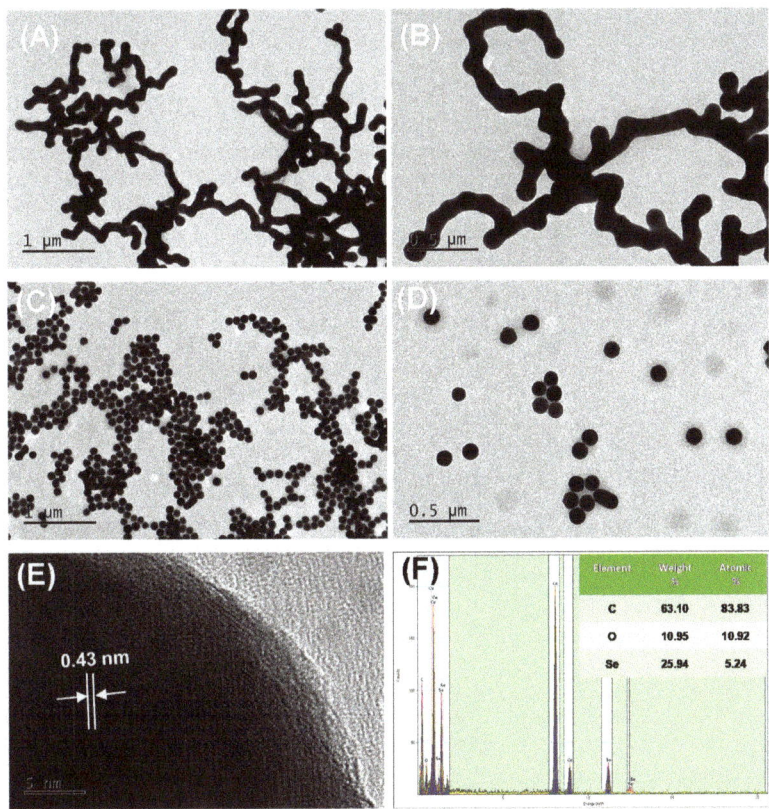

Figure 3. TEM images of SeNPs (**A,B**) and PSP-SeNPs in the presence of 0.1 mg/mL PSP (**C,D**). HRTEM of an individual PSP-SeNPs (**E**) and typical EDX from HRTEM (**F**).

3.4. The Stability of SeNPs and PSP-SeNPs

Stability is an important factor influencing the functionality and applications of nanomaterials. In this study, the effect of pH, temperature, and ionic strength on the stability of PSP-SeNPs was investigated. As shown in Figure 4A, the average size of PSP-SeNPs significantly decreased from 1262 to 186 nm when pH was increased from 2 to 3. It could be observed that no obvious changes occurred in the average size at pH range from 4 to 10. Similar results were also described previously on the stability of Polyporus umbellatus polysaccharide (PUP) coated SeNPs [34]. This might be ascribed to the protonation of PSP at pH 2 that weakened the electrostatic interactions between SeNPs and PSP, leading to the aggregation of nanoparticles. Moreover, the ζ-potential of PSP-SeNPs kept increasing with pH increased and reached the highest value of -32.6 mV at pH 7. A further increase in pH did not significantly affect the ζ-potential of PSP. It has been reported that the ζ-potential of nanoparticles was highly associated with the pKa value of the polysaccharides. The pH value higher than the pKa of polysaccharides resulted in more deprotonated characteristic groups, contributing to the increase in ζ-potential [27]. The average size of PSP-SeNPs increased from 113 to 191 nm, accompanied by the temperature increase from 25 °C to 90 °C with a constant ζ-potential at around -31 mV (Figure 4B). The result indicated that heating could increase the chances and strength of collisions, resulting in a larger size [29]. As shown in Figure 4C, the particle size of PSP-SeNPs exhibited a slight increase in 10 and 50 mM NaCl with decreased ζ-potential, and steeply increased to 882 nm in a high concentration of NaCl at 100 mM. High ion strength could remarkably reduce the surface charge of nanoparticles due to the electrostatic interaction between positive charged Na$^+$ and negatively charged PSP-SeNPs, resulting in the decrease of the electrostatic repulsion among nanoparticles [35]. It was observed that PSP-SeNPs were stable at about 113 nm for at least 20 days of storage (Figure 4D). The stability of PSP-SeNPs was higher than that of SeNPs decorated with a hyperbranched polysaccharide from Lignosus rhinocerotis 14. It should be pointed out that SeNPs in the absence of PSP precipitated after 1-day storage (Figure 1). Moreover, the particle size of PSP-SeNPs only increased from 113 to 123 nm after 30 days of storage and the ζ-potential of PSP-SeNPs presented at around −30 mV during the storage time, suggesting that PSP-SeNPs had better stability.

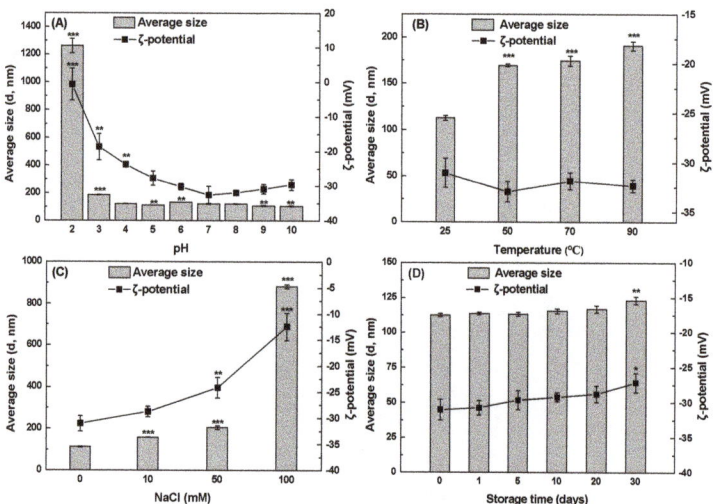

Figure 4. Effect of pH (**A**), temperature (**B**), ion strength (**C**), and storage time (**D**) on the average size and ζ-potential of PSP-SeNPs. Values marked with *: $p < 0.05$, **: $p < 0.01$, and ***: $p < 0.001$ indicated significant differences when compared to the conditions of pH: 7, temperature: 25 °C, NaCl: 0 mM, or storage time: 0 day.

3.5. Characterization and Possible Stabilizing Mechanism of PSP-SeNPs

The UV-vis spectra of PSP and PSP-SeNPs in the range of 190 to 800 nm were presented in Figure 5A. It was shown that no characteristic absorption peaks were observed on the UV-vis spectra of PSP at the concentration of 0.01 mg/mL. The PSP-SeNPs exhibited wide absorption bands with a maximum absorption peak at about 288 nm. The characteristic absorption peak corresponded to a localized surface plasmon response (LSPR), further demonstrating the formation of nanoparticles [36].

FTIR spectra were performed to clarify the interaction between PSP and SeNPs. In the spectrum of PSP (Figure 5B), the broad absorption band at nearly 3390 cm^{-1} was assigned to the O-H stretching vibration. The peak presented at 2927 cm^{-1} was attributed to the C-H stretching vibration. The signals that occurred in the region of 1200–1000 cm^{-1} were associated with the C-O stretching vibration, indicating the existence of a pyranose ring [37]. The FTIR spectrum of PSP-SeNPs was similar to that of the pure PSP, indicating the presence of PSP on the surface of SeNPs. In addition, the O-H stretching vibration occurred red-shift from 3390 cm^{-1} to 3376 cm^{-1}, suggesting the formation of hydrogen bonds between SeNPs and the PSP chains [38]. Based on the above results, we proposed that the interaction mechanism was similar to the combination of arabinogalactans/and SeNPs as described previously [36]. Briefly, the SeO$_3^{2-}$ reacted with the -OH group in the PSP molecule to form special chain-shaped intermediates first, then reduced to the element Se by ascorbic acid. The Se atom further aggregated into the nucleus to form SeNPs as the reaction processed and the -OH groups of PSP were bound to the surface of SeNPs to prevent the aggregation of nanoparticles.

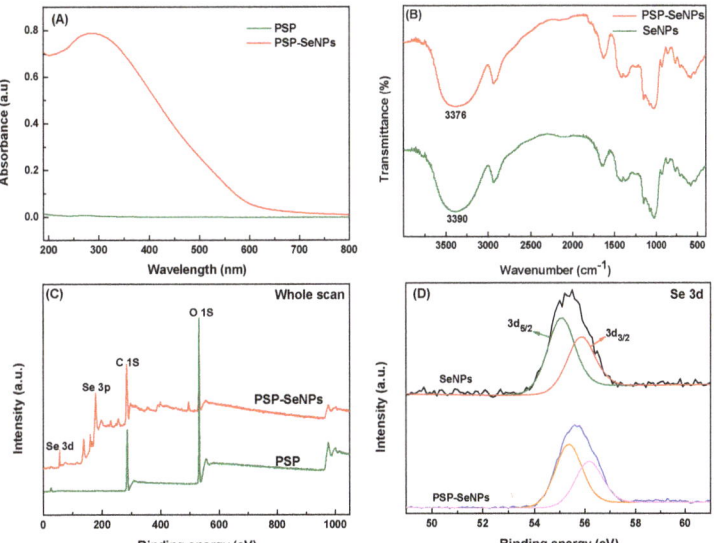

Figure 5. UV-vis spectra (**A**), FTIR spectra (**B**), XPS spectra (**C**), and XPS spectra of Se 3d (**D**) of PSP and PSP-SeNPs.

The XPS spectra were further used to analyze the valence state of selenium. The peaks of Se 3d and 3p orbitals at the binding energy of 55.6 and 179.3 eV (Figure 5C) indicated the zero-valent state of Se within the PSP-SeNPs [10]. As shown in Figure 5D, the peaks of Se $3d_{5/2}$ and Se $3d_{3/2}$ were up-shifted from 55.1 and 55.9 (SeNPs) to 55.4 and 56.2 (PSP-SeNPs), respectively. The results indicate that the Se 3d orbit participated in the formation of PSP-SeNPs [39], confirming that PSP was successfully conjugated to the SeNPs. Meanwhile, no peak was found at 59.5 eV, which represented the typical Se 3d signal of Se (IV), suggesting that Se (IV) was completely reduced to elemental selenium [40].

3.6. Antioxidant Assays

The DPPH and ABTS radical scavenging activity were measured in our study to evaluate the antioxidant activity of PSP, SeNPs, and PSP-SeNPs. As shown in Figure 6A, PSP exhibited a low DPPH radical scavenging ability at the tested concentrations. Both SeNPs and PSP-SeNPs had a concentration-dependent DPPH radical scavenging effect at 0.01–1.0 mg/mL. PSP-SeNPs showed a higher scavenging ability than SeNPs. The scavenging effect of PSP-SeNPs reached 59% at the concentration of 1.0 mg/mL, whereas SeNPs could only scavenge 43% DPPH radical at the same concentration. This might be attributed to the enhanced hydrogen-donating ability of PSP-SeNPs to form a stable DPPH-H molecule [41]. Compared to the DPPH radical, all the tested samples performed more efficiently in scavenging ABTS radical (Figure 6B). Similar to the DPPH scavenging assay, the ABTS radical scavenging capacity of PSP-SeNPs was significantly stronger than that of PSP and SeNPs. At 1.0 mg/mL, the scavenging effects of PSP, SeNPs, and PSP-SeNPs were 20%, 62% and 89%, respectively. It has been reported that the DPPH scavenging ability of gum arabic-selenium nanocomposites was lower than 60% at 1.0 mg/mL [42]. The ABTS radical scavenging activity of SeNPs functionalized with a polysaccharide from *Rosa roxburghii* fruit only reached about 50% at 1.0 mg/mL 15. The free radical scavenging ability of PSP-SeNPs synthesized in our study was higher than the above nanoparticles. Moreover, the results showed that the surface decoration of SeNPs with PSP could remarkably improve the antioxidant activity of SeNPs and PSP. PSP-SeNPs with a smaller size could provide more radical reactive sites due to their larger specific surface area, resulting in higher antioxidant activity [29,43]. However, barely SeNPs were easily aggregated with a decreased active surface to react with the free radicals, further reducing their biological activities [43].

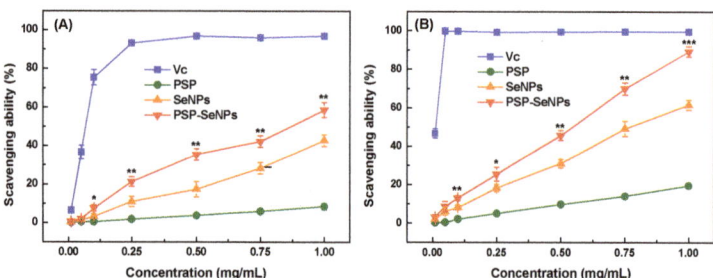

Figure 6. Antioxidant activities of PSP, SeNPs, and PSP-SeNPs in vitro. (**A**) DPPH radical scavenging activity. (**B**) ABTS radical scavenging activity. Ascorbic acid (Vc) is used as a positive control. Values marked with *: $p < 0.05$, **: $p < 0.01$, and ***: $p < 0.001$ indicated significant differences when compared to SeNPs at the same concentration.

3.7. Effects of PSP-SeNPs on H_2O_2-Induced PC-12 Cells Toxicity

Although the free radical scavenging assays proved the excellent antioxidant activity of PSP-SeNPs, the antioxidant assays based on chemical reactions may not necessarily reflect the behavior of antioxidants in biological systems [16]. Thus, the effect of different selenium species on oxidative stress-induced damage to PC-12 cells was further investigated by MTT assay. As depicted in Figure 7A, the cell viability was higher than 90% when incubated with SeNPs and PSP-SeNPs at the concentration of 1–20 µg/mL. However, the cell viability dramatically decreased to 67% after treatment with 20 µg/mL Na_2SeO_3, suggesting that both SeNPs and PSP-SeNPs showed lower cytotoxicity than Na_2SeO_3.

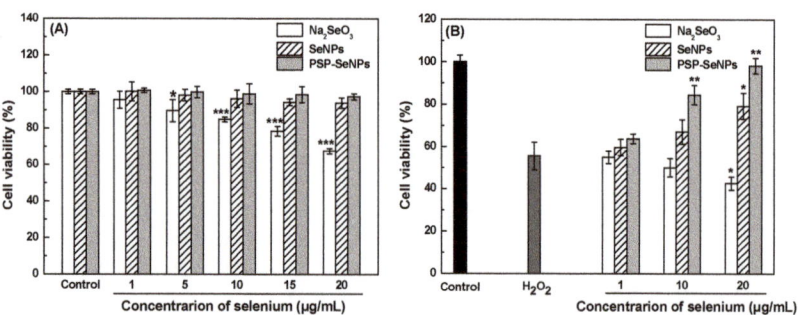

Figure 7. Effects of sodium selenite (Na_2SeO_3), SeNPs, and PSP-SeNPs on the viability of PC-12 cells (**A**). Values marked with *: $p < 0.05$ and ***: $p < 0.001$ indicated significant differences when compared to the control group. The protective effect against H_2O_2 (0.5 µM)-induced PC-12 cells toxicity by MTT assay (**B**). Values marked with *: $p < 0.05$ and **: $p < 0.01$ indicated significant differences when compared to the H_2O_2 treated group.

The overproduction of reactive oxygen species (ROS) is considered to be the main cause of oxidative damage [44]. Herein, exogenous H_2O_2 was used as an inducer of cell damage in our model. PC-12 cells incubated with 500 µM H_2O_2 showed a remarkable decrease of cell viability reaching 56% (Figure 7B). However, the viability of PC-12 cells decreased to 55%, 50%, and 43% when pretreated with Na_2SeO_3 at concentrations of 1,

10, and 20 µg/mL, respectively. Interestingly, compared with the H_2O_2-induced oxidative stress model group, cells pretreated with SeNPs or PSP-SeNPs alleviated the H_2O_2-induced toxicity on PC-12 cells in a concentration-dependent manner, as reflected by the increase in cell viability. The viability of PC-12 cells pretreated with 20 µg/mL SeNPs or PSP-SeNPs significantly increased to 79% and 98%, respectively. In addition, the protective effect of PSP-SeNPs on H_2O_2-induced oxidative damage on PC-12 cells was better than that of SeNPs. The results confirmed that PSP-SeNPs had excellent antioxidant activity in cells, which may be associated with the free radical scavenging ability.

4. Conclusions

Our present study provided a facile approach for the synthesis of size-controlled SeNPs by using PSP as a stabilizer in the redox system of sodium selenite and ascorbic acid. The synthesized PSP-SeNPs presented a monodisperse spherical structure with zero-valent Se. The interaction between the hydroxyl groups of PSP chains and the surface of SeNPs contributed to the stable structure of PSP-SeNPs. Furthermore, PSP-SeNPs exhibited stronger free radical scavenging ability and a higher protective effect against H_2O_2-induced PC-12 cell death than SeNPs. Our findings not only provide the foundations for the utilization of PSP in the development of stable SeNPs but also emphasize the potential application of PSP-SeNPs as an antioxidant in food additives, dietary supplements, and nutraceuticals.

Author Contributions: Conceptualization, W.C. and H.C.; methodology, W.C.; software, W.C.; validation, W.C.; formal analysis, W.C. and H.C.; investigation, W.C.; resources, W.X.; data curation, H.C.; writing—original draft preparation, W.C.; writing—review and editing, H.C.; visualization, W.X.; supervision, W.X.; project administration, W.C. and H.C.; funding acquisition, W.C. and H.C. All authors have read and agreed to the published version of the manuscript.

Funding: This research was funded by the China Postdoctoral Science Foundation (Grant No. 2021M691288) and the National Natural Science Foundation of China (Grant No. 32101939).

Institutional Review Board Statement: Not applicable.

Informed Consent Statement: Not applicable.

Data Availability Statement: The data presented in this study are available in this manuscript.

Conflicts of Interest: The authors declare no conflict of interest.

References

1. Hariharan, S.; Dharmaraj, S. Selenium and selenoproteins: It's role in regulation of inflammation. *Inflammopharmacology* **2020**, *28*, 667–695. [CrossRef] [PubMed]
2. Kursvietiene, L.; Mongirdiene, A.; Bernatoniene, J.; Sulinskiene, J.; Staneviciene, I. Selenium anticancer properties and impact on cellular redox status. *Antioxidants* **2020**, *9*, 80. [CrossRef] [PubMed]
3. Dawood, M.A.O.; El Basuini, M.F.; Yilmaz, S.; Abdel-Latif, H.M.R.; Kari, Z.A.; Razab, M.; Ahmed, H.A.; Alagawany, M.; Gewaily, M.S. Selenium nanoparticles as a natural antioxidant and metabolic regulator in aquaculture: A review. *Antioxidants* **2021**, *10*, 1364. [CrossRef] [PubMed]
4. Zhu, X.K.; Jiang, M.D.; Song, E.Q.; Jiang, X.J.; Song, Y. Selenium deficiency sensitizes the skin for UVB-induced oxidative damage and inflammation which involved the activation of p38 MAPK signaling. *Food Chem. Toxicol.* **2015**, *75*, 139–145. [CrossRef] [PubMed]
5. Zhang, L.Q.; Song, H.X.; Guo, Y.B.; Fan, B.; Huang, Y.T.; Mao, X.F.; Liang, K.H.; Hu, Z.Q.; Sun, X.D.; Fang, Y.; et al. Benefit-risk assessment of dietary selenium and its associated metals intake in China (2017–2019): Is current selenium-rich agro-food safe enough? *J. Hazard. Mater.* **2020**, *398*, 123224. [CrossRef] [PubMed]
6. Jin, Y.X.; Cai, L.Q.; Yang, Q.; Luo, Z.Y.; Liang, L.; Liang, Y.X.; Wu, B.L.; Ding, L.; Zhang, D.D.; Xu, X.J.; et al. Anti-leukemia activities of selenium nanoparticles embedded in nanotube consisted of triple-helix beta-D-glucan. *Carbohydr. Polym.* **2020**, *240*, 116329. [CrossRef] [PubMed]
7. Huang, Y.Y.; Su, E.Z.; Ren, J.S.; Qu, X.G. The recent biological applications of selenium-based nanomaterials. *Nano Today* **2021**, *38*, 101205. [CrossRef]
8. Ye, X.G.; Chen, Z.Z.; Zhang, Y.Y.; Mu, J.J.; Chen, L.Y.; Li, B.; Lin, X.R. Construction, characterization, and bioactive evaluation of nano-selenium stabilized by green tea nano-aggregates. *LWT-Food Sci. Technol.* **2020**, *129*, 109475. [CrossRef]
9. Yan, J.K.; Qiu, W.Y.; Wang, Y.Y.; Wang, W.H.; Yang, Y.; Zhang, H.N. Fabrication and stabilization of biocompatible selenium nanoparticles by carboxylic curdlans with various molecular properties. *Carbohydr. Polym.* **2018**, *179*, 19–27. [CrossRef]

10. Liu, L.; Xiao, Z.; Niu, S.; He, Y.; Wang, G.; Pei, X.; Tao, W.; Wang, M. Preparation, characteristics and feeble induced-apoptosis performance of non-dialysis requiring selenium nanoparticles@chitosan. *Mater. Des.* **2019**, *182*, 108024. [CrossRef]
11. Ayadi, F.; Bayer, I.S.; Marras, S.; Athanassiou, A. Synthesis of water dispersed nanoparticles from different polysaccharides and their application in drug release. *Carbohydr. Polym.* **2016**, *136*, 282–291. [CrossRef] [PubMed]
12. Shi, X.D.; Tian, Y.Q.; Wu, J.L.; Wang, S.Y. Synthesis, characterization, and biological activity of selenium nanoparticles conjugated with polysaccharides. *Crit. Rev. Food Sci. Nutr.* **2021**, *61*, 2225–2236. [CrossRef] [PubMed]
13. Chen, W.W.; Li, Y.F.; Yang, S.; Yue, L.; Jiang, Q.; Xia, W.S. Synthesis and antioxidant properties of chitosan and carboxymethyl chitosan-stabilized selenium nanoparticles. *Carbohydr. Polym.* **2015**, *132*, 574–581. [CrossRef]
14. Cai, W.F.; Hu, T.; Bakry, A.M.; Zheng, Z.M.; Xiao, Y.D.; Huang, Q.L. Effect of ultrasound on size, morphology, stability and antioxidant activity of selenium nanoparticles dispersed by a hyperbranched polysaccharide from Lignosus rhinocerotis. *Ultrason. Sonochem.* **2018**, *42*, 823–831. [CrossRef] [PubMed]
15. Lei, W.; Chao, L.; Qiang, H.; Xiong, F. Biofunctionalization of selenium nanoparticles with a polysaccharide from *Rosa roxburghii* fruit and their protective effect against H_2O_2-induced apoptosis in INS-1 cells. *Food Funct.* **2019**, *10*, 539–553. [CrossRef]
16. Zhang, W.; Zhang, J.; Ding, D.; Zhang, L.; Muehlmann, L.A.; Deng, S.-E.; Wang, X.; Li, W.; Zhang, W. Synthesis and antioxidant properties of *Lycium barbarum* polysaccharides capped selenium nanoparticles using tea extract. *Artif. Cells Nanomed. Biotechnol.* **2018**, *46*, 1463–1470. [CrossRef]
17. Xie, J.H.; Jin, M.L.; Morris, G.A.; Zha, X.Q.; Chen, H.Q.; Yi, Y.; Li, J.E.; Wang, Z.J.; Gao, J.; Nie, S.P.; et al. Advances on bioactive polysaccharides from medicinal plants. *Crit. Rev. Food Sci. Nutr.* **2016**, *56*, S60–S84. [CrossRef]
18. Li, X.J.; Chen, Q.; Liu, G.K.; Xu, H.R.; Zhang, X. Chemical elucidation of an arabinogalactan from rhizome of *Polygonatum sibiricum* with antioxidant activities. *Int. J. Biol. Macromol.* **2021**, *190*, 730–738. [CrossRef]
19. Huang, S.; Yuan, H.Y.; Li, W.Q.; Liu, X.Y.; Zhang, X.J.; Xiang, D.X.; Luo, S.L. *Polygonatum sibiricum* polysaccharides protect against MPP-induced neurotoxicity via the Akt/mTOR and Nrf2 pathways. *Oxid. Med. Cell. Longev.* **2021**, *2021*, 8843899. [CrossRef]
20. Wang, J.; Lu, C.-S.; Liu, D.-Y.; Xu, Y.-T.; Zhu, Y.; Wu, H.-H. Constituents from *Polygonatum sibiricum* and their inhibitions on the formation of advanced glycosylation end products. *J. Asian Nat. Prod. Res.* **2016**, *18*, 697–704. [CrossRef]
21. Cui, X.; Wang, S.; Cao, H.; Guo, H.; Li, Y.; Xu, F.; Zheng, M.; Xi, X.; Han, C. A Review: The bioactivities and pharmacological applications of *Polygonatum sibiricum* polysaccharides. *Molecules* **2018**, *23*, 1170. [CrossRef] [PubMed]
22. Liu, J.; Li, T.Y.; Chen, H.Y.; Yu, Q.; Yan, C.Y. Structural characterization and osteogenic activity in vitro of novel polysaccharides from the rhizome of *Polygonatum sibiricum*. *Food Funct.* **2021**, *12*, 6626–6636. [CrossRef]
23. Chen, Z.; Liu, J.; Kong, X.; Li, H. Characterization and immunological activities of polysaccharides from *Polygonatum sibiricum*. *Biol. Pharm. Bull.* **2020**, *43*, 959–967. [CrossRef] [PubMed]
24. Sun, T.T.; Zhang, H.; Li, Y.; Liu, Y.; Dai, W.; Fang, J.; Cao, C.; Die, Y.; Liu, Q.; Wang, C.L.; et al. Physicochemical properties and immunological activities of polysaccharides from both crude and wine-processed *Polygonatum sibiricum*. *Int. J. Biol. Macromol.* **2020**, *143*, 255–264. [CrossRef]
25. Ma, W.; Wei, S.; Peng, W.; Sun, T.; Huang, J.; Yu, R.; Zhang, B.; Li, W. Antioxidant effect of *Polygonatum sibiricum* polysaccharides in D-galactose-induced heart aging mice. *Biomed. Res. Int.* **2021**, *2021*, 6688855. [CrossRef] [PubMed]
26. Wang, G.J.; Fu, Y.W.; Li, J.J.; Li, Y.N.; Zhao, Q.H.; Hu, A.L.; Xu, C.D.; Shao, D.L.; Chen, W.J. Aqueous extract of *Polygonatum sibiricum* ameliorates ethanol-induced mice liver injury via regulation of the Nrf2/ARE pathway. *J. Food Biochem.* **2021**, *45*, 11. [CrossRef] [PubMed]
27. Song, X.; Chen, Y.; Sun, H.; Liu, X.; Leng, X. Physicochemical stability and functional properties of selenium nanoparticles stabilized by chitosan, carrageenan, and gum Arabic. *Carbohydr. Polym.* **2021**, *255*, 117379. [CrossRef]
28. Wang, T.; Zhao, H.; Bi, Y.; Fan, X. Preparation and antioxidant activity of selenium nanoparticles decorated by polysaccharides from *Sargassum fusiforme*. *J. Food Sci.* **2021**, *86*, 977–986. [CrossRef]
29. Chen, W.W.; Yue, L.; Jiang, Q.X.; Liu, X.L.; Xia, W.S. Synthesis of varisized chitosan-selenium nanocomposites through heating treatment and evaluation of their antioxidant properties. *Int. J. Biol. Macromol.* **2018**, *114*, 751–758. [CrossRef]
30. Li, H.Y.; Liu, D.D.; Li, S.H.; Xue, C.H. Synthesis and cytotoxicity of selenium nanoparticles stabilized by alpha-D-glucan from *Castanea mollissima* Blume. *Int. J. Biol. Macromol.* **2019**, *129*, 818–826. [CrossRef]
31. Ren, L.R.; Wu, Z.C.; Ma, Y.; Jian, W.J.; Xiong, H.J.; Zhou, L.N. Preparation and growth-promoting effect of selenium nanoparticles capped by polysaccharide-protein complexes on tilapia. *J. Sci. Food Agric.* **2021**, *101*, 476–485. [CrossRef] [PubMed]
32. Chen, W.W.; Yue, L.; Jiang, Q.X.; Xia, W.S. Effect of chitosan with different molecular weight on the stability, antioxidant and anticancer activities of well-dispersed selenium nanoparticles. *IET Nanobiotechnol.* **2019**, *13*, 30–35. [CrossRef]
33. Xiao, Y.D.; Huang, Q.L.; Zheng, Z.M.; Guan, H.; Liu, S.Y. Construction of a *Cordyceps sinensis* exopolysaccharide-conjugated selenium nanoparticles and enhancement of their antioxidant activities. *Int. J. Biol. Macromol.* **2017**, *99*, 483–491. [CrossRef] [PubMed]
34. Gao, X.; Li, X.; Mu, J.; Ho, C.-T.; Su, J.; Zhang, Y.; Lin, X.; Chen, Z.; Li, B.; Xie, Y. Preparation, physicochemical characterization, and anti-proliferation of selenium nanoparticles stabilized by *Polyporus umbellatus* polysaccharide. *Int. J. Biol. Macromol.* **2020**, *152*, 605–615. [CrossRef] [PubMed]
35. Evageliou, V.I.; Ryan, P.M.; Morris, E.R. Effect of monovalent cations on calcium-induced assemblies of kappa carrageenan. *Food Hydrocoll.* **2019**, *86*, 141–145. [CrossRef]

36. Tang, S.; Wang, T.; Jiang, M.; Huang, C.; Lai, C.; Fan, Y.; Yong, Q. Construction of arabinogalactans/selenium nanoparticles composites for enhancement of the antitumor activity. *Int. J. Biol. Macromol.* **2019**, *128*, 444–451. [CrossRef]
37. Wang, Y.; Liu, N.; Xue, X.; Li, Q.; Sun, D.; Zhao, Z. Purification, structural characterization and in vivo immunoregulatory activity of a novel polysaccharide from *Polygonatum sibiricum*. *Int. J. Biol. Macromol.* **2020**, *160*, 688–694. [CrossRef]
38. Zhou, L.; Song, Z.; Zhang, S.; Li, Y.; Xu, J.; Guo, Y. Construction and antitumor activity of selenium nanoparticles decorated with the polysaccharide extracted from *Citrus limon* (L.) Burm. f. (Rutaceae). *Int. J. Biol. Macromol.* **2021**, *188*, 904–913. [CrossRef]
39. Liu, W.; Li, X.; Wong, Y.-S.; Zheng, W.; Zhang, Y.; Cao, W.; Chen, T. Selenium nanoparticles as a carrier of 5-fluorouracil to achieve anticancer synergism. *ACS Nano* **2012**, *6*, 6578–6591. [CrossRef]
40. Chunyue, Z.; Xiaona, Z.; Guanghua, Z.; Fazheng, R.; Xiaojing, L. Synthesis, characterization, and controlled release of selenium nanoparticles stabilized by chitosan of different molecular weights. *Carbohydr. Polym.* **2015**, *134*, 158–166. [CrossRef]
41. Cheng, Y.; Xiao, X.; Li, X.; Song, D.; Lu, Z.; Wang, F.; Wang, Y. Characterization, antioxidant property and cytoprotection of exopolysaccharide-capped elemental selenium particles synthesized by *Bacillus paralicheniformis* SR14. *Carbohydr. Polym.* **2017**, *178*, 18–26. [CrossRef] [PubMed]
42. Kong, H.; Yang, J.; Zhang, Y.; Fang, Y.; Nishinari, K.; Phillips, G.O. Synthesis and antioxidant properties of gum arabic-stabilized selenium nanoparticles. *Int. J. Biol. Macromol.* **2014**, *65*, 155–162. [CrossRef] [PubMed]
43. Tang, L.; Luo, X.; Wang, M.; Wang, Z.; Guo, J.; Kong, F.; Bi, Y. Synthesis, characterization, in vitro antioxidant and hypoglycemic activities of selenium nanoparticles decorated with polysaccharides of *Gracilaria lemaneiformis*. *Int. J. Biol. Macromol.* **2021**, *193*, 923–932. [CrossRef] [PubMed]
44. Qi, Y.; Yi, P.; He, T.; Song, X.; Liu, Y.; Li, Q.; Zheng, J.; Song, R.; Liu, C.; Zhang, Z.; et al. Quercetin-loaded selenium nanoparticles inhibit amyloid-beta aggregation and exhibit antioxidant activity. *Colloid Surf. A-Physicochem. Eng. Asp.* **2020**, *602*, 125058. [CrossRef]

Article

Improving Physicochemical Stability of Quercetin-Loaded Hollow Zein Particles with Chitosan/Pectin Complex Coating

Muhammad Aslam Khan [1,2], Chufan Zhou [1,2], Pu Zheng [3], Meng Zhao [4] and Li Liang [1,2,*]

[1] State Key Laboratory of Food Science and Technology, Jiangnan University, Wuxi 214122, China; aslamkhan255@yahoo.com (M.A.K.); chufan.zhou@mail.mcgill.ca (C.Z.)
[2] School of Food Science and Technology, Jiangnan University, Wuxi 214122, China
[3] School of Biotechnology, Jiangnan University, Wuxi 214122, China; zhengpu@jiangnan.edu.cn
[4] State Key Laboratory of Biobased Material and Green Papermaking, Qilu University of Technology, Shandong Academy of Sciences, Jinan 250353, China; 2001zhaomeng@163.com
* Correspondence: liliang@jiangnan.edu.cn

Abstract: Hollow nanoparticles are preferred over solid ones for their high loading capabilities, sustained release and low density. Hollow zein particles are susceptible to aggregation with a slight variation in the ionic strength, pH and temperature of the medium. This study was aimed to fabricate quercetin-loaded hollow zein particles with chitosan and pectin coating to improve their physicochemical stability. Quercetin as a model flavonoid had a loading efficiency and capacity of about 86–94% and 2.22–5.89%, respectively. Infrared and X-ray diffraction investigations revealed the interaction of quercetin with zein and the change in its physical state from crystalline to amorphous upon incorporation in the composite particles. The chitosan/pectin coating improved the stability of quercetin-loaded hollow zein particles against heat treatment, sodium chloride and in a wide range of pH. The complex coating protected quercetin that was encapsulated in hollow zein particles from free radicals in the aqueous medium and enhanced its DPPH radical scavenging ability. The entrapment of quercetin in the particles improved its storage and photochemical stability. The storage stability of entrapped quercetin was enhanced both at 25 and 45 °C in hollow zein particles coated with chitosan and pectin. Therefore, composite hollow zein particles fabricated with a combination of polysaccharides can expand their role in the encapsulation, protection and delivery of bioactive components.

Keywords: hollow zein particle; chitosan; pectin; quercetin; coating

1. Introduction

Hollow zein particles have been fabricated by wrapping sodium carbonate (Na_2CO_3) nanoprecipitate as sacrificial templet with zein in ethanol–water binary mixture followed by antisolvent precipitation [1]. Hollow particles in the loading and controlled release of bioactive components were preferred over their solid counterpart for more surface area and low density [2], but proteinaceous nature limits their utility as an efficient delivery system due to destabilization around pI (5–6.5), presence of counterion and high temperature-induced denaturation [3,4]. Numerous strategies have been adopted to overcome instability issues of zein particles, for instance, coating with proteins [5,6], polysaccharides [7] and lipids [8]. Pectin is an anionic biodegradable polymer found in the plant cell wall and mainly made up of methyl esterified 1-4 linked α-D-galacturonic acid and 1-2-linked α-L-rhamnopyranose. Composite hollow zein particles with casein and pectin were developed by heating at 80 °C and 6.2 pH for 1 h to attain outstanding stability under simulated gastrointestinal conditions. Still, these composite particles were limited only to encapsulate and deliver heat-sensitive bioactives [9].

Chitosan-coated solid zein particles were fabricated through hydrophobic, hydrogen and van der Waals interactions at pH 4, improving the entrapment, photo/thermal protection and controlled release of bioactive components [10–12]. Chitosan, a N-acetyl-D-glucosamine and D-glucosamine β1-4 linked cationic polymer, is obtained by deacetylation of chitin and considerably used for the stabilization of delivery systems [13]. However, the deprotonation of the amine groups of chitosan above pk_a (~pH 6.5) reduces charge density, and chitosan competes with counterions with the increase in ionic strength, heading towards destabilization [12,14]. It has been reported that pectin imparted good pH and heating stability to zein core–shell nanoparticles, but the endurance for increasing ionic strength was extremely weak [15,16]. Chitosan and pectin could form polyelectrolyte complexes via electrostatic interaction [17], which was used to improve the physiochemical stability of nanoliposome [18]. Therefore, a combination of chitosan and pectin may synergistically and resourcefully bear variation in pH, temperature and counterions.

Quercetin is a flavonoid with antioxidant, anticarcinogenic, antiviral and anti-inflammatory properties. Its low solubility in water and chemical instability have been addressed through biopolymer-based nano/micro-delivery systems for the application in functional foods [19,20]. In the current work, composite hollow zein particles were fabricated with chitosan-pectin complex coating for the encapsulation and protection of quercetin. The particles were characterized for size, ζ-potential and loading efficiency of quercetin. The lyophilized samples of quercetin-loaded composite particles were analyzed with infrared and X-ray diffraction techniques. Moreover, the particle dispersions were subject to varying pH, ionic strength and temperature conditions to assess physical stability. Finally, the antioxidant activity, photochemical and storage stability of quercetin encapsulated in composite hollow zein particles were examined to judge the protective effects of the particles. This study focused on the fabrication of composite hollow zein particles with improved physical stability and better protective effect on flavonoids through tailoring a complex polysaccharide interfacial layer.

2. Materials and Methods

2.1. Materials

Zein from corn (~98%) was purchased from J&K Chemical Ltd. (Shanghai, China). Sodium carbonate (Na_2CO_3, ~99.8%) was purchased from Sinopharm Chemical Reagent Co., Ltd. (Shanghai, China). Pectin from citrus peel was purchased from Shanghai Sangong Bioengineering Co., Ltd. (Shanghai, China). Chitosan (low molecular weight, 50–190 KDa) and quercetin (≥95%, HPLC) were purchased from Sigma-Aldrich Co. (Shanghai, China). Ultra-pure water obtained using a Milli-Q direct water purification system equipped with Quantum TEX column (Molsheim, France) was used throughout all the experiments.

2.2. Preparation of Blank and Quercetin-Loaded Hollow Zein, Hollow Zein-Chitosan, and Hollow Zein-Chitosan/Pectin Particles

Hollow zein (HZ) particles were fabricated with sodium carbonate sacrificial templet by mixing 1.75 mL of absolute ethanol and 2.5 mL of 50 mg/mL zein in 70% (v/v) ethanol–water binary mixture with 0.75 mL of 0.5, 1 and 2 (w/v%) Na_2CO_3 aqueous solution under magnetic stirring at 1000 rpm for 1 min followed by adding to a 20 mL ultra-pure water [1,21]. The HZ particles were mixed with 0.5, 1 and 2 mg/mL chitosan in 1% (v/v) acetic acid at a 1:1 volume ratio under stirring at 1000 rpm for 30 min. Ethanol in the particle dispersion was removed under vacuum with a rotary evaporator RE-52C (Shanghai Tianheng Instrument Co. Ltd., Shanghai, China) at 35 °C for 30 min. Unstable particles were separated by centrifugation the particle dispersion at 2000× g for 15 min; the supernatant was then centrifuged at 15,000× g for 30 min to remove unabsorbed chitosan, and the precipitate was redispersed in an equal volume of distilled water to obtained chitosan-coated hollow zein (HZ-chi) particles. The aqueous solutions of pectin at 0.01, 0.025, 0.05 and 0.1 mg/mL were added to the dispersion of HZ-chi particles under stirring for 30 min at pH 4–4.5 to obtain pectin- and chitosan-coated hollow zein (HZ-chi/pec) particles. The quercetin-loaded hollow particles were prepared by adding 2, 3, 4 and 5 mg of quercetin in 50 mg/mL zein stock solution in ethanol–water binary mixture corresponding to 100, 150, 200 and 250 µg/mL of quercetin in the final particle dispersion.

2.3. Characterization of Particles

2.3.1. Particle Size and ζ-Potential

Samples were diluted by 200 folds with distilled water and measured at 25 °C and analyzed on a NanoBrook Omini particle size analyzer (Brookhaven Instrument, New York, NY, USA) at a scattering angle of 90°. NNLS function was used to acquire the size distribution, while phase analysis light scattering (PALS) was employed to estimate the ζ-potential. Samples were prepared in triplicates for each measurement.

2.3.2. Loading Efficiency and Loading Capacity of Quercetin

Quercetin-loaded hollow zein and hollow zein-chitosan particles were centrifuged at $2000 \times g$ for 15 min to remove unencapsulated quercetin, and the supernatant and samples were diluted 50-fold in ethanol for the measurement of quercetin absorption at a λ_{max} of 373 nm using a UV1800 UV-Vis spectrophotometer (Shimadzu Corporation, Tokyo, Japan) with a standard curve constructed from 1 to 20 µg/mL of quercetin dissolved in ethanol (a correlation coefficient of 0.999, Figure S4), adopting the method of Wang et al. [22]. Samples were prepared in triplicates for each measurement. Loading efficiency and capacity of quercetin in the particles were determined by the following equations.

$$\text{Loading efficiency (\%)} = \frac{\text{Quercetin in supernatant}}{\text{Total added quercetin}} \times 100 \quad (1)$$

$$\text{Loading capacity (\%)} = \frac{\text{Quercetin entrapped in particles (µg)}}{\text{Zein and chitosan (µg)}} \times 100 \quad (2)$$

2.3.3. Microstructural Analysis

Lyophilized particles were mounted to the surface of double-sided carbon tape and coated with a thin layer of gold. The morphology of particles was observed with a Hitachi SU8010 FE-SEM (Hitachi, Co., Tokyo, Japan) operated at an accelerating voltage of 8 kV.

2.3.4. Infrared Spectroscopy (IR)

Samples were freeze-dried and pressed into a transparent pellet with KBr, Infrared spectra of particles and raw materials were collected in the range of 400–4000 cm^{-1} on Nicolet™ iS™ 10 FT-IR Spectrometer (Thermo Fisher Scientific, Waltham, MA, USA) at a 4 cm^{-1} resolution and 16 scans.

2.3.5. X-ray Diffraction (XRD)

A Bruker D2 PHASER (Brucker, Odelzhausen, Germany) X-ray diffractometer, operated at 30 kV, 10 mA was used to obtain X-ray diffractograms of quercetin, zein, chitosan, pectin and quercetin-loaded hollow zein particles coated with chitosan and pectin. The data were collected over an angular range from 5° to 40° 2θ in continuous mode using step size and time of 0.02° and 5 s, respectively.

2.4. Stability Assessment of Particles under Stressed Condition

2.4.1. Sodium Chloride Stability

Freshly prepared quercetin-loaded HZ, HZ-chi and HZ-chi/pec particles were exposed to 50, 100, 200, 300, 400, 500 mM sodium chloride under continuously mixing for 30 min followed by a 5 min rest [8]. Particle size and ζ-potential were measured, as mentioned above in Section 2.3.1.

2.4.2. pH Stability

The pH values of quercetin-loaded HZ, HZ-chi and HZ-chi/pec particles were adjusted from 2 to 9 with 0.1 mM hydrochloric acid and sodium hydroxide and then continuously stirred for 30 min [8]. Particle size and ζ-potential were analyzed by diluting them in pH-adjusted Milli-Q water [23].

2.4.3. Temperature Durability

Quercetin-loaded HZ, HZ-chi and HZ-chi/pec particles were incubated in a water bath at 30, 40, 50, 60, 70, 80 and 90 °C for 30 min and evaluated for particle size and ζ-potential [8].

2.5. Antioxidant Activity

ABTS and DPPH assays of quercetin-loaded HZ and HZ-chi/pec particles were estimated according to the method of Dong et al. and Pan et al. [24,25]. In brief, equal volumes of potassium persulfate (2.6 mM) and ABTS (7.4 mM) were mixed and allowed to react and generate ABTS$^+$ for 12 h in the dark. After the diluted ABTS$^+$ solution with an absorbance of 0.7 at 734 nm was mixed with samples at a volume ratio of 2:1 in the dark for 6 min, the absorbance was measured at 734 nm on Synergy H1 Microplate Reader (BioTek Instruments, Inc., Winooski, VT, USA). Likewise, after 0.1 mM DPPH in ethanol was added to samples in equal volumes and allowed to react for 30 min in the dark, the absorbance was recorded at 517 nm. The ABTS$^+$ and DPPH scavenging capacity was calculated with the help of the following equation,

$$\text{Free radical scavenging capacity (\%)} = (A_{Control} - A_{Sample})/A_{Control} \times 100 \quad (3)$$

where $A_{Control}$ and A_{Sample} are the absorbances of the free radical solution without and with samples, respectively.

2.6. Functional Characteristics of Particles

2.6.1. Photochemical Stability of Quercetin

Photochemical stability of pristine and encapsulated quercetin was evaluated by the procedure presented by Sun et al. [26]. Quercetin dispersed in water and encapsulated in HZ and HZ-chi/pec particles at 5 μg/mL were irradiated up to 120 min with a 365 nm ultraviolet lamp (VWR International Inc., West Chester, PA, USA). Samples were collected at 0, 15, 30, 60, 90 and 120 min, and the content of quercetin was analyzed with the help of UV-Vis spectrophotometer mentioned above in Section 2.3.2.

2.6.2. Storage Stability of Quercetin and Particles

Samples were stored at 25 and 45 °C for 28 days inside an LRH-250F incubator (Yiheng Scientific Instrument Co., Ltd. Shanghai, China). The stability of quercetin-loaded particles was analyzed in terms of particle size and ζ-potential during storage. The retention of quercetin was expressed as percent retention and calculated by using the following equation:

$$\text{Quercetin retention (\%)} = Q_t/Q_i \times 100 \quad (4)$$

where Q_i and Q_t are the content of quercetin at the beginning and specific time intervals during storage.

2.7. Statistical Analysis

All experiments were done in triplicates. The results were expressed in mean plus standard deviation and analyzed for a significant difference ($p < 0.05$) with IBM SPSS statistics 20.0 software package (IBM, Armonk, NY, USA).

3. Results and Discussion
3.1. Characterization of Hollow Zein Particles
3.1.1. Effect of Na₂CO₃ and Chitosan on Hollow Zein Particles

The size of hollow zein particles is greatly influenced by the concentration of Na_2CO_3 used for the preparation of sacrificial templet [1,27]. The smallest particles of 76 nm were fabricated with 1% Na_2CO_3 (Figure 1A). At 0.5 and 2% Na_2CO_3, the size of HZ particles was 145 and 210 nm, respectively. The reason for the bigger particles is due to the formation of a thicker zein shell around all the available sodium carbonate nuclei formed in ethanolic conditions at a Na_2CO_3 concentration of 0.5% but the formation of bigger Na_2CO_3 nanocrystal in size due to excessive aggregation and crystal growth at 2% [27]. The pH values of the HZ particle dispersions prepared with 0.5, 1 and 2% Na_2CO_3 were 9.04, 10.31 and 10.77, respectively, which is above the pI of zein [28]. The ζ-potential of HZ particles was −8, −26 and −24 mV when prepared with 0.5, 1 and 2% Na_2CO_3 (Figure 1B), receptively. There was a slight increase in the particle size of HZ particles upon the addition of chitosan (Figure 1A), except that precipitation was observed at 0.05% chitosan and 2% Na_2CO_3. The increase in the particle size was the most obvious when the chitosan concentration was 0.1% at 1% and 2% Na_2CO_3. In the presence of chitosan, the pH of particle dispersions shifted to 4.0–4.5. Chitosan interacts with proteins below their pI through hydrophobic, electrostatic, van der Waals and hydrogen bonding [12,29]. The ζ-potential of HZ-chi particles ranged between +30 and +58 mV (Figure 1B). The amine groups in chitosan contribute to the positive surface charge of composite zein particles [30,31]. These results suggest the formation of a chitosan shell.

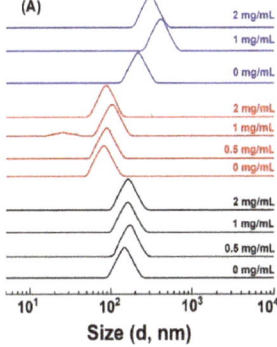 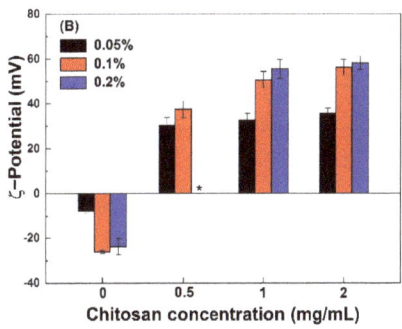

Figure 1. Particle size distribution (**A**) and ζ−potential (**B**) of hollow zein particles fabricated with 0.5% (black), 1% (red) and 2% (blue) Na2CO3 and coated with chitosan at various concentrations (Mark at the right). * indicates unstable particles.

3.1.2. Fabrication of Pectin-Coated Composite Particles

HZ-chi particles prepared with 1% Na_2CO_3 were further coated with pectin to form a second layer. The addition of pectin did not influence particle PDI (Table 1). When the concentration of chitosan was 0.5 mg/mL, the size of HZ-chi particles increased from 87 to 170 and 180 nm upon adding 0.01 and 0.025 mg/mL of pectin, respectively. When the concentration of chitosan was 2 mg/mL, the size of HZ-chi particles increased from 84 to 194 and 192 nm upon adding 0.01 and 0.025 mg/mL of pectin, respectively. Meanwhile, a reduction in the ζ-potential positive values was observed, showing insufficient pectin to fully cover the surface of particles. With further increasing the pectin concentration to 0.05 and 0.1 mg/mL in the presence of 0.5 mg/mL chitosan and to at 0.05 mg/mL pectin in the presence of 2 mg/mL, the particle dispersion became unstable with the formation of precipitates, due to charge neutralization [7]. The neutralization was observed at 0.01 and 0.025 mg/mL pectin in the presence of 1 mg/mL chitosan (Table 1). However, in the presence of 1 mg/mL chitosan, an increment in size to 248 and 219 nm was observed at 0.05 and 0.1 mg/mL pectin, respectively, and ζ-potentials changed to negative values, suggesting the formation of pectin surface layer. An absolute value of ζ-potential above 20 mV is so high enough to ensure the physical stability of biopolymer-based particles [32]. Therefore, the hollow zein particles prepared with 1 mg/mL chitosan and 0.1 mg/mL pectin were used for further study on their physicochemical and functional attributes.

Table 1. Effect of pectin on size, PDI and ζ-potential of hollow zein particles coated with chitosan.

Chitosan Concentration (mg/mL)	Parameter	Pectin Concentration (mg/mL)				
		0	0.01	0.025	0.05	0.1
0.5	Size (nm)	87 ± 2 [Aa]	170 ± 1 [Ab]	180 ± 2 [Ac]	–	–
	PDI	0.21 ± 0.02 [Aa]	0.20 ± 0.02 [Aa]	0.21 ± 0.02 [Aa]	–	–
	ζ-Potential (mV)	+37 ± 4 [Aa]	+33 ± 1 [Aa]	+26 ± 0 [Ab]	–	–
1	Size (nm)	82 ± 2 [Aa]	–	–	248 ± 4 [Ac]	219 ± 1 [Ab]
	PDI	0.12 ± 0.01 [Ba]	–	–	0.18 ± 0.08 [Aa]	0.12 ± 0.013 [Aa]
	ζ-Potential (mV)	+51 ± 4 [Ba]	–	–	−18 ± 0 [Ac]	−28 ± 1 [Ab]
2	Size (nm)	84 ± 2 [Aa]	194 ± 5 [Ab]	192 ± 4 [Ab]	–	300 ± 6 [Bc]
	PDI	0.18 ± 0.01 [Aa]	0.14 ± 0.05 [Aa]	0.11 ± 0.05 [Ba]	–	0.13 ± 0.07 [Aa]
	ζ-Potential (mV)	+56 ± 4 [Ba]	+21 ± 1 [Bb]	+22 ± 2 [Bb]	–	−32 ± 2 [Bc]

Values with different letters (upper case A and B for the concentration of chitosan; lower case a, b and c for the concentration of pectin) are significantly different in rows ($p < 0.05$); – represents the aggregation followed by precipitation of nanoparticles.

3.1.3. Encapsulation of Quercetin

Quercetin was used as a model bioactive flavonoid to be encapsulated in hollow zein particles. The loading efficiencies of quercetin were 72.71–79.86% in HZ particles, while the loading efficiencies increased to 90.58–93.86% at the flavonoid concentrations of 100–200 µg/mL and 85.84% at 250 µg/mL in HZ-chi particles (Table 2). The higher loading efficiency of quercetin in the presence of chitosan is attributed to its interaction with chitosan through electrostatic and hydrogen bonding, in addition to hydrophobic interaction with zein [33,34]. The loading capacity of quercetin in the absence and presence of chitosan remained similar and increased from about 2% to 6%. This is different from a decreasing trend that was commonly observed in composite protein particles with increasing the content of the wall materials [35,36].

Table 2. Loading efficiency and loading capacity of quercetin in hollow zein and zein-chitosan particles.

Quercetin Concentration (µg/mL)	Loading Efficiency (%)		Loading Capacity	
	Zein	Zein-Chitosan	Zein	Zein-Chitosan
100	79.07 ± 4.12 [Aa]	93.86 ± 1.19 [Ab]	2.59 ± 0.56 [Ca]	2.22 ± 0.20 [Ca]
150	75.72 ± 2.61 [Aa]	91.36 ± 0.80 [Ab]	3.92 ± 0.85 [BCa]	3.76 ± 0.24 [Ba]
200	79.86 ± 5.18 [Aa]	90.58 ± 2.02 [Ab]	5.69 ± 0.74 [ABa]	5.53 ± 0.47 [Aa]
250	72.71 ± 3.82 [Aa]	85.84 ± 4.16 [Bb]	6.29 ± 0.98 [Aa]	5.89 ± 1.08 [Aa]

Values with different letters (upper case A, B and C for the concentration of quercetin in column; lower case a and b in the row for the type of particles) are significantly different ($p < 0.05$).

SEM images showed that quercetin-loaded HZ, HZ-chi and HZ-chi/pec were spherical with a smooth surface (Figure 2). The loading of quercetin increased the size of HZ particles without and with chitosan and/or pectin coating (Figure S1A). A similar trend was previously reported for the encapsulation of resveratrol and curcumin in composite hollow zein particles [11,37]. The loading of quercetin did not affect the ζ-Potential of HZ particles but increased ζ-potential absolute values of HZ-chi and HZ-chi/pec particles (Figure S1C), possibly due to rearrangement and exposure of more charged groups upon flavonoid inclusion [33].

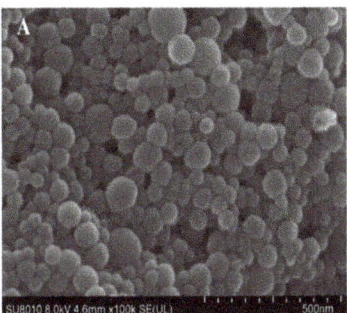 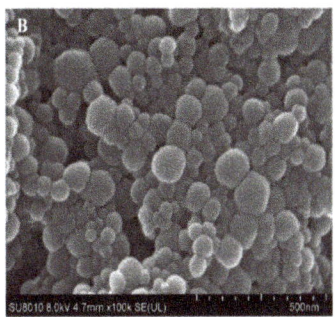 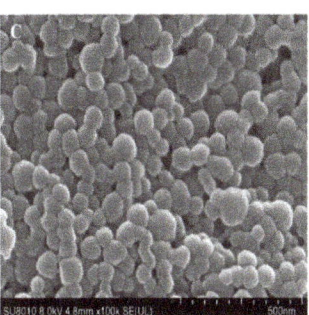

Figure 2. SEM images of quercetin-loaded hollow zein (qHZ, (**A**)), zein-chitosan (qHZ-chi, (**B**)) and zein-chitosan/pectin (qHZ-chi/pec, (**C**)) particles. The concentrations of chitosan and pectin were 1 and 0.1 mg/mL, respectively.

3.2. Physical Characterization

3.2.1. XRD

Quercetin is a crystalline solid in its pristine form [38], as evident from sharp X-ray diffraction patterns at 8, 10, 12, 13.01, 15, 17 and 27 of diffraction angle (2θ) (Figure 3). The distinct peaks of quercetin with lower intensities were apparent in its mixture of zein, chitosan and pectin. In the diffractograms of qHZ, qHZ-chi and qHZ-chi/pec particles, the crystalline peaks of quercetin disappeared, advocating the transformation from crystalline to an amorous state upon encapsulation in particles [10,39]. Pure zein and chitosan showed mild crystallinity with moderate peaks at 9.56 and 20.24 2θ because of respective α-helixes and crystal lattice structure [40,41]. When zein was structured into HZ, HZ-chi, HZ-chi/pec, qHZ, qHZ-chi and qHZ-chi/pec particles, the diffraction pattern raised from α-helixes in zein and crystal lattice of chitosan became flattened due to interaction between the polymers during the formation of the composite particles [42,43].

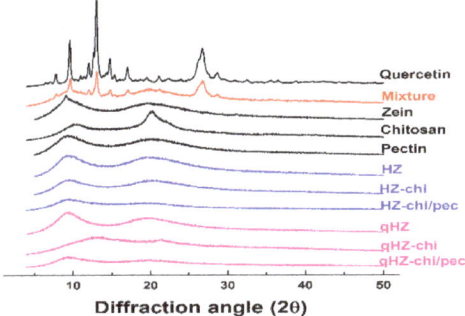

Figure 3. XRD diffraction patterns of quercetin, mixture (quercetin, zein, chitosan and pectin), zein, chitosan, pectin, HZ, HZ-chi, HZ-chi/pec, qHZ, qHZ-chi, qHZ-chi/pec.

3.2.2. IR Spectroscopy

OH stretching of zein and HZ particles was at 3318 cm^{-1} but changed to 3316 and 3405 cm^{-1} in HZ-chi and HZ-chi/pec particles, indicating hydrogen bonding among zein, chitosan and pectin (Figure 4B,C). Likewise, amide I and II absorption band of zein changed from 1658 and 1541 cm^{-1} to 1656 and 1536 cm^{-1} in HZ particles, to 1658 and 1542 cm^{-1} in HZ-chi particles, and to 1654 and 1635 cm^{-1} in HZ-chi/pec particles. The variations of stretching and bending vibrations of C=O and N-H indicate that hydrophobic and electrostatic interactions occurred during the fabrication of HZ, chitosan and chitosan/pectin coated HZ particles [29]. The distinct absorption peaks of quercetin IR spectrum at 3386, 1655, 1599, 1373, 1257 and 1160 cm^{-1} (Figure 4A) signify O-H, C=O, C=C, C-OH, C-O-C and C-OH (B ring), respectively [34,44]. The absorption bands were clearly seen with lower intensities in the physical mixture of quercetin, zein, chitosan and pectin (Figure 4A). However, the characteristic absorption bands of quercetin were invisible in HZ, HZ-chi and HZ-chi/pec particles, indicating the entrapment of quercetin in the particles and limited starching and bending of various bonds [42]. The encapsulation of quercetin resulted in the variation of OH stretching of HZ, HZ-chi and HZ-chi/pec particles (Figure 4A,B), suggesting the flavonoid interaction with wall materials [45]. Upon encapsulation of quercetin in hollow zein particle, the amide I peak of zein was unchanged, whereas the amide II showed a redshift to 1537, 1543 and 1539 cm^{-1} in HZ, HZ-chi and HZ-chi/pec particles (Figure 4A). These shifts in the amide II result from hydrophobic, electrostatic and hydrogen bonding of quercetin with wall materials [46,47].

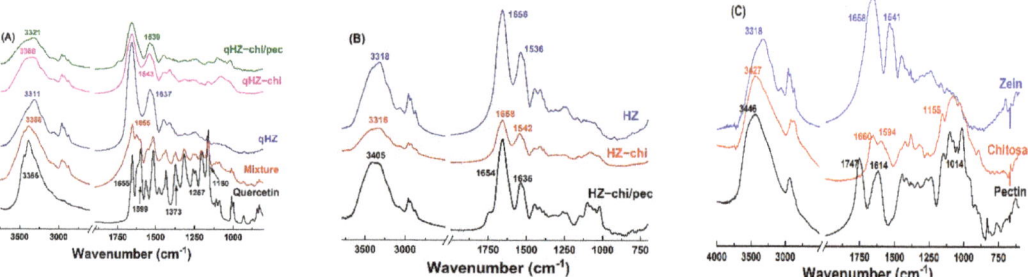

Figure 4. IR spectra of quercetin, mixture (quercetin, zein, chitosan and pectin), quercetin-loaded HZ−chi and HZ−chi/pec particles (**A**), HZ, HZ−chi and HZ−chi/pec blank particles (**B**), and raw materials zein, chitosan and pectin (**C**).

3.3. Stability of Hollow Particles

3.3.1. Salt Endurance

Colloidal stability against salt was assessed by exposing particles to various concentrations of NaCl. Quercetin-loaded hollow zein particles without and with chitosan were highly unstable and precipitated out even at 50 mM NaCl (Figure S2A,B), attributed to the screening effect and charge neutralization by counterions [7,43]. Quercetin-loaded HZ-chi/pec showed a remarkable endurance to precipitation up to 500 mM NaCl (Figure S2C). The size of quercetin-loaded HZ-chi/pec particles kept a single peak and increased gradually as the concentration of NaCl increased up to 200 mM (Figure 5A). Meanwhile, their ζ-potential changed to −29 mV (Figure 5B). It is possible that the NaCl ions screen the repulsion among the polysaccharide chain, leading to a greater adsorption of pectin on the particle surface [48,49]. These results are different from the sedimentation of pectin-coated zein particle at 70 mM NaCl reported by Huang et al. [16]. It can be thus speculated that the complex coating of chitosan and pectin provided better stability against salt than did by pectin alone. The size of quercetin-loaded HZ-chi/pec particles became bigger and had two peaks (Figure 5A), and their ζ-potential absolute values significantly decreased upon further increasing the concentration of NaCl. These results suggest the salt at high concentrations reduced the electrostatic repulsion between particles, resulting in their aggregation [46].

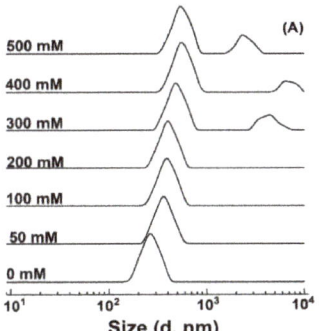
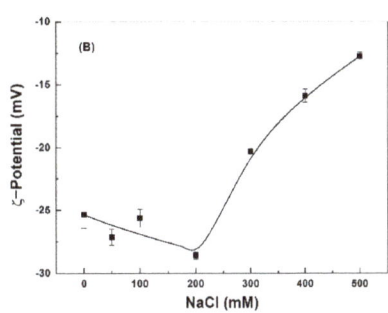

Figure 5. Size distribution (**A**) and ζ−potential (**B**) of quercetin−loaded hollow zein particles coated with chitosan and pectin as a function of NaCl concentration.

3.3.2. pH Stability

Protein-based particles are liable to aggregation around pI [6], limiting their application. Quercetin-loaded hollow zein particles were sable with similar size distribution at pH 2, 3 and 4 (Figure 6A) but precipitated at pH 5 around pI, due to charge neutralization and lack of repulsive forces [50,51]. It is difficult to re-disperse the precipitated zein particles above the pI of zein. Quercetin-loaded hollow zein-chitosan particles showed two size distributions above and below pH 4 (Figure 6A). The change in the particle size below pH 4 is attributed to swelling and dissolution of the polymers with higher charge density [18], while the formation of bigger aggregates is at pH 5, close to the pI of zein [7]. The particles with chitosan were unstable at pH 6–9 due to the deprotonation of chitosan amine groups at pk_a and above [52,53]. It is noted that quercetin-loaded HZ-chi/pec particles showed good stability across the investigated pH range of 2–5 (Figure 6A). The ζ-potential of quercetin-loaded HZ-chi/pec was the highest at pH 5 and decreased as pH decreased (Figure 6B), due to deionization of the carboxylic groups of pectin below its pk_a (3.5) [16]. At pH 6–9, the size of quercetin-loaded HZ-chi/pec particles was around 220 nm and less than those at lower pH. Their ζ-potential also decreased slightly as the pH increased from pH 5. Karim et al. also reported a decline in the ζ-potential and size of pectin/chitosan-coated nanoliposome at pH 8 compared with that of pH 5 [18]. These changes might be due to the detachment of loosely adsorbed pectin molecules from the surface, since chitosan possesses less charged groups as the pH increases [54]. These findings suggest that the double coating with chitosan and pectin provides excellent protection against aggregation, especially around the pI of zein and pk_a of chitosan.

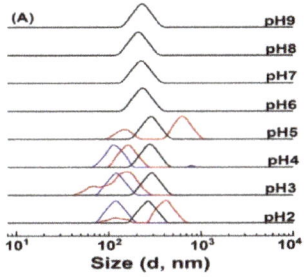
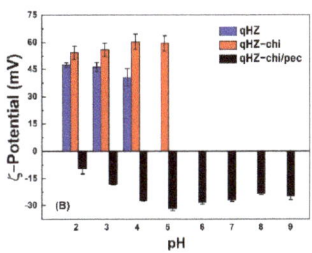

Figure 6. Size distribution (**A**) and ζ−potential (**B**) of quercetin−loaded hollow zein particles without (blue) and with chitosan (red) and chitosan/pectin (black) coating at various pH values.

3.3.3. Temperature Stability

Bioactive-component-loaded carrier systems for food application might be exposed to different temperature treatments during the production cycles. Figure 7 shows the effect of temperature on the size distribution and ζ-potential of quercetin-loaded HZ, HZ-chi and HZ-chi/pec particles. Quercetin-loaded HZ particles had ζ-potential values between +41–+48 mV (Figure 7B). Size of quercetin-loaded HZ particles increased as temperature increased and showed two peaks above 40 °C (Figure 7A), possibly attributed to rearrangement of zein molecules and exposure of nonpolar groups followed by a collapse of hollow structure [55]. This is different from solid zein particles, those were colloidally stable when heated at pH 4 and 80 °C for 120 min [43]. Likewise, Figure 7A shows that size of quercetin-loaded HZ-chi particles increased with increasing temperature, but two peaks were observed above 70 °C, suggesting that the chitosan coating improves the particle stability against heat treatment. Their ζ-potential was +32 mV at 30 °C and increased to a range of +45 and +48 mV at 40–80 °C and to +56 mV at 90 °C. The changes might be due to that the realignment of zein and chitosan at higher temperatures possibly leads to inter/intramolecular interactions [5,56]. The quercetin-loaded HZ-chi/pec particles were stable when the temperature was increased to 40 °C (Figure 7A). Then, their size gradually increased upon further increase in temperature and showed two peaks at 90 °C. Their ζ-potential was kept between −25 and −32 mV at all the investigated temperatures. These results indicate that the pectin coating further improves the colloidal stability of quercetin-loaded HZ-chi particles. It is presumed that steric stabilization of pectin coating inhibits the collision of zein particles with more exposed reactive functional groups at the higher temperature, thus preventing increment in size and particle aggregation [57].

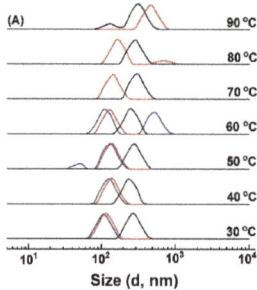
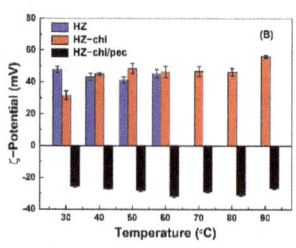

Figure 7. Effect of temperature on size distribution qHZ (blue), qHZ−chi (red) and qHZ−chi/pec (black) (**A**) and ζ−potential (**B**) of quercetin−loaded composite hollow zein particles.

3.3.4. Storage Stability

After storage at 25 and 45 °C for 7 days, quercetin-loaded HZ and HZ-chi particles precipitated, while no precipitation was observed for quercetin-loaded HZ-chi/pec particles (Figure S3). Figure 8 shows size distribution and ζ-potential of quercetin-loaded HZ-chi/pec particles during storage. Their size distribution remained a single peak during storage at 45 °C and increased to around 325 nm after 6 days. At 25 °C, the size distribution kept a single peak around 220 nm after 6 days and then showed two peaks. These results indicate quercetin-loaded HZ-chi/pec particles are more stable at 45 °C than 25 °C, possibly due to the high temperature facilitates adsorption of pectin to the particles more effectively and improves its complex formation capacity [58,59]. At 25 °C, a smaller size, around 90 nm was observed after storage for 13 days (Figure 8), probably due to the detachment, depolymerization and hydrolysis of galacturonic acid glycan chains of pectin at pH < 5 [60]. At 25 °C, a larger size around 900–1500 nm was observed after 20 days (Figure 8). The ζ-potential of HZ-chi/pec particles became more negative over time, both at 25 and 45 °C. These changes indicate the realignment of pectin during storage and exposure of more charged COO$^-$ groups at the interface, leading to an upsurge in ζ-potential [18,54]. Pectin coating substantially enhances the colloidal stability of quercetin-loaded HZ-chi/pec compared with that of qHZ and qHZ-chi (Figure S3A,B) by preventing aggregation of particles through electrostatic repulsions and steric stabilization as well [61].

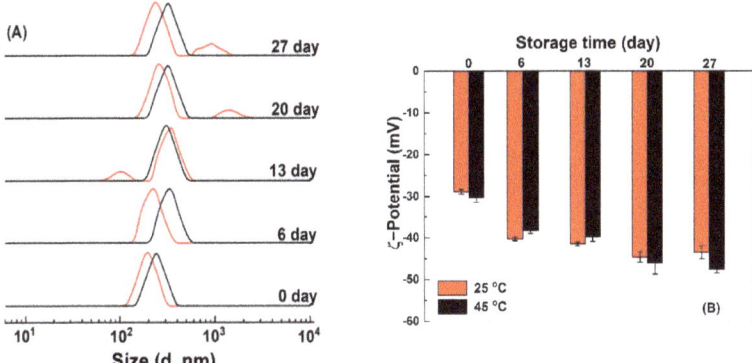

Figure 8. Size distribution (**A**) and ζ−potential (**B**) of quercetin−loaded hollow zein particles coated with chitosan and pectin at 25 °C (red) and 45 °C (black) during storage.

3.4. Antioxidant Activity

The antioxidant activity of encapsulated quercetin was investigated by ABTS and DPPH radical scavenging assays (Figure 9). The ABTS$^+$ scavenging capacity of quercetin-loaded HZ was lower than the quercetin dispersed in water and ethanol. The reason for the lower scavenging capacity of encapsulated quercetin is being embedded in the hydrophobic pockets of protein and inaccessible to ABTS$^+$ [30]. Furthermore, the hydroxyl groups in the B ring of quercetin are the main contributor of H$^+$ and are involved in hydrogen bonding in the composite particles, thus unavailable to scavenge the free radicals. Earlier, quercetin encapsulated in SPI and solid zein particles showed a similar reduction in the ABTS$^+$ savaging capacity. On the other hand, the interfacial chitosan-pectin coating in quercetin-loaded HZ-chi/pec particles may increase the accessibility of hydrophilic ABTS$^+$ to quercetin, demonstrating higher scavenging of ABTS$^+$ compared with that of qHZ (Figure 9). DPPH scavenging capacity of quercetin was improved upon encapsulation in HZ and HZ-chi/pec particles by compared with that of dispersed in ethanol and water (Figure 9). Earlier, it has been reported that both zein and DPPH being soluble in ethanol–water binary medium facilitate scavenging of DPPH by hydrophobic antioxidants encapsulated in the zein particles [62,63]. The ABTS and DPPH radical scavenging assay revealed that by encapsulating quercetin in composite hollow zein particles, it could be better protected from free reactive radicals in the surrounding medium along with an improved or sustained antioxidant activity.

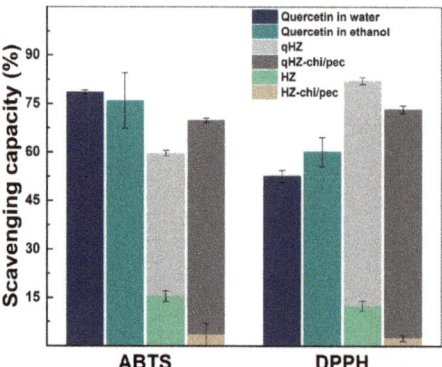

Figure 9. ABTS+ and DPPH scavenging capacity of quercetin, blank and quercetin-loaded HZ and HZ-chi/pec particles.

3.5. Stability of Quercetin

3.5.1. Photochemical Stability

Quercetin is prone to degradation upon exposure to ultraviolet light due to oxidative decarboxylation of the C ring [22]. The retention of unencapsulated quercetin sharply decreased to 25% after 120 min of irradiation (Figure 10A). The quercetin loaded in HZ and HZ-chi/pec particles was more stable, with retention of about 80% after 120 min of irradiation. These results suggesting the hollow particle provide the excellent stability of quercetin against ultraviolet light. The protection results from the physical barrier and light-scattering effect of the particles [61]. The similar retention of quercetin in HZ and HZ-chi/pec particles (Figure 10A) indicates that the protection was fundamentally attributed to zein. A comparable protective effect has previously been reported by encapsulating quercetin in WPI/lotus root amylopectin, pea protein-isolated and zein/soluble soybean polysaccharide composite nanoparticles due to hydrogen and hydrophobic interaction between quercetin and proteins [39,43,64].

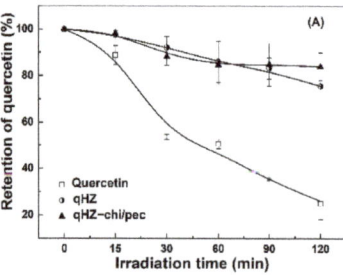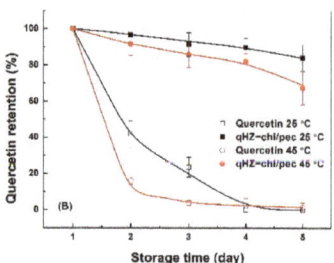

Figure 10. Retention of quercetin free and encapsulated in hollow zein particles coated without (qHZ) and with (qHZ-chi/pec) 1 mg/mL chitosan and 0.1 mg/mL pectin under irradiation (**A**) and during storage at 25 °C and 45 °C (**B**).

3.5.2. Storage Stability

Liu and coworkers reported a drastic decrease in the content of quercetin to around 38% in 3 days of storage at room temperature [39]. The degradation of quercetin was faster at 45 °C than 25 °C, with the retention of quercetin being 42% and 16% after 2 days (Figure 10B), respectively. Its complete loss was observed at 25 °C and 45 °C after 4 days. The degradation of quercetin is attributed to its auto-oxidation in the aqueous medium, which is much pronounced at elevated temperature [19,65]. The retention of quercetin was significantly greater in HZ-chi/pec particles during storage, being 84% and 67% at 25 and

45 °C after 27 days (Figure 10B), respectively. The ABTS$^+$ scavenging capacity of quercetin-loaded HZ-chi/pec particles was lower than quercetin alone (Figure 9), indicating that the entrapment of quercetin reduces its accessibility to ABTS$^+$ [30]. It is thus speculated that the entrapment in HZ-chi/pec particles (Table 2) provides a physical barrier between quercetin and environment-sensitive factors, resulting in the improved stability of quercetin (Figure 10B). Moreover, the hydrophobic interaction and hydrogen bond of quercetin with excipient biopolymer (Figure 4) may inhibit the flavonoid autoxidation and prolong its storage life.

4. Conclusions

The hollow zein particles coated with chitosan and pectin were prepared with 1% Na_2CO_3 as a sacrificial template. The hollow particles coated with 1 mg/mL chitosan and 0.1 mg/mL pectin had a size of 219 nm and ζ-potential of -28 mV. Chitosan coating improved the loading efficiency of quercetin in hollow zein particles. The coating of chitosan/pectin improved the stability of quercetin-loaded hollow zein particles against heat treatment, pH variation and salt. The entrapment in the hollow particles improved the photostability and storage stability of quercetin. The storage stability was better at 25 °C for entrapped quercetin but at 45 °C for hollow zein particles coated with chitosan and pectin. These findings will extend the application of composite hollow zein particles for the incorporation of bioactive components in functional products.

Supplementary Materials: The following are available online at https://www.mdpi.com/article/10.3390/antiox10091476/s1, Figure S1: Size distribution (A) and ζ-potential (B) of hollow zein (HZ) particles coated with chitosan (HZ-chi) and with chitosan and pectin (HZ-chi/pec) without and with quercetin. Figure S2: Visual appearance of quercetin-loaded hollow zein particles (A) coated with chitosan (B) and with chitosan and pectin (C) at 50–500 mM NaCl. Figure S3: Visual appearance of qHZ, qHZ-chi, qHZ-chi/pec at 25 °C (A), 45 °C (B). Figure S4: Calibration curve of quercetin.

Author Contributions: M.A.K. and C.Z.; Conceptualization, Investigation, Writing—original draft preparation, Writing—review and editing, P.Z. and M.Z.; Writing—review and editing, L.L.; Writing—review and editing, Reviewing, Supervision. All authors have read and agreed to the published version of the manuscript.

Funding: This work received support from the National Natural Science Foundation of China (NSFC Project 31571781).

Institutional Review Board Statement: Not applicable.

Informed Consent Statement: Not applicable.

Data Availability Statement: The data presented in this study are available in this manuscript.

Conflicts of Interest: The authors declare no conflict of interest.

References

1. Xu, H.; Jiang, Q.; Reddy, N.; Yang, Y. Hollow nanoparticles from zein for potential medical applications. *J. Mater. Chem.* **2011**, *21*, 18227–18235. [CrossRef]
2. Bentz, K.C.; Savin, D.A. Hollow polymer nanocapsules: Synthesis, properties, and applications. *Polym. Chem.* **2018**, *9*, 2059–2081. [CrossRef]
3. Chuacharoen, T.; Sabliov, C.M. Stability and controlled release of lutein loaded in zein nanoparticles with and without lecithin and pluronic F127 surfactants. *Colloids Surf. A Physicochem. Eng. Asp.* **2016**, *503*, 11–18. [CrossRef]
4. Luo, Y.; Teng, Z.; Wang, Q. Development of zein nanoparticles coated with carboxymethyl chitosan for encapsulation and controlled release of vitamin D3. *J. Agric. Food Chem.* **2012**, *60*, 836–843. [CrossRef] [PubMed]
5. Chen, J.; Zheng, J.; McClements, D.J.; Xiao, H. Tangeretin-loaded protein nanoparticles fabricated from zein/β-lactoglobulin: Preparation, characterization, and functional performance. *Food Chem.* **2014**, *158*, 466–472. [CrossRef] [PubMed]
6. Patel, A.R.; Bouwens, E.C.M.; Velikov, K.P. Sodium caseinate stabilized zein colloidal particles. *J. Agric. Food Chem.* **2010**, *58*, 12497–12503. [CrossRef]
7. Hu, K.; McClements, D.J. Fabrication of biopolymer nanoparticles by antisolvent precipitation and electrostatic deposition: Zein-alginate core/shell nanoparticles. *Food Hydrocoll.* **2015**, *44*, 101–108. [CrossRef]

8. Dai, L.; Zhou, H.; Wei, Y.; Gao, Y.; McClements, D.J. Curcumin encapsulation in zein-rhamnolipid composite nanoparticles using a pH-driven method. *Food Hydrocoll.* **2019**, *93*, 342–350. [CrossRef]
9. Chang, C.; Wang, T.; Hu, Q.; Zhou, M.; Xue, J.; Luo, Y. Pectin coating improves physicochemical properties of caseinate/zein nanoparticles as oral delivery vehicles for curcumin. *Food Hydrocoll.* **2017**, *70*, 143–151. [CrossRef]
10. Chen, S.; Ma, X.; Han, Y.; Wei, Y.; Guo, Q.; Yang, S.; Zhang, Y.; Liao, W.; Gao, Y. Effect of chitosan molecular weight on zein-chitosan nanocomplexes: Formation, characterization, and the delivery of quercetagetin. *Int. J. Biol. Macromol.* **2020**, *164*, 2215–2223. [CrossRef]
11. Khan, M.A.; Chen, L.; Liang, L. Improvement in storage stability and resveratrol retention by fabrication of hollow zein-chitosan composite particles. *Food Hydrocoll.* **2021**, *113*, 106477. [CrossRef]
12. Li, M.F.; Chen, L.; Xu, M.Z.; Zhang, J.L.; Wang, Q.; Zeng, Q.Z.; Wei, X.C.; Yuan, Y. The formation of zein-chitosan complex coacervated particles: Relationship to encapsulation and controlled release properties. *Int. J. Biol. Macromol.* **2018**, *116*, 1232–1239. [CrossRef]
13. Luo, Y.; Wang, Q. Recent development of chitosan-based polyelectrolyte complexes with natural polysaccharides for drug delivery. *Int. J. Biol. Macromol.* **2014**, *64*, 353–367. [CrossRef]
14. Wu, Y.; Yang, W.; Wang, C.; Hu, J.; Fu, S. Chitosan nanoparticles as a novel delivery system for ammonium glycyrrhizinate. *Int. J. Pharm.* **2005**, *295*, 235–245. [CrossRef]
15. Huang, X.; Huang, X.; Gong, Y.; Xiao, H.; McClements, D.J.; Hu, K. Enhancement of curcumin water dispersibility and antioxidant activity using core-shell protein-polysaccharide nanoparticles. *Food Res. Int.* **2016**, *87*, 1–9. [CrossRef] [PubMed]
16. Huang, X.; Liu, Y.; Zou, Y.; Liang, X.; Peng, Y.; McClements, D.J.; Hu, K. Encapsulation of resveratrol in zein/pectin core-shell nanoparticles: Stability, bioaccessibility, and antioxidant capacity after simulated gastrointestinal digestion. *Food Hydrocoll.* **2019**, *93*, 261–269. [CrossRef]
17. Ji, F.; Li, J.; Qin, Z.; Yang, B.; Zhang, E.; Dong, D.; Wang, J.; Wen, Y.; Tian, L.; Yao, F. Engineering pectin-based hollow nanocapsules for delivery of anticancer drug. *Carbohydr. Polym.* **2017**, *177*, 86–96. [CrossRef] [PubMed]
18. Karim, N.; Shishir, M.R.I.; Chen, W. Surface decoration of neohesperidin-loaded nanoliposome using chitosan and pectin for improving stability and controlled release. *Int. J. Biol. Macromol.* **2020**, *164*, 2903–2914. [CrossRef] [PubMed]
19. Wang, W.; Sun, C.; Mao, L.; Ma, P.; Liu, F.; Yang, J.; Gao, Y. The biological activities, chemical stability, metabolism and delivery systems of quercetin: A review. *Trends Food Sci. Technol.* **2016**, *56*, 21–38. [CrossRef]
20. Srinivas, K.; King, J.W.; Howard, L.R.; Monrad, J.K. Solubility and solution thermodynamic properties of quercetin and quercetin dihydrate in subcritical water. *J. Food Eng.* **2010**, *100*, 208–218. [CrossRef]
21. Xu, H.; Shen, L.; Xu, L.; Yang, Y. Controlled delivery of hollow corn protein nanoparticles via non-toxic crosslinking: In vivo and drug loading study. *Biomed. Microdevices* **2015**, *17*, 8. [CrossRef]
22. Wang, Y.; Wang, X. Binding, stability, and antioxidant activity of quercetin with soy protein isolate particles. *Food Chem.* **2015**, *188*, 24–29. [CrossRef] [PubMed]
23. Yi, J.; Fan, Y.; Zhang, Y.; Wen, Z.; Zhao, L.; Lu, Y. Glycosylated α-lactalbumin-based nanocomplex for curcumin: Physicochemical stability and DPPH-scavenging activity. *Food Hydrocoll.* **2016**, *61*, 369–377. [CrossRef]
24. Dong, H.; Yin, X.; Wusigale; Cheng, H.; Choijilsuren, N.; Chen, X.; Liang, L. Antioxidant activity and stability of α-tocopherol, resveratrol and epigallocatechin-3-gallate in mixture and complexation with bovine serum albumin. *Int. J. Food Sci. Technol.* **2021**, *56*, 1788–1800. [CrossRef]
25. Pan, K.; Luo, Y.; Gan, Y.; Baek, S.J.; Zhong, Q. PH-driven encapsulation of curcumin in self-assembled casein nanoparticles for enhanced dispersibility and bioactivity. *Soft Matter* **2014**, *10*, 6820–6830. [CrossRef] [PubMed]
26. Sun, C.; Dai, L.; Gao, Y. Interaction and formation mechanism of binary complex between zein and propylene glycol alginate. *Carbohydr. Polym.* **2017**, *157*, 1638–1649. [CrossRef] [PubMed]
27. Hu, S.; Wang, T.; Fernandez, M.L.; Luo, Y. Development of tannic acid cross-linked hollow zein nanoparticles as potential oral delivery vehicles for curcumin. *Food Hydrocoll.* **2016**, *61*, 821–831. [CrossRef]
28. Shukla, R.; Cheryan, M. Zein: The industrial protein from corn. *Ind. Crop. Prod.* **2001**, *13*, 171–192. [CrossRef]
29. Chen, S.; Han, Y.; Jian, L.; Liao, W.; Zhang, Y.; Gao, Y. Fabrication, characterization, physicochemical stability of zein-chitosan nanocomplex for co-encapsulating curcumin and resveratrol. *Carbohydr. Polym.* **2020**, *236*, 116090. [CrossRef]
30. Pauluk, D.; Padilha, A.K.; Khalil, N.M.; Mainardes, R.M. Chitosan-coated zein nanoparticles for oral delivery of resveratrol: Formation, characterization, stability, mucoadhesive properties and antioxidant activity. *Food Hydrocoll.* **2019**, *94*, 411–417. [CrossRef]
31. Ren, X.; Hou, T.; Liang, Q.; Zhang, X.; Hu, D.; Xu, B.; Chen, X.; Chalamaiah, M.; Ma, H. Effects of frequency ultrasound on the properties of zein-chitosan complex coacervation for resveratrol encapsulation. *Food Chem.* **2019**, *279*, 223–230. [CrossRef]
32. Pfeiffer, C.; Rehbock, C.; Hühn, D.; Carrillo-Carrion, C.; De Aberasturi, D.J.; Merk, V.; Barcikowski, S.; Parak, W.J. Interaction of colloidal nanoparticles with their local environment: The (ionic) nanoenvironment around nanoparticles is different from bulk and determines the physico-chemical properties of the nanoparticles. *J. R. Soc. Interface* **2014**, *11*, 20130931. [CrossRef] [PubMed]
33. Ma, J.J.; Yu, Y.G.; Yin, S.W.; Tang, C.H.; Yang, X.Q. Cellular Uptake and Intracellular Antioxidant Activity of Zein/Chitosan Nanoparticles Incorporated with Quercetin. *J. Agric. Food Chem.* **2018**, *66*, 12783–12793. [CrossRef]

34. Yan, L.; Wang, R.; Wang, H.; Sheng, K.; Liu, C.; Qu, H.; Ma, A.; Zheng, L. Formulation and characterization of chitosan hydrochloride and carboxymethyl chitosan encapsulated quercetin nanoparticles for controlled applications in foods system and simulated gastrointestinal condition. *Food Hydrocoll.* **2018**, *84*, 450–457. [CrossRef]
35. Chen, S.; Han, Y.; Sun, C.; Dai, L.; Yang, S.; Wei, Y.; Mao, L.; Yuan, F.; Gao, Y. Effect of molecular weight of hyaluronan on zein-based nanoparticles: Fabrication, structural characterization and delivery of curcumin. *Carbohydr. Polym.* **2018**, *201*, 599–607. [CrossRef]
36. Chen, S.; Han, Y.; Huang, J.; Dai, L.; Du, J.; McClements, D.J.; Mao, L.; Liu, J.; Gao, Y. Fabrication and Characterization of Layer-by-Layer Composite Nanoparticles Based on Zein and Hyaluronic Acid for Codelivery of Curcumin and Quercetagetin. *ACS Appl. Mater. Interfaces* **2019**, *11*, 16922–16935. [CrossRef] [PubMed]
37. Li, X.; Maldonado, L.; Malmr, M.; Rouf, T.B.; Hua, Y.; Kokini, J. Development of hollow kafirin-based nanoparticles fabricated through layer-by-layer assembly as delivery vehicles for curcumin. *Food Hydrocoll.* **2019**, *96*, 93–101. [CrossRef]
38. Patel, A.R.; Heussen, P.C.M.; Hazekamp, J.; Drost, E.; Velikov, K.P. Quercetin loaded biopolymeric colloidal particles prepared by simultaneous precipitation of quercetin with hydrophobic protein in aqueous medium. *Food Chem.* **2012**, *133*, 423–429. [CrossRef]
39. Liu, K.; Zha, X.Q.; Shen, W.; Li, Q.M.; Pan, L.H.; Luo, J.P. The hydrogel of whey protein isolate coated by lotus root amylopectin enhance the stability and bioavailability of quercetin. *Carbohydr. Polym.* **2020**, *236*, 116009. [CrossRef]
40. Deng, L.; Zhang, X.; Li, Y.; Que, F.; Kang, X.; Liu, Y.; Feng, F.; Zhang, H. Characterization of gelatin/zein nanofibers by hybrid electrospinning. *Food Hydrocoll.* **2018**, *75*, 72–80. [CrossRef]
41. Epure, V.; Griffon, M.; Pollet, E.; Avérous, L. Structure and properties of glycerol-plasticized chitosan obtained by mechanical kneading. *Carbohydr. Polym.* **2011**, *83*, 947–952. [CrossRef]
42. Fan, Y.; Liu, Y.; Gao, L.; Zhang, Y.; Yi, J. Improved chemical stability and cellular antioxidant activity of resveratrol in zein nanoparticle with bovine serum albumin-caffeic acid conjugate. *Food Chem.* **2018**, *261*, 283–291. [CrossRef]
43. Li, H.; Wang, D.; Liu, C.; Zhu, J.; Fan, M.; Sun, X.; Wang, T.; Xu, Y.; Cao, Y. Fabrication of stable zein nanoparticles coated with soluble soybean polysaccharide for encapsulation of quercetin. *Food Hydrocoll.* **2019**, *87*, 342–351. [CrossRef]
44. Baranović, G.; Šegota, S. Infrared spectroscopy of flavones and flavonols. Reexamination of the hydroxyl and carbonyl vibrations in relation to the interactions of flavonoids with membrane lipids. *Spectrochim. Acta-Part. A Mol. Biomol. Spectrosc.* **2018**, *192*, 473–486. [CrossRef]
45. Souza, M.P.; Vaz, A.F.M.; Correia, M.T.S.; Cerqueira, M.A.; Vicente, A.A.; Carneiro-da-Cunha, M.G. Quercetin-Loaded Lecithin/Chitosan Nanoparticles for Functional Food Applications. *Food Bioprocess Technol.* **2014**, *7*, 1149–1159. [CrossRef]
46. Chen, Y.; Xia, G.; Zhao, Z.; Xue, F.; Chen, C.; Zhang, Y. Formation, structural characterization, stability and: In vitro bioaccessibility of 7,8-dihydroxyflavone loaded zein-/sophorolipid composite nanoparticles: Effect of sophorolipid under two blending sequences. *Food Funct.* **2020**, *11*, 1810–1825. [CrossRef] [PubMed]
47. Wang, T.X.; Li, X.X.; Chen, L.; Li, L.; Janaswamy, S. Carriers Based on Zein-Dextran Sulfate Sodium Binary Complex for the Sustained Delivery of Quercetin. *Front. Chem.* **2020**, *8*, 662. [CrossRef]
48. Liu, Y.; Yang, J.; Zhao, Z.; Li, J.; Zhang, R.; Yao, F. Formation and characterization of natural polysaccharide hollow nanocapsules via template layer-by-layer self-assembly. *J. Colloid Interface Sci.* **2012**, *379*, 130–140. [CrossRef]
49. Ye, S.; Wang, C.; Liu, X.; Tong, Z. Multilayer nanocapsules of polysaccharide chitosan and alginate through layer-by-layer assembly directly on PS nanoparticles for release. *J. Biomater. Sci. Polym. Ed.* **2005**, *16*, 909–923. [CrossRef] [PubMed]
50. Chen, Y.; Zhao, Z.; Xia, G.; Xue, F.; Chen, C.; Zhang, Y. Fabrication and characterization of zein/lactoferrin composite nanoparticles for encapsulating 7,8-dihydroxyflavone: Enhancement of stability, water solubility and bioaccessibility. *Int. J. Biol. Macromol.* **2020**, *146*, 179–192. [CrossRef]
51. Wang, L.; Zhang, Y. Heat-induced self-assembly of zein nanoparticles: Fabrication, stabilization and potential application as oral drug delivery. *Food Hydrocoll.* **2019**, *90*, 403–412. [CrossRef]
52. Xu, W.; Tang, Y.; Yang, Y.; Wang, G.; Zhou, S. Establishment of a stable complex formed from whey protein isolate and chitosan and its stability under environmental stresses. *Int. J. Biol. Macromol.* **2020**, *165*, 2823–2833. [CrossRef] [PubMed]
53. Ding, L.; Huang, Y.; Cai, X.X.; Wang, S. Impact of pH, ionic strength and chitosan charge density on chitosan/casein complexation and phase behavior. *Carbohydr. Polym.* **2019**, *208*, 133–141. [CrossRef]
54. Birch, N.P.; Schiffman, J.D. Characterization of self-Assembled polyelectrolyte complex nanoparticles formed from chitosan and pectin. *Langmuir* **2014**, *30*, 3441–3447. [CrossRef] [PubMed]
55. Yu, Y.B.; Wang, C.; Chen, T.T.; Wang, Z.W.; Yan, J.K. Enhancing the colloidal stabilities of zein nanoparticles coated with carboxylic curdlans. *Lwt* **2020**, *137*, 110475. [CrossRef]
56. Dai, L.; Sun, C.; Wang, D.; Gao, Y. The interaction between zein and lecithin in ethanol-Water solution and characterization of zein-Lecithin composite colloidal nanoparticles. *PLoS ONE* **2016**, *11*, e0167172. [CrossRef]
57. Joye, I.J.; Nelis, V.A.; McClements, D.J. Gliadin-based nanoparticles: Fabrication and stability of food-grade colloidal delivery systems. *Food Hydrocoll.* **2015**, *44*, 86–93. [CrossRef]
58. Shamsara, O.; Jafari, S.M.; Muhidinov, Z.K. Fabrication, characterization and stability of oil in water nano-emulsions produced by apricot gum-pectin complexes. *Int. J. Biol. Macromol.* **2017**, *103*, 1285–1293. [CrossRef]
59. Luo, Y.; Pan, K.; Zhong, Q. Casein/pectin nanocomplexes as potential oral delivery vehicles. *Int. J. Pharm.* **2015**, *486*, 59–68. [CrossRef] [PubMed]

60. Shao, P.; Wang, P.; Niu, B.; Kang, J. Environmental stress stability of pectin-stabilized resveratrol liposomes with different degree of esterification. *Int. J. Biol. Macromol.* **2018**, *119*, 53–59. [CrossRef]
61. Joye, I.J.; Davidov-Pardo, G.; McClements, D.J. Encapsulation of resveratrol in biopolymer particles produced using liquid antisolvent precipitation. Part 2: Stability and functionality. *Food Hydrocoll.* **2015**, *49*, 127–134. [CrossRef]
62. Tai, K.; He, X.; Yuan, X.; Meng, K.; Gao, Y.; Yuan, F. A comparison of physicochemical and functional properties of icaritin-loaded liposomes based on different surfactants. *Colloids Surf. A Physicochem. Eng. Asp.* **2017**, *518*, 218–231. [CrossRef]
63. Khan, M.A.; Fang, Z.; Wusigale; Cheng, H.; Gao, Y.; Deng, Z.; Liang, L. Encapsulation and protection of resveratrol in kafirin and milk protein nanoparticles. *Int. J. Food Sci. Technol.* **2019**, *54*, 2998–3007. [CrossRef]
64. Cuevas-Bernardino, J.C.; Leyva-Gutierrez, F.M.A.; Vernon-Carter, E.J.; Lobato-Calleros, C.; Román-Guerrero, A.; Davidov-Pardo, G. Formation of biopolymer complexes composed of pea protein and mesquite gum – Impact of quercetin addition on their physical and chemical stability. *Food Hydrocoll.* **2018**, *77*, 736–745. [CrossRef]
65. Makris, D.P.; Rossiter, J.T. Heat-induced, metal-catalyzed oxidative degradation of quercetin and rutin (quercetin 3-O-rhamnosylglucoside) in aqueous model systems. *J. Agric. Food Chem.* **2000**, *48*, 3830–3838. [CrossRef] [PubMed]

Article

Whey Protein Isolate-Xylose Maillard-Based Conjugates with Tailored Microencapsulation Capacity of Flavonoids from Yellow Onions Skins

Ștefania Adelina Milea, Iuliana Aprodu, Elena Enachi, Vasilica Barbu, Gabriela Râpeanu, Gabriela Elena Bahrim and Nicoleta Stănciuc *

Faculty of Food Science and Engineering, Dunarea de Jos University of Galati, 111 Domnească Street, 800201 Galați, Romania; adelina.milea@ugal.ro (Ș.A.M.); iuliana.aprodu@ugal.ro (I.A.); elena.ionita@ugal.ro (E.E.); vasilica.barbu@ugal.ro (V.B.); gabriela.rapeanu@ugal.ro (G.R.); Gabriela.Bahrim@ugal.ro (G.E.B.)
* Correspondence: Nicoleta.Stanciuc@ugal.ro

Citation: Milea, Ș.A.; Aprodu, I.; Enachi, E.; Barbu, V.; Râpeanu, G.; Bahrim, G.E.; Stănciuc, N. Whey Protein Isolate-Xylose Maillard-Based Conjugates with Tailored Microencapsulation Capacity of Flavonoids from Yellow Onions Skins. *Antioxidants* **2021**, *10*, 1708. https://doi.org/10.3390/antiox10111708

Academic Editors: Li Liang and Hao Cheng

Received: 1 October 2021
Accepted: 26 October 2021
Published: 27 October 2021

Publisher's Note: MDPI stays neutral with regard to jurisdictional claims in published maps and institutional affiliations.

Copyright: © 2021 by the authors. Licensee MDPI, Basel, Switzerland. This article is an open access article distributed under the terms and conditions of the Creative Commons Attribution (CC BY) license (https://creativecommons.org/licenses/by/4.0/).

Abstract: The objective of this study is to encapsulate flavonoids from yellow onion skins in whey protein isolates (WPI) and xylose (X), by Maillard-based conjugates, as an approach to improve the ability to entrap flavonoids and to develop powders with enhanced antioxidant activity. WPI (0.6%, w/v) was conjugated to X (0.3%, w/v) through the Maillard reaction at 90 °C for 120 min, in the presence of a flavonoid-enriched extract. Two variants of powders were obtained by freeze-drying. The glycation of WPI allowed a better encapsulation efficiency, up to $90.53 \pm 0.29\%$, corresponding to a grafting degree of $30.38 \pm 1.55\%$. The molecular modelling approach was used to assess the impact of X interactions with α-lactalbumin and β-lactoglobulin on the ability of these proteins to bind the main flavonoids from the yellow onion skins. The results showed that X might compete with quercetin glucosides to bind with α-lactalbumin. No interference was found in the case of β-lactoglobulin. The microstructural appearance of the powders revealed finer spherosomes in powder with WPI–X conjugates via the Maillard reaction. The powders were added to nachos, followed by a phytochemical characterization, in order to test their potential added value. An increase in antioxidant activity was observed, with no significant changes during storage.

Keywords: glycation; flavonoids; microencapsulation; onion skins; antioxidant activities

1. Introduction

The recent trends in food intakes have experienced a transition to a more healthy oriented nutrition, which shifted eating habits to natural foods, paving the way for the extraction and identification of new biologically active compounds, with beneficial or even therapeutic functions, with a considerable emphasis put on well-being and the prevention of disease. Therefore, obtaining and incorporating bioactive-enriched plant extracts in food may significantly contribute to lowering the risk of specific illnesses [1]. Certain advantages may result from the use of bioactive-enriched plant extracts when compared with individual or synthetic compounds, particularly in terms of the synergistic actions of different molecules [2].

Onion (*Allium cepa* L.) is cultivated around the world, being the second most grown horticultural crop after tomatoes. It has been estimated that more than 550,000 tonnes of onion skin bio-waste was generated by the use of the 89 million tonnes onion harvest [3]. The onion waste represents an environmental problem since it is not suitable for animal feeding and so is usually sent to landfill. Onion skins and the outer layers contain significant quantities of fiber and phenolic compounds, such as flavonoids, glucosides, phenolic acids, and organosulfur compounds [3]. In particular, the onion solid waste is rich in quercetin, quercetin glucosides, quercetin polymers, ferulic acid, gallic acid, and kaempferol, with significant beneficial effects [4] associated with biological activities such as: antidiabetic, antioxidant, anti-inflammatory, anticancer, antimicrobial, and enzyme inhibitory effects [5]. Thus, it can be appreciated that onion has nutritional complexity and holds suitable potential for functional food development, as a source of antioxidant, antimicrobial, anticancer, and antibrowning compounds [6].

Nowadays, the food industry is focusing on implementing methods for the valorization of onion solid waste, as a natural resource with a high amount of value-added ingredients, into eco-friendly functional foods [7]. However, adding polyphenols in a free form in foods may lead to chemical instability due to the unsaturated bonds contained in their molecular structures. The stability is affected by the presence of oxidants, heat, light, and enzymes during storage [8]. Suitable techniques to protect phenolic compounds from chemical damage before their industrial application carry out microencapsulation using different methods, such as freeze-drying [9], which may overcome the drawbacks of their instability, improve their bioavailability as well as shelf life [10] and widen the industrial applications in the food, pharmaceutical and cosmetics industries [11]. In our recent study, different delivery systems were developed for extracts enriched in onion skin flavonoids using a unique combination of whey protein isolates, whey proteins hydrolysates, pectin, and maltodextrin as coating materials [12]. The coating materials should have thermal or mechanical stability to protect the core materials from external factors. Since proteins have amphiphilic properties, they can correlate with the interaction of various chemical groups. However, when using proteins as coatings, some limitations should be considered, given by several external factors, including pH variation, ionic strength and in vitro proteolysis by pepsin, which lead to the degradation of protective walls, causing the release and degradation of bioactives during digestion [13]. These authors tested various structural designs, such as Maillard-based conjugation, to modify the structure and properties of whey proteins and to produce more stable delivery systems with excellent properties. The functional and physico-chemical properties gained with glycation reaction refer to significantly improved emulsifying properties, thermal stability, antioxidant properties, antibacterial activity, and water solubility [14], simultaneously with enhancing the thermal stability of proteins over a wide range of pH and thermal aggregation values [13]. Numerous studies are focused on whey protein as an encapsulating material, but, to the best of our knowledge, there are no studies using whey protein isolates (WPI) in conjugate form with xylose (X) as an encapsulation material for flavonoids. Therefore, the aim of this study was to test the possibility of using WPI–X Maillard-based conjugates as coating materials for flavonoids extracted from yellow onions skins. Flavonoids were isolated by means of solid–liquid extraction in combination with ultrasound-assisted extraction using ethanol as solvent. The WPI–X conjugates were generated via heating in an alkali environment, whereas the flavonoid microcapsules were generated using freeze-drying. Two powders were obtained, using WPI-X conjugates with and without heating, and the resulting freeze-dried powders were characterized in terms of their encapsulation efficiency, phytochemical content and antioxidant activity. Structural and morphological particularities of the samples were analyzed using confocal laser electron microscopy. In order to test the added value, the powders were added to the recipe of a food product (nachos), followed by phytochemical characterization. The results obtained in this study could bring certain benefits in terms of exploiting the bioactive potential of phytochemicals and glycation reaction for developing formulas with improved functional properties.

2. Materials and Methods

2.1. Chemicals

Whey protein isolate (WPI) (protein content of 95%) was purchased from Fonterra (New Zealand). Xylose (X) (about 99% purity) and the reagents used to determine the total phenolic compounds, total flavonoid content, 2,2-diphenyl-1-picrylhydrazyl (DPPH), and 6-hydroxy-2,5,7,8-tetramethylchroman 2-carboxylic acid (Trolox) were purchased from Sigma-Aldrich Corp. (St. Louis, MO, USA). All other reagents were of analytical grade.

2.2. Ethanolic Ultrasound-Assisted Extraction of Flavonoids from Onion Skins

Yellow onions were purchased from a local market (Galati, Romania) in June 2020. The outer layers of onions were collected, cleaned with distilled water and dried. Before extraction, in order to obtain a homogeneous batch of particle size, the onion skins were ground to sizes smaller than 0.5 mm × 0.5 mm and used for further extraction. The flavonoidic extract was obtained by mixing 50 g of the onion skins with 450 mL of 70% ethanol solution and glacial acetic acid (ratio 9:1, *v/v*). The extraction was performed using a sonication bath at 40 °C for 30 min. In order to obtain flavonoids-enriched extracts, the extraction was repeated three times, while the supernatants were centrifuged at 5000× *g* for 10 min at 4 °C, collected and concentrated under reduced pressure at 40 °C. The obtained extract was characterized in terms of selected phytochemicals and used for microencapsulation experiments.

2.3. Preparation of WPI–X–Flavonoid Conjugates

WPI–X conjugates were prepared according to the method described by Jia et al. [13] with minor modifications. An amount of 60 mg/mL of WPI was first dissolved in 100 mL ultrapure water under gentle stirring, followed by the addition of 30 mg/mL of X (mass ratio of 2:1). The mixture was allowed to hydrate for 14 h at room temperature (25 °C). After complete hydration, about 750 mg of concentrated flavonoid-enriched extract was added and the mixture was allowed to dissolve by ultrasonication for 1 h at 35 °C. The resulting solution was divided into two: variant 1 (coded V1) and variant 2 (coded V2). The pH of V2 solution was adjusted at 9.0 and placed in a sealed screw-top glass tube. In order to promote the Maillard conjugation, Variant 2 was heated to 90 °C in water bath for 3 h. After heating, the temperature of Variant 2 was lowered to 25 °C in an ice bath. Both variants were freeze-dried (CHRIST Alpha 1–4 LD plus, Osterode am Harz Germany) at −42 °C under a pressure of 10 Pa for 48 h and stored at −4 °C.

2.4. Characterization of the Extract and of Microencapsulated Powders

Both the extract and the powders were characterized for flavonoids, total phenolic compound contents and antioxidant activity using the aluminum chloride method, Folin–Ciocâlteu and DPPH method, respectively, as described by Milea et al. [12]. The total flavonoid contents are expressed in mg quercetin equivalents (QE)/g dry weight (DW), whereas the total polyphenol contents are expressed in mg gallic acid equivalents (GAE)/g DW. In each case, the concentrations of bioactives and antioxidant activity were expressed through selected standard calibration curves.

The encapsulation efficiency of the powders was calculated as described by Saénz et al. [15]. In brief, the microencapsulation efficiency was determined by assessing the surface flavonoid contents (SFC) and total flavonoid contents (TFC) of the powders, expressed as mg QE/g DW. In order to quantify the SFC, 50 mg of powders was mixed with 5 mL of ethanol and methanol (ratio 1:1, *v/v*). These dispersions were stirred at room temperature for 1 min and then centrifuged at 4000× *g* and 4 °C for 10 min. For TFC, 50 mg of powder was accurately weighed and dispersed in 5 mL of a mixture of ethanol, acetic acid, and water (50:8:42, *v/v/v*). The resulting dispersion was vortexed (1 min), followed by ultrasonication for 30 min at 40 ± 1.0 °C, to break the microcapsules. The supernatant was centrifuged at 14,000× *g* for 10 min and then filtered. The content of flavonoids in the resulting supernatants was measured by the aluminum chloride method, as explained by Milea et al. [12]. The microencapsulation efficiency (ME, %) was calculated using Equation (1):

$$\text{ME}\ (\%) = \frac{\text{TFC} - \text{SFC}}{\text{TFC}} \times 100 \quad (1)$$

The antioxidant activity was assessed using the DPPH method. Briefly, 0.1 mL of supernatant resulted from TFC determination was added to 3.9 mL of DPPH stock solution. The DPPH stock solution was prepared mixing 3 g of DPPH with 100 mL of methanol. Simultaneously, a control sample was prepared by adding 0.1 mL methanol to 3.9 mL

DPPH. The absorbance for both samples was read at 515 nm after 1.5 h. The results were calculated using a calibration curve and are expressed in mMol Trolox/g DW.

2.5. Browning Index and Grafting Degree Measurement of the Powders

The browning intensity was measured using a spectrophotometric method with minor modifications [16]. A volume of 5 mL of a mixture consisting of equal volume of acetic acid (2% v/v) and formaldehyde (1% v/v) was added to 0.1 mL of 1 mg/mL of samples and centrifuged for 10 min at 4500× g. The resulting supernatant (5 mL) was mixed with 5 mL of ethanol and the mixture was centrifuged again. Absorbances of the supernatant at 420 and 600 nm were measured. The difference between the two absorbance values was used to evaluate the browning index.

The grafting degree was determined using the o-phthalaldehyde (OPA) method, according to Jia et al. [13]. A volume of 4 mL of OPA was added to 0.2 mL of the diluted samples (0.1 mg/mL) in test tubes. Upon homogenization, all tubes were placed in a water bath at 35 °C for 1 min. The absorbance was measured at 340 nm. A blank sample was also made using the same volume of ultrapure water. Grafting degree was determined using Equation (2):

$$\text{Grafting Degree (\%)} = 100 \times \frac{A_0 - A_s}{A_0} \quad (2)$$

where A_0 and A_s are the absorbance of blank sample and absorbance of tested samples, respectively.

2.6. In Silico Investigations

In agreement with the experimental approach, the molecular modeling tools were used to simulate X binding by the two major whey proteins. The three-dimensional molecular models of the α-lactalbumin (α-LA, PDB ID: 1F6S) [17] and β-lactoglobulin monomers (β-LG, PDB ID: 4DQ3) [18] from the RCSB Protein Data Bank were optimized and relaxed at 25 °C using GROMACS 4.6 software [19], in agreement with the protocol previously described by Aprodu et al. [20]. The equilibrated protein models were used as receptors for the X binding. The PatchDock algorithm [21], which is very efficient for performing protein–small ligand docking, was used to identify the most probable binding site of X molecules to the WP. Matching the receptor and ligand molecules was carried out through rigid body docking, which is based on the shape complementarity principles. The algorithm employed involves the following major stages: the surface of the receptor was first segmented to identify the so-called hot spot residues on the concave, convex or flat geometric patches, which were selected for a further surface patch matching step with the ligand. The resulting complexes were filtered to disqualify the solutions, involving steric clashes, and finally ranked based on the geometric shape complementarity score. The best three WPI X fits were selected based on the binding energy values among the potential docking models generated by the PatchDock algorithm [21]. An in-depth analysis of the binding pockets was carried out using the PDBePISA [22] tools and DoGSiteScorer web server [23] to identify the extent to which protein glycation affects flavonoids binding.

2.7. Confocal Laser Microscope Spectroscopy

A confocal laser scanning microscopy analysis of the samples was employed to assess the structural appearance of the microencapsulated powders. The CLSM images were captured with a Zeiss confocal laser scanning system (LSM710) equipped with several types of lasers such as a diode laser (405 nm), Ar laser (458, 488, 514 nm), DPSS laser (diode pumped solid state—561 nm) and HeNe laser (633 nm). The powders were observed with a 20× apochromatic objective, at zoom values of 1 and 0.6, respectively. The obtained 3D images were rendered and processed by ZEN 2012 SP1 software (Black Edition).

2.8. Formulation of a Value-Added Food Product

To support the multifunctional properties, the powders were added as an ingredient in a nacho recipe at a ratio of 3%. The recipe involved mixing of corn (250 g) and wheat flour (100 g) in a ratio of 2.5:1, onion (100 mg), pepper (10 mg), oil (20 mL), salt (10 mg), powders (3%) and water (150 mL). The corresponding samples were coded as N1 and N2. The control sample was nachos without the addition of microencapsulated powder (C). Samples were homogenized and allowed to stand for 1 h at room temperature to equilibrate. After homogenization, the nachos were formed and cooked for 6 min in an oven (3 min on each side) at 200 °C. The storage stability of bioactives was tested at 0 days and after 28 days at 25 °C.

2.9. Statistical Analyses

All analyses were performed in triplicate and data are reported as mean ± standard deviation (SD). After running the normality and homoscedasticity tests, experimental data were subjected to one-way analysis of variance (ANOVA) in order to identify significant differences. The Tukey method with a 95% confidence interval was employed for post hoc analysis; $p < 0.05$ was considered to be statistically significant. The statistical analysis was carried out using Minitab 18 software.

3. Results

3.1. Phytochemical Characterization of the Yellow Onion Skin Extract

The solid–liquid ethanolic ultrasound-assisted method applied in this study allowed us to obtain a bioactive-enriched extract, containing flavonoids of 228.7 ± 3.0 mg QE/g DW, with a total polyphenolic content of 96.1 ± 2.7 mg GAE/g DW, and yielding an antioxidant activity of 495.9 ± 2.4 mM TEAC/g DW. In our previous study, different extraction techniques were tested in order to select the most suitable method to obtain flavonoid-enriched extracts from yellow onion skins [24]. The results showed a satisfactory content in phytochemicals, when comparing the ultrasound-assisted technique versus the conventional solid–liquid extraction. However, the selection of the ultrasound-assisted extraction in this study was based on the reduction time and protection of thermolabile compounds. Therefore, Constantin et al. [24] reported similar values for flavonoid contents in ultrasound-assisted extracts of 230.6 ± 8.4 mg QE/g DW. Additionally, Milea et al. [12] extracted the biologically active compounds from yellow onion skins using a similar method and reported flavonoids of 97.3 ± 3.0 mg QE/g DW, polyphenols of 55.3 ± 2.5 mg GAE/g DW and an antioxidant activity of 345.0 ± 2.7 mM TE/g DW. Singh et al. [25] used an ultrasound-assisted method to extract the bioactive compounds from onions. The extraction with 70% ethanol showed similar values for flavonoid extraction of 212.3 ± 14.6 mg QE/g and a higher amount of phenolic compounds (418.0 ± 34.4 mg GAE/g). On the other hand, Pobłocka-Olech et al. [26] extracted flavonoids from different varieties of yellow onion skins using only methanol and obtained a lower level compared with the current results, between 2.4 and 12.2 mg QE/g. Benito-Román et al. [27] performed a comparative study of polyphenols from onion wastes between conventional and ultrasound-assisted extraction. They reported smaller values for flavonoid contents, ranging from 7.7 ± 0.1 to 23.8 mg QE/g dry onion skins (DOS). The different values could be explained by the distinct selected parameters, different origin of raw materials or by the method of expressing final results (DW/DOS). Likewise, the experimental conditions allowed us to extract a significant amount of polyphenols (73.3 ± 1.8 mg GAE/g DOS), which were further increased to 102.1 ± 5.1 mg GAE/g DOS.

The difference between results is due to the extraction method. As is known, ultrasound-assisted method simplifies and accelerates the extraction because the high-intensity ultrasounds increase pressure and temperature, causing a disruption of the cell wall of the matrix, with the subsequent release of polyphenols. Moreover, this technique offers the advantages of lower extraction times and temperatures compared to conventional extraction techniques [28].

3.2. Correlation between Browning Intensity, Grafting Degree and Microencapsulation Efficiency

An effective method to improve the functional properties of proteins, including the ability to include and protect low molecular bioactives, is based on the interaction with polysaccharides and smaller carbohydrates, via Maillard conjugation [29]. The Maillard reaction is a complex reaction occurring between amines and carbonyls [30], which involves, first, to consumption of the free amino group by the carbonylation reaction, mainly coming from the free amino group on the side chain such as lysine and arginine, or the free amino group on the N-terminus of the peptide chain of the protein molecule. Therefore, the carbonyl condensation between the reducing sugar and nucleophilic amino group could be analyzed by the loss of amino acids in the reaction [31]. Ideally, as reported by Jiménez-Castanõ et al. [32], to produce a glycol conjugate destined for incorporation into food, to avoid the formation of the highly colored, insoluble, nitrogen-containing polymeric compounds, referred to as melanoidins, the Maillard reaction needs to be performed under carefully controlled conditions to prevent the later stage changes. As reported by Jia et al. [13], protein glycation by the Maillard reaction might be favored in alkali conditions, while glycation is inhibited by the partial denaturation of the protein in acid conditions. These authors suggested a higher grafting degree and lower browning intensity at pH 9.0 after heating for 3 h. Therefore, these parameters were selected in our study to promote Maillard-based conjugates between WPI and X.

The brown-colored pigment formation in foods is caused by the Maillard reactions or caramelization [16]; therefore, the browning index is generally accepted as an indicator of the Maillard reaction. The brown pigment formation in the microencapsulated powders was evaluated by absorbance measurements at 420 nm and 600 nm, respectively. As expected, the browning intensity was higher (0.12 ± 0.01) for the heat-treated variant (Variant 2) than for the untreated variant (0.09 ± 0.01) (Variant 1). A proportional increase was observed between the browning intensity and antioxidant activity of the powders, indicating the strong antioxidant potency of the glycated variant due to the heating process. The powders showed significant differences in antioxidant activity ($p < 0.05$), with values of 179.7 ± 4.5 mMol TE/g DW for V1 and 184.4 ± 0.7 mMol TE/g DW for V2. Suminar et al. [33] explained that this was probably caused by reducing sugar reacting more easily with amino acids in heating conditions and producing antioxidant activity. The formation of the conjugate was also confirmed by grafting degree, which is able to reflect the level of glycation [13]. In the present study, a correlation between the grafting degree and encapsulation efficiency can be observed in both variants. Thereby, a grafting degree of $22.6 \pm 2.5\%$ and an encapsulation efficiency of $86.7 \pm 1.4\%$ were observed for V1. A significantly higher ($p < 0.05$) values were estimated for Variant 2, with a grafting degree of $30.4 \pm 1.6\%$ and an encapsulation efficiency of $90.5 \pm 0.3\%$. Therefore, it can be appreciated that the Maillard-based conjugates showed a higher ability to entrap the flavonoids from yellow onion skin extract. These results indicate that the glycated form of the powder has a positive effect on the encapsulation efficiency.

In the conditions applied in our study, the Maillard-based conjugates between WPI and X caused structural changes in proteins that improved the ability to entrap flavonoids, in good agreement with reports of Liu et al. [34], Xu et al. [35] and Liu et al. [36]. The glycation degree can be correlated with the decrease in available -NH_2 groups. For example, Shang et al. [37] suggested a dramatic loss in Lys and Arg, whereas a significant decrease in Tyr and Cys was also found, due to the formation of the dehydroalpropyl side chain. These authors also reported a transition toward a higher molecular weight distribution of WPI heated in the presence of X, at 90 °C and 95 °C and pH 9.0, whereas the contents of protein polymers larger than 40 kDa increased with the reaction time, thus indicating a protein crosslinking phenomenon. The heat-induced glycation reaction between WPI and X molecules might induce the formation of hydrogen bonds, thus weakening the interaction between molecules, and result in a reduction in the β-sheet and β-turns but an increase in the random coil [37].

In another study, lycopene was encapsulated in whey protein isolate and xylo-oligosaccharides conjugates and presented values ranging from 10.0 ± 0.4% to 27.0 ± 0.5%, depending on the other parameters of the reaction (temperature, time, pH). Muhoza et al. [38] evaluated the possibility of glycating the casein by the Maillard reaction with dextran for delivering coenzyme Q_{10}. These authors reported that when the reaction time was less than 8 h, the grafting degree of the mixture about 20%. Ghatak and Iyyaswami [39] encapsulated quercetin from dry onion peels and obtained an encapsulation yield between 40.4% and 96.4%. They found that the highest encapsulation yield was achieved under the following process conditions: casein concentration of pH 7.09 for the crude extract containing the quercetin concentration of 16.27 M. Akdeniz et al. [40] reported the encapsulation efficiency values of phenolic from onion skins as being between 55.6 and 89.2% for different coating material combinations, with a maximum value found for maltodextrin:casein in a ratio of 6:4.

3.3. Molecular Modeling

Details on how protein glycation with X impacts further flavonoids binding were collected by means of an in silico approach. The potential binding sites for X were predicted for both α-LA and β-LG molecules by performing molecular docking simulations. The best three fits involving α-LA and β-LG as receptors, decided based on the binding energy values and the interface area, were further analyzed in detail (Table 1). In the case of α-LA, mainly two binding sites located on the protein surface appeared to accommodate the ligands used in the study. The highest affinity of the receptor for its ligand was estimated based on the lowest binding energy. Thus, α-LA exhibited the highest affinity toward X molecules when bound to the cavity involving residues of the α-LA core (Phe^{53}, Gln^{54}, Tyr^{103}, Trp^{104}) and the amino-terminal section of the Leu^{105}-Leu^{110} helix. In addition to the hydrophobic contacts, three hydrogen bonds of 3.68 Å, 2.37 Å, and 2.83 Å involving Thr^{33}, Leu^{105} and Ala^{106}, respectively, contributed to the attraction between α-LA and the X molecule hosted within this cavity. Two different relative binding positions of the X molecules with respect to the α-LA receptor, sharing common amino acid residues in contact with the ligand, were predicted with high scores. None of the two X binding modes appeared to affect the attachment of the main flavonoids prevailing in the onion skin extract, namely quercetin-4'-O-monoglucoside (QMG) and quercetin-3,4'-O-diglucoside (QDG) [41], to the α-LA. In agreement with Horincar et al. [41], the α-LA molecule accommodates, with high specificity, the same binding site of both QDG and QMG ligands (Figure 1). The amino acids establishing direct contacts with QMG are Leu^3, Glu^{11}, Leu^{12}, Lys^{13}, Asp^{14}, Thr^{38}, Leu^{52}, Leu^{85}, Thr^{86}, Asp^{88}, Ile^{89}, Met^{90} and Lys^{93}, whereas the residues responsible for the QDG binding are Glu^1, Leu^3, Arg^{10}, Glu^{11}, Leu^{12}, Lys^{13}, Thr^{38}, Leu^{52}, Asp^{83}, Leu^{85}, Thr^{86}, Asp^{88} and Ile^{89} [41].

In addition, this wide pocket with a volume of 435.8 Å^3 appears to be able to accommodate an X molecule, which overlaps the QMG without interfering with QDG binding. It should be noted that α-LA shows a better affinity towards QMG and QDG (binding energy of −24.21 kcal/mol and −32.01 kcal/mol, respectively) with respect to X (binding energy of −7.48 kcal/mol).

On the other hand, the β-LG molecule is able to accommodate the X molecule in three different pockets with volumes ranging from 137.86 to 217.15 Å (Table 1), without interfering with QMG or QDG binding (Figure 1).

Table 1. Molecular details on the interactions between the main whey proteins (α-lactalbumin (α-LA) and β-lactoglobulin monomer (β-LG)) equilibrated at 25 °C, xylose (X) and the major flavonoids from onion skins (quercetin-4′-O-monoglucoside (QMG) and quercetin-3,4′-O-diglucoside (QDG) [41].

	α-LA—X			α-LA—QMG	α-LA—QDG
	Complex 1	Complex 2	Complex 3	Horincar et al. [41]	
Interaction descriptors					
Amino acids interacting with ligands	Thr^{33}, Glu^{49}, Phe^{53}, Gln^{54}, Tyr^{103}, Trp^{104}, Leu^{105}, Ala^{106}	Thr^{33}, Val^{42}, Asn^{44}, Glu^{49}, Phe^{53}, Gln^{54}, Tyr^{103}, Trp^{104}, Leu^{105}, Ala^{106}	Glu^{11}, Leu^{12}, Asp^{14}, Leu^{15}, Thr^{38}, Leu^{85}, Thr^{86}, Ile^{89}, Met^{90}, Lys^{93}	Leu^3, Glu^{11}, Leu^{12}, Lys^{13}, Asp^{14}, Thr^{38}, Leu^{52}, Leu^{85}, Thr^{86}, Asp^{88}, Ile^{89}, Met^{90}, Lys^{93}	Glu^1, Leu^3, Arg^{10}, Glu^{11}, Leu^{12}, Lys^{13}, Thr^{38}, Leu^{52}, Asp^{83}, Leu^{85}, Thr^{86}, Asp^{88}, Ile^{89}
Binding energy, kcal/mol	−13.00	−10.52	−7.48	−24.41	−32.01
Interface area, Å^2	153.5	147.5	159.8	625.20	541.20
Pocket descriptors					
Volume, Å^3	339.58	339.58	381.50	435.84	
Depth, Å	15.72	15.72	13.18	13.94	
Enclosure	0.16	0.16	0.20	0.30	
	β-LG—X			β-LG—QMG	β-LG—QDG
	Complex 1	Complex 2	Complex 3	Horincar et al. [41]	
Interaction descriptors					
Amino acids interacting with ligands	Tyr^{20}, Tyr^{42}, Val^{43}, Glu^{44}, Gln^{59}, Cys^{66}, Pro^{126}, Leu^{156}, Glu^{157}, Glu^{158}, Gln^{159}, Cys^{160}, His^{161}	Ala^{23}, Met^{24}, Ala^{25}, Leu^{133}, Phe^{136}, Asp^{137}, Leu^{140}, Arg^{148}, Leu^{149}, Ser^{150}	Ile^2, Leu^{93}, Glu^{108}, Ala^{111}, Gln^{115}, Leu^{117}	Ala^{37}, Pro^{38}, Leu^{39}, Val^{41}, Leu^{58}, Lys^{60}, Ile^{71}, Ile^{84}, Asp^{85}, Ala^{86}, Leu^{87}, Asn^{88}, Asn^{90}, Met^{107}, Glu^{108}, Asn^{109}	Pro^{38}, Leu^{39}, Val^{41}, Leu^{58}, Lys^{60}, Glu^{62}, Ala^{67}, Lys^{69}, Ile^{71}, Asn^{90}, Met^{107}, Glu^{108}, Asn^{109}
Binding energy, kcal/mol	−13.37	−16.41	−13.00	−35.05	−34.37
Interface area, Å^2	169.1	152.8	157.6	502.10	585.90
Pocket descriptors					
Volume, Å^3	217.15	141.31	137.86	229.70	
Depth, Å	9.42	7.68	10.16	13.08	
Enclosure	0.12	0.19	0.19	0.33	

β-LG binds the X molecules more tightly compared to α-LA; the X binding energy by the β-LG monomer varies between −16.41 and −13.00 kcal/mol. In addition, the binding of two X molecules to the β-LG pockets involves hydrogen bonds established with Glu^{158} in the case of complex 1 and Met^{24}, Asp^{137} and Leu^{149} in the case of complex 2, as presented in Table 1. The in silico results successfully complement the experimental findings, adding valuable details on how whey protein glycation with X further impacts flavonoid binding. These atomic level observations indicate that, upon glycation, the β-LG molecule might play a major role in flavonoid biding.

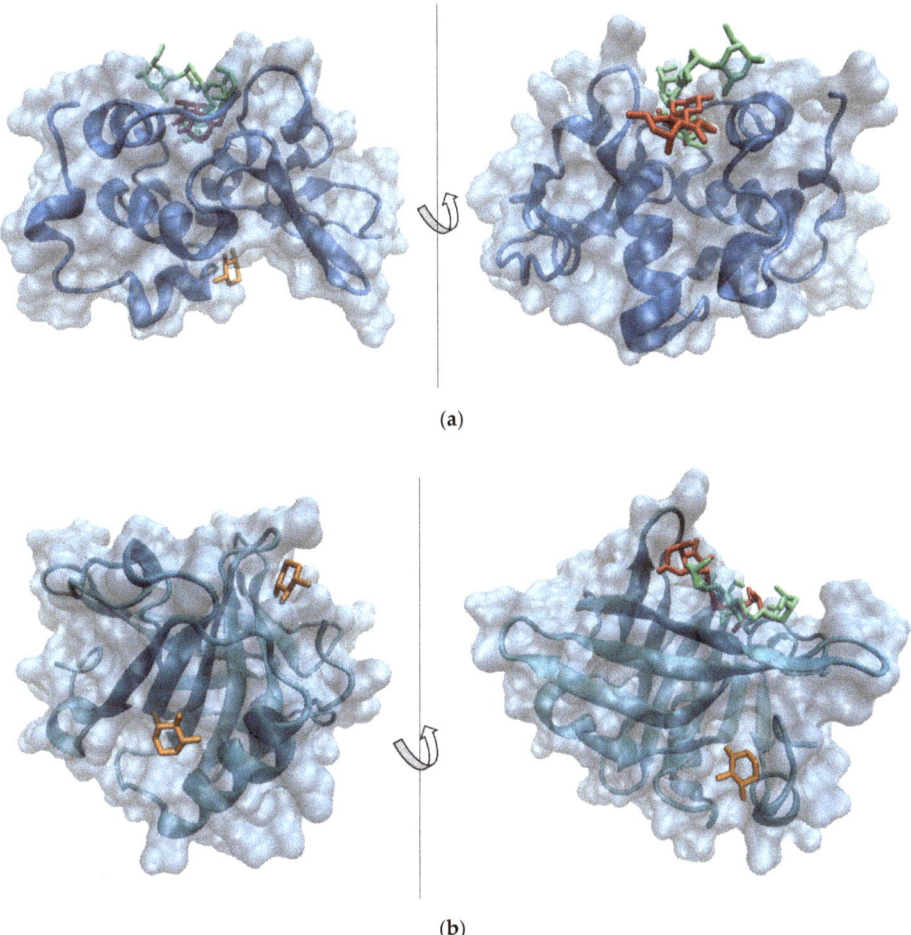

Figure 1. Superposition of the models showing the most probable complexes formed between (**a**) α-lactalbumin and (**b**) β-lactoglobulin (represented in blue in surf style) and xylose (represented in orange in licorice style—models 1 and 2), quercetin-4'-O-monoglucoside (represented in red in licorice style) and quercetin-3,4'-O-diglucoside (represented in green in licorice style). Images were prepared using VMD software [42].

3.4. Phytochemical Profile of the Powders

The two powders showed a total flavonoid content of 97.7 ± 3.7 mg QE/g DW in V1 and 120.0 ± 1.6 mg QE/g DW in V2. Total polyphenolic content showed no significant ($p < 0.05$) differences of 45.1 ± 0.6 mg GAE/g DW and 46.9 ± 1.5 mg GAE/g DW in V1 and V2, yielding a corresponding antioxidant activity of 179.7 ± 4.5 mMol TE/g DW and 184.4 ± 0.7 mMol TE/g DW, respectively. Therefore, the glycation of WPI allowed a better encapsulation of flavonoids, yielding a powder with a higher antioxidant activity. To the best of our knowledge, no other studies are available that exploit the potential of WPI conjugates as biopolymeric wall materials used in the microencapsulation of flavonoids.

In a previous study, Milea et al. [43] encapsulated flavonoids from yellow onion skins using maltodextrin, pectin and whey protein hydrolysates as coating materials in different ratios. The concentration of flavonoids, polyphenols and the antioxidant activity of the freeze-dried variants showed comparable levels, as flavonoids varied from 98.1 ± 0.5 to 103.7 ± 0.6 mg QE/g DW, whereas significant higher polyphenol contents (varying from 53.5 ± 1.7 to 69.3 ± 1.0 mg GAE/g DW) and antioxidant activities (varying from 280.6 ± 3.1 to 337.6 ± 0.9 mM TE/g DW) were reported for different variants. Horincar et al. [41] used different combinations of biopolymeric coatings based on whey protein isolate and chitosan, maltodextrin and pectin as adjuvants for encapsulation. These authors obtained two variants of freeze-dried powder with different profiles. Therefore, lower values for total flavonoid content of 5.8 ± 0.2 mg QE/g DW and antioxidant activity of 175.9 ± 1.5 mM TE/g DW were suggested in coatings with WPI-chitosan. When using a more complex biopolymeric wall material, including WPI-maltodextrin-pectin, these authors obtained a powder with significant higher flavonoid content and antioxidant activity of 104.9 ± 5.0 mg QE/g DW and 269.2 ± 3.6 mM TE/g DW, respectively.

3.5. Structure and Morphology of the Powder

The confocal laser scanning microscopy technique allows the simultaneous identification of several compounds, at the surface of the particles using specific samples that usually discharge light at different wavelengths. This type of analysis also permits the visualization of the internal morphology of any type of particle by different fluorophores labels, hence displaying the compositional evolution of the targeted molecules, which represents the main aim of any study that regards the protection of valuable molecules such as antioxidant compounds, mainly polyphenols, through an encapsulation process. Therefore, it can be applied as a nondestructive visualization technique for microparticles.

By using a confocal laser scanning Carl Zeiss 710 microscope with the ZEN 2012 SP1 software (Black Edition), the images of V1 and V2 powders (Figure 2) were captured, both in the native state without any other additional dye added (Figure 2a,c) and stained with Congo Red (Figure 2b,d, respectively). In the native state, the powders showed an autofluorescence (in the range of 520–580 nm) due to the rich content of polyphenols among which quercetin predominates [39]. The biologically active compounds were captured in the WPI–X matrix (with a displayed autofluorescence showed in blue). Several irregular scaly formations could be observed with larger dimensions for the V2 (199.6–253.3 µm), compared to V1 (94.5–104.8 µm), probably due to the WPI–X conjugates. The microstructure was rather similar to that reported by Horincar et al. [41], who used different polymers as the encapsulating matrices such as chitosan, maltodextrin and pectin.

Through the fluorescent staining with Congo Red, a fluorophore usually used to highlight the fluorescence of proteins, several spherosomes were revealed, highlighting the encapsulated flavonoids in the WPI–X matrix. Nonetheless, the displayed formations were larger (up to 51.5 µm) and less numerous in the V1 sample where a significant amount of nonencapsulated flavonoids was visualized (Figure 2b).

The interaction of the WPI with the small carbohydrates via the Maillard reaction favored a better incorporation of the biologically active compounds from the onion extract into finer spherosomes (approximately 20 µm) or in the form of scales with digitiform extensions. The fluorophore bound to the conjugated proteins and generated a fine, orange wall with a fluorescent emission between 600 and 620 nm around the bioactives.

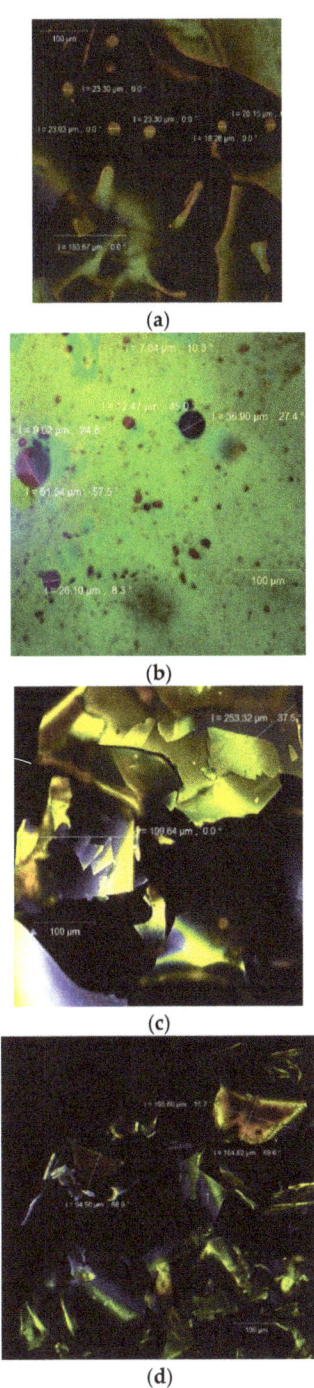

Figure 2. Confocal laser scanning microscopy images of the V1 sample ((**a**)—native state and (**b**)—stained with fluorophores) and V2 sample ((**c**)—native state and (**d**)—stained with fluorophores).

3.6. Characterization of New Formulated Food Product

To test selected functionality, the powders were added to a recipe of nachos in a ratio of 3%. Therefore, three variants of nachos were obtained according to the two variants, coded as N1 (3% addition of variant V1) and N2 (3% addition of variant V2) and a blank without powder (C). The obtained food products were analyzed in terms of bioactive stability for 28 days over storage at 25 °C. As expected, the differences in bioactives and antioxidant activity between samples correlated with the added quantity of powder (Table 2).

Table 2. Phytochemical profile of added-value nachos and stability during 28 days of storage.

Selected Phytochemicals	Control		Samples with 3% Addition of Variant 1		Samples with 3% Addition of Variant 2	
	Time 0	After 28 Days	Time 0	After 28 Days	Time 0	After 28 Days
Total flavonoids, mg QE/g DW	0.66 ± 0.02 [a]	0.63 ± 0.02 [a]	1.04 ± 0.05 [a]	0.99 ± 0.01 [a]	1.08 ± 0.05 [a]	0.98 ± 0.04 [b]
Total polyphenols, mg GAE/g DW	0.73 ± 0.04 [a]	0.75 ± 0.02²	1.24 ± 0.03 [a]	1.22 ± 0.05 [a]	1.37 ± 0.19 [a]	1.08 ± 0.019 [a]
Antioxidant activity, mMol TE/g DW	156.07 ± 2.57 [b]	197.97 ± 1.74 [a]	157.89 ± 1.41 [b]	199.49 ± 0.81 [a]	158.19 ± 0.48 [b]	198.57 ± 0.35 [a]

Means on the same row that do not share letter (a, b) are significantly different, based on Tukey method and 95% confidence.

During the storage test, no significant decrease ($p > 0.05$) was found in the flavonoid contents of N1, in contrast with N2, where a significant decrease of 9% was observed ($p < 0.05$). From Table 2, a significant increase in antioxidant activity values for both variants, during storage, can be observed, probably due to the release of some other compounds, apart from flavonoids, such as phenolics from microcapsules. Therefore, an increase in antioxidant activity was found in both variants, at approximatively 26%. Milea et al. [12] reported a decrease of 43% for flavonoids, 35% for polyphenols and 8% for antioxidant activity, in the case of a new formulated soft cheese with the addition of 1% microencapsulated powder, and 47% for flavonoids and 31% and 9% for polyphenols and antioxidant activity in the case of a soft cheese with 2% microencapsulated powder.

4. Conclusions

In this study, flavonoid-loaded microcapsules using whey protein isolates glycated with xylose via the Maillard reaction were successfully obtained. The liquid–solid, ultrasound-assisted extraction method was applied to obtain a flavonoid-enriched extract. The extract showed a significant content of flavonoids and a satisfactory antioxidant activity. Whey protein isolates and xylose were used, in nonglycated and glycated forms, as possible candidates for the microencapsulation of flavonoid-enriched onion skin extracts by freeze-drying. Both powders were characterized, showing significant amounts of polyphenols, flavonoids and a remarkable antioxidant capacity. A positive correlation was found between the browning index and antioxidant activity, and consecutively between the grafting degree and microencapsulation efficiency. The confocal laser scanning microscopy confirmed the higher ability of the whey protein isolate–xylose conjugates to entrap flavonoids. The molecular docking studies allowed the identification of the potential zones from α-lactalbumin and β-lactoglobulin surfaces involved in the interaction with xylose molecules. Xylose appeared to attach with high affinity to the α-lactalbumin protein pocket involved in flavonoid binding. In the case of β-lactoglobulin, the tested ligands docked to different sites; smaller cavities located on the protein surface are preferred by xylose. The powder was added to nachos, and a slightly decrease in phytochemicals was found during storage. However, the antioxidant activity of the added-value products increased, probably due to the release of some other bioactives from microcapsules. Based on the reported results, the protein–monosaccharide Maillard-type conjugates are a good alternative for food ingredient carriers and promising attractive methods of delivery.

Author Contributions: Conceptualization, N.S.; methodology, N.S.; software, Ș.A.M.; validation, I.A. and N.S.; formal analysis, Ș.A.M., E.E. and V.B.; resources, G.E.B. and G.R.; writing—original draft preparation, Ș.A.M., I.A. and N.S.; writing—review and editing, I.A. and N.S.; supervision, N.S.; project administration, N.S.; funding acquisition, G.E.B. and G.R. All authors have read and agreed to the published version of the manuscript.

Funding: This work was supported by a grant of the Romanian Ministry of Research and Innovation, CCCDI-UEFISCDI, project number PN-III-P1-1.2-PCCDI-2017-0569-PRO-SPER (10PCCI), within the PNCDI programme.

Institutional Review Board Statement: Not applicable.

Informed Consent Statement: Not applicable.

Data Availability Statement: Data are contained within the article.

Acknowledgments: The Integrated Centre for Research, Expertise and Technological Transfer in Food Industry is acknowledged for providing technical support.

Conflicts of Interest: The authors declare no conflict of interest.

References

1. Mark, R.; Lyu, X.; Lee, J.J.L.; Parra-Saldívar, R.; Chen, W.N. Sustainable production of natural phenolics for functional food applications. *J. Funct. Foods* **2019**, *57*, 233–254. [CrossRef]
2. Conti, V.M.; Guzzetti, L.; Panzeri, D.; De Giuseppe, R.; Coccetti, P.; Labra, M.; Cena, H. Bioactive compounds in legumes: Implications for sustainable nutrition and health in the elderly population. *Trends Food Sci. Technol.* **2021**, in press. [CrossRef]
3. Choi, I.; Cho, E.; Moon, J.; Bae, H. Onion skin waste as a valorization resource for the by-products quercetin and biosugar. *Food Chem.* **2015**, *188*, 537–542. [CrossRef]
4. Abouzed, T.K.; Contreras, M.D.M.; Sadek, K.M.; Shukry, M.; Abdelhady, D.H.; Gouda, W.M.; Abdo, W.; Nasr, N.E.; Mekky, R.H.; Segura-Carretero, A.; et al. Red onion scales ameliorated streptozotocin-induced diabetes and diabetic nephropathy in Wistar rats in relation to their metabolite fingerprint. *Diabetes Res. Clin. Pract.* **2018**, *140*, 253–264. [CrossRef]
5. Sharma, K.; Mahato, N.; Nile, S.H.; Lee, Y.R. Economical and environment friendly approaches for usage of onion (*Allium cepa* L.) wastes. *Food Funct.* **2016**, *7*, 3354–3369. [CrossRef] [PubMed]
6. Ma, Y.L.; Zhu, D.Y.; Thakur, K.; Wang, C.H.; Wang, H.; Ren, Y.F.; Zhang, J.G.; Wei, Z.J. Antioxidant and antibacterial evaluation of polysaccharides sequentially extracted from onion (*Allium cepa* L.). *Int. J. Biol. Macromol.* **2018**, *111*, 92–101. [CrossRef] [PubMed]
7. Kim, S.W.; Ko, M.J.; Chung, M.S. Extraction of the flavonol quercetin from onion waste by combined treatment with intense pulsed light and subcritical water extraction. *J. Clean. Prod.* **2019**, *231*, 1192–1199. [CrossRef]
8. Cheynier, V. Polyphenols in foods are more complex than often thought. *Am. J. Clin. Nutr.* **2005**, *81*, 223–229. [CrossRef]
9. Saikia, S.; Mahnot, N.K.; Mahanta, C.L. Optimisation of phenolic extraction from Averrhoa carambola pomace by response surface methodology and its microencapsulation by spray and freeze-drying. *Food Chem.* **2015**, *171*, 144–152. [CrossRef]
10. Cam, M.; İçyer, N.C.; Erdoğan, F. Pomegranate peel phenolics: Microencapsulation, storage stability and potential ingredient for functional food development. *LWT-Food Sci. Technol.* **2014**, *55*, 117–123. [CrossRef]
11. Munin, A.; Edwards-Lévy, F. Encapsulation of natural polyphenolic compounds; a review. *Pharmaceutics* **2011**, *3*, 793–829. [CrossRef]
12. Milea, A.S.; Aprodu, I.; Vasile, A.M.; Barbu, V.; Râpeanu, G.; Bahrim, G.E.; Stănciuc, N. Widen the functionality of flavonoids from yellow onion skins through extraction and microencapsulation in whey proteins hydrolysates and different polymers. *J. Food Eng.* **2019**, *251*, 29–35. [CrossRef]
13. Jia, C.; Caoa, D.; Jia, S.; Lina, W.; Zhanga, X.; Muhoza, B. Whey protein isolate conjugated with xylo-oligosaccharides via maillard reaction: Characterization, antioxidant capacity, and application for lycopene microencapsulation. *LWT-Food Sci. Technol.* **2020**, *118*, 108837. [CrossRef]
14. Fathi, M.; Donsi, F.; McClements, D.J. Protein-based delivery systems for the nanoencapsulation of food ingredients. *Compr. Rev. Food Sci. Food Saf.* **2018**, *17*, 920–936. [CrossRef]
15. Saénz, C.; Tapia, S.; Chávez, J.; Robert, P. Microencapsulation by spray drying of bioactive compounds from cactus pear (*Opuntia ficus-indica*). *Food Chem.* **2009**, *114*, 616–622. [CrossRef]
16. Coklar, H.; Akbulut, M. The control of Maillard reaction in white grape molasses by the method of reducing reactant concentration. *Food Sci. Technol.* **2020**, *40*, 179–189. [CrossRef]
17. Chrysina, E.; Brew, K.; Acharya, K. Crystal structures of apo- and holo-bovine α-lactalbumin at 2.2-Å resolution reveal an effect of calcium on inter-lobe interactions. *J. Biol. Chem.* **2000**, *275*, 37021–37029. [CrossRef]
18. Loch, J.I.; Bonarek, P.; Polit, A.; Riès, D.; Dziedzicka-Wasylewska, M.; Lewiński, K. Binding of 18-carbon unsaturated fatty acids to bovine β-lactoglobulin-structural and thermodynamic studies. *Int. J. Biol. Macromol.* **2013**, *57*, 226–231. [CrossRef]
19. Hess, B.; Kutzner, C.; van Der Spoel, D.; Lindahl, E. GROMACS 4: Algorithms for highly efficient, load-balanced, and scalable molecular simulation. *J. Chem. Theory Comput.* **2008**, *4*, 435–447. [CrossRef]

20. Aprodu, I.; Ursache, F.M.; Turturică, M.; Râpeanu, G.; Stănciuc, N. Thermal stability of the complex formed between carotenoids from sea buckthorn (*Hippophae rhamnoides* L.) and bovine β-lactoglobulin. *Spectrochim. Acta A Mol. Biomol. Spectrosc.* **2017**, *173*, 562–571. [CrossRef] [PubMed]
21. Schneidman-Duhovny, D.; Inbar, Y.; Nussinov, R.; Wolfson, H.J. PatchDock and SymmDock: Servers for rigid and symmetric docking. *Nucleic Acids Res.* **2005**, *33*, W363–W367. [CrossRef]
22. Krissinel, E. Crystal contacts as nature's docking solutions. *J. Comput. Chem.* **2010**, *31*, 133–143. [CrossRef] [PubMed]
23. Volkamer, A.; Kuhn, D.; Rippmann, F.; Rarey, M. DoGSiteScorer: A web server for automatic binding site prediction, analysis and druggability assessment. *Bioinformatics* **2012**, *28*, 2074–2075. [CrossRef] [PubMed]
24. Constantin, O.E.; Milea, A.Ș.; Bolea, C.A.; Mihalcea, L.; Enachi, E.; Copolovici, D.M.; Copolovici, L.; Munteanu, F.; Bahrim, G.E.; Râpeanu, G. Onion (*Allium cepa* L.) peel extracts characterization by conventional and modern methods. *Int. J. Food Eng.* **2021**, *17*, 485–493. [CrossRef]
25. Singh, V.; Krishan, P.; Shri, R. Extraction of antioxidant phytoconstituents from onion waste. *J. Pharmacog. Phytochem.* **2016**, *6*, 502–505.
26. Pobłocka-Olech, L.; Głód, D.; Żebrowska, E.M.; Sznitowska, M.; Baranowska, M.K. TLC determination of flavonoids from different cultivars of *Allium cepa* and *Allium ascalonicum*. *Acta Pharmacol.* **2016**, *66*, 543–554. [CrossRef]
27. Benito-Román, O.; Blanco, B.; Sanz, M.T.; Beltrán, S. Subcritical Water Extraction of Phenolic Compounds from Onion Skin Wastes (*Allium cepa* cv. Horcal): Effect of Temperature and Solvent Properties. *Antioxidants* **2020**, *9*, 1233. [CrossRef]
28. Galanakis, C.M. *Polyphenols: Properties, Recovery, and Applications*; Woodhead Publishing: Sawston, UK, 2018.
29. Oliver, C.M.; Melton, L.D.; Stanley, R.A. Creating proteins with novel functionality via the Maillard reaction: A review. *Crit. Rev. Food Sci. Nutr.* **2006**, *46*, 337–350. [CrossRef] [PubMed]
30. Liu, S.-C.; Yang, D.-J.; Jin, S.-J.; Hsu, C.-S.; Chen, S.-L. Kinetics of color development, pH decreasing, and anti-oxidative activity reduction of Maillard reaction in galactose/glycine model systems. *Food Chem.* **2008**, *108*, 533–541. [CrossRef]
31. Li, Y.; Zhong, F.; Ji, W.; Yokoyama, W.; Shoemaker, C.F.; Zhu, S.; Xia, W. Functional properties of Maillard reaction products of rice protein hydrolysates with mono-, oligo- and polysaccharides. *Food Hydrocoll.* **2013**, *30*, 53–60. [CrossRef]
32. Jiménez-Castaño, L.; Villamiel, M.; Lòpez-Fandiño, R. Glycosylation of individual whey proteins by Maillard reaction using dextran of different molecular mass. *Food Hydrocoll.* **2007**, *21*, 433–443. [CrossRef]
33. Suminar, A.T.; Al-Baarri, A.N.; Legowo, A.M. Demonstration of physical phenomena's and scavenging activity from d-psicose and methionine Maillard reaction products. *Potravin. Slovak J. Food Sci.* **2017**, *11*, 417–424.
34. Liu, L.; Li, X.; Zhu, Y.; Massounga Bora, A.F.; Zhao, Y.; Du, L.; Li, D.; Bi, W. Effect of microencapsulation with Maillard reaction products of whey proteins and isomaltooligosaccharide on the survival of *Lactobacillus rhamnosus*. *LWT* **2016**, *73*, 37–43. [CrossRef]
35. Xu, Y.; Pitkänen, L.; Maina, N.H.; Coda, R.; Katina, K.; Tenkanen, M. Interactions between fava bean protein and dextrans produced by *Leuconostoc pseudomesenteroides* DSM 20193 and *Weissella cibaria* Sj 1b. *Carbohydr. Polym.* **2018**, *190*, 315–323. [CrossRef]
36. Liu, Q.; Cui, H.; Muhoza, B.; Hayat, K.; Hussain, S.; Tahir, M.U.; Zhang, X.; Ho, C.-T. Whey protein isolate-dextran conjugates: Decisive role of glycation time dependent conjugation degree in size control and stability improvement of colloidal nanoparticles. *LWT* **2021**, *148*, 111766. [CrossRef]
37. Shang, J.; Zhong, F.; Zhu, S.; Wang, J.; Huang, D.; Li, Y. Structure and physiochemical characteristics of whey protein isolate conjugated with xylose through Maillard reaction at different degrees. *Arab. J. Chem.* **2020**, *13*, 8051–8059. [CrossRef]
38. Muhoza, B.; Xia, S.; Cai, J.; Zhang, X.; Su, J.; Li, L. Time effect on coenzyme Q 10 loading and stability of micelles based on glycosylated casein via Maillard reaction. *Food Hydrocoll.* **2017**, *72*, 271–280. [CrossRef]
39. Ghatak, D.; Iyyaswami, R. Selective encapsulation of quercetin from dry onion peel crude extract in reassembled casein particles. *Food Bioprod. Process.* **2019**, *115*, 100–109. [CrossRef]
40. Akdeniz, N.; Sumnu, G.; Sahin, S. Microencapsulation of phenolic compounds extracted from onion (*Allium cepa*) skin. *J. Food Process. Preserv.* **2018**, *42*, e13648. [CrossRef]
41. Horincar, G.; Aprodu, I.; Barbu, V.; Râpeanu, G.; Bahrim, G.E.; Stănciuc, N. Interactions of flavonoids from yellow onion skins with whey proteins: Mechanisms of binding and microencapsulation with different combinations of polymers. *Spectrochim Acta A Mol. Biomol. Spectrosc.* **2019**, *215*, 158–167. [CrossRef]
42. Humphrey, W.; Dalke, A.; Schulten, K. VMD-Visual Molecular Dynamics. *J. Mol. Graph.* **1996**, *14*, 33–38. [CrossRef]
43. Milea, Ș.A.; Vasile, M.A.; Crăciunescu, O.; Prelipcean, A.M.; Bahrim, G.E.; Râpeanu, G.; Oancea, A.; Stănciuc, N. Co-microencapsulation of flavonoids from yellow onion skins and lactic acid bacteria lead to a multifunctional ingredient for foods and pharmaceutics applications. *Pharmaceutics* **2020**, *12*, 1053. [CrossRef] [PubMed]

Article

Resveratrol Stabilization and Loss by Sodium Caseinate, Whey and Soy Protein Isolates: Loading, Antioxidant Activity, Oxidability

Xin Yin [1,2], Hao Cheng [1,2], Wusigale [3,4], Huanhuan Dong [1,2], Weining Huang [1,2] and Li Liang [1,2,*]

[1] State Key Lab of Food Science and Technology, Jiangnan University, Wuxi 214122, China; 7190112057@stu.jiangnan.edu.cn (X.Y.); haocheng@jiangnan.edu.cn (H.C.); 17865917062m@sina.cn (H.D.); wnhuang@jiangnan.edu.cn (W.H.)
[2] School of Food Science and Technology, Jiangnan University, Wuxi 214122, China
[3] Key Laboratory of Dairy Biotechnology and Engineering, Ministry of Education, Inner Mongolia Agricultural University, Hohhot 010018, China; wusigale@imau.edu.cn
[4] Inner Mongolia Key Laboratory of Dairy Biotechnology and Engineering, Hohhot 010018, China
* Correspondence: liliang@jiangnan.edu.cn; Tel.: +86-(510)-8519-7367

Abstract: The interaction of protein carrier and polyphenol is variable due to their environmental sensitivity. In this study, the interaction between resveratrol and whey protein isolate (WPI), sodium caseinate (SC) and soy protein isolate (SPI) during storage were systematically investigated from the aspects of polyphenol loading, antioxidant activity and oxidability. It was revealed that resveratrol loaded more in the SPI core and existed both in the core of SC micelles and on the particle surface, while WPI and resveratrol mainly formed in complexes. The loading capacity of the three proteins ranked in order SC > SPI > WPI. ABTS assay showed that the antioxidant activity of the protein carriers in the initial state was SC > SPI > WPI. The results of sulfhydryl, carbonyl and amino acid analysis showed that protein oxidability was SPI > SC > WPI. WPI, with the least oxidation, improved the storage stability of resveratrol, and the impact of SC on resveratrol stability changed from a protective to a pro-degradation effect. Co-oxidation occurred between SPI and resveratrol during storage, which refers to covalent interactions. The data gathered here suggested that the transition between the antioxidant and pro-oxidative properties of the carrier is the primary factor to investigate its protective effect on the delivered polyphenol.

Keywords: protein; resveratrol; loading; antioxidant activity; oxidability; stability

1. Introduction

Protein-based assemblies including molecular complexes, nano-/micro-particles, and their stabilized emulsions and emulsion gels have been expected to protect antioxidants [1,2]. Even though the stabilization mechanism of polyphenols in proteins is not fully clear, there is a hypothesis that proteins express a protective effect by shielding the environmental accessibility of polyphenols and scavenging the free radical [3]. However, amphiphilic and hydrophilic polyphenols cannot be completely encapsulated in a carrier, and a portion of polyphenols are still in the free form. Meanwhile, it is worth noting that antioxidants can be converted into pro-oxidants under certain conditions and proteins may generate reactive oxidative species [4]. The imbalance between pro-oxidation and anti-oxidation in the physiological system eventually leads to the oxidation of biomolecules, the so-called oxidative stress [5]. The interaction between proteins and polyphenols might also be complicated and changeable, since they are both environment sensitive.

It has been reported that bioactive components including vitamins, polyunsaturated fatty acids and polyphenols may also affect the delivery carriers. The photo-decomposition of folic acid caused the indirect oxidation of the whey protein isolate (WPI), which enhanced the protein antioxidant activity, leading to increased protection for the folic acid [6].

β-Carotene, one of the major carotenoids, produced free radicals that can accelerate the oxidation of WPI in oil-in-water emulsions [7], while phenolic anthocyanins provided protection against the oxidation of Trp [8]. The oxidation of fish protein rich in polyunsaturated fatty acids was promoted by lipid oxidation products, especially secondary oxidation products, and the oxidation of protein and lipids occurred in parallel, showing a good correlation [9]. Although protein–polyphenol interaction has been investigated for the modification of protein structure and colloidal stability [10,11], its impact on the stability of polyphenols has rarely been reported until now. It is thus necessary to clarify the mechanism of proteins on the stability of polyphenols in more depth.

Resveratrol (*trans*-3,5,4'-trihydroxy-stilbene) is known as a polyphenolic compound with antioxidant activity. However, resveratrol is prone to oxidation, which limits its application in commercial products. Various proteins (e.g., zein, gliadin and ovalbumin) have been reported to stabilize resveratrol [12,13], but bovine serum albumin (BSA) accelerates the degradation of resveratrol [14]. Proteins in the molecular level and in the form of micelles might provide a different microenvironment and unique carrying properties for targeted antioxidants [15]. β-casein in the molecular level improved the storage stability of both *cis*- and *trans*-resveratrol better than β-casein micelles, although β-casein micelles could inhibit the transformation of resveratrol from *trans*-isomer to *cis*-isomer to a certain extent [16]. The stabilization effect on resveratrol is dependent on the type and concentration of protein carriers [17], but lacking systematic comparison study. Therefore, there is a growing demand for clarifying the theoretical basis to select suitable proteins as carrier materials for resveratrol.

Sodium caseinate (SC) and WPI are major milk proteins, and SC has a disordered structure and is more hydrophobic properties than WPI, while WPI contains two major globular proteins β-lactoglobulin and α-lactalbumin [18]. Soy protein isolate (SPI) is mainly composed of 7S and 11S globulins. WPI, SC and SPI are generally recognized as safe (GRAS), and their assemblies are commonly used to protect antioxidants against oxidation and degradation [19,20]. In the present study, WPI, SC, and SPI at various concentrations were used to investigate their effect on the storage stability of resveratrol and the polyphenol impact on the composition of proteins. The data gathered here should help guide the shelf life of the protein–polyphenol system used in commercial products.

2. Materials and Methods

2.1. Materials

WPI (≥92%) was obtained from Davisco International Inc (Le Sueur, MN, USA). SPI (≥90%) was from Shandong Xiya Chemical Industry Co., Ltd. (Linshu, Shandong, China). SC, resveratrol (trans-isomer, >99%) and polydatin (HPLC grade, >95%) were purchased from Sigma-Aldrich Co. (St. Louis, MO, USA). 2,2'-azino-bis-3- ethylbenzthiazoline-6-sulphonic acid (ABTS) was purchased from Aladdin Bio-Chem Technology Co., Ltd. (Shanghai, China). Other agents were of analytical grade and purchased from SinoPharm CNCM Ltd. (Shanghai, China).

2.2. Sample Preparation

WPI, SPI or SC powder was dissolved in ultrapure water. The solutions were adjusted to pH 12 with 2 M NaOH and hydrated fully with magnetic stirring for 1 h, and then neutralized the pH to 7 with 2 M HCl under agitation for another 1 h. Stock solutions of proteins were 0.02%, 0.2% and 2.0% (w/v). Stock solution of resveratrol was prepared at a concentration of 2 mM by dissolving in 70% (v/v) ethanol. The resveratrol solution was added into protein solutions and diluted with water at pH 7 under stirring for 30 min. The final concentrations were 0.01%, 0.10% and 1.00% for proteins and 25, 50 and 100 μM for resveratrol. The 0.02% (w/v) sodium azide was added to solutions as an antimicrobial agent.

2.3. Fluorescence Spectroscopy

Fluorescence of pyrene as a probe was measured on a Cary Eclipse fluorescence spectrophotometer (Agilent Co., Ltd., New York, NY, USA) equipped with 10 mm quartz cuvettes. The spectral resolution was 2.5 nm for both excitation and emission. Pyrene in acetone was added into samples with its final concentration of 1 μM under stirring for at least 12 h before measurement. Fluorescence emission spectra were scanned from 350 to 600 nm with the excitation wavelength of 335 nm and the ratio of the intensity of the first and third bands (I_1/I_3) was calculated [21].

Fluorescence emission spectra of resveratrol in the absence or presence of proteins were recorded from 330 to 600 nm at an excitation wavelength of 320 nm. Slit widths with a nominal band-pass of 5 nm were used for both excitation and emission. Background of proteins was subtracted from raw spectra.

2.4. Particle Size and ζ-Potential

Size distribution by the intensity and ζ-potential were determined by a NanoBrooker Omni Particle size Analyzer (Brookhaven Instruments Ltd., New York, NY, USA) with a He/Ne laser (λ = 633 nm) at a scattering angle of 173°. They were obtained using an NNLS model and Smoluchowski model through phase analysis light-scattering (PALS) measurement, respectively.

2.5. Color Evaluation

The color parameters of protein-resveratrol solutions before and after storage at 45 °C for 30 days were measured using a ColorQuest XE colorimeter (ColorQuest XE, Hunter Lab, Reston, VA, USA) and calculated using the Hunter Lab color scale (L*a*b*). L* represents the lightness (black = 0 to white = 100), a* varies from red (positive) to green (negative), and b* varies from yellow (positive) to blue (negative). The total color difference (ΔE) was calculated from the tristimulus color coordinates using the following equation:

$$\Delta E = \left[(L^* - L_i^*)^2 + (a^* - a_i^*)^2 + (b^* - b_i^*)^2 \right]^{1/2} \qquad (1)$$

where, L_i^*, a_i^*, b_i^* are the initial values of the CIE L*a*b* color coordinates of freshly-prepared samples, and L*, a*, b* are the color coordinates of samples after 30 days. Additionally, the difference in chroma (ΔC*) value, which represents the color intensity of samples, was analyzed by the following equation [22]:

$$\Delta C^* = \left[(a^* - a_i^*)^2 + (b^* - b_i^*)^2 \right]^{1/2} \qquad (2)$$

2.6. Resveratrol Quantification

An exactly 0.5 mL sample was mixed with 0.5 mL polydatin (internal standard, 50 μM) in methanol and then added into 4 mL methanol under vortexing for 60 s. After the mixture was centrifuged at 15,000× g for 60 min, the supernatant was measured on the Alliance HPLC system equipped with a 2695 separation module and 2998 PDA detector (Waters, Milford, MA, USA). The mobile phase was a mixture of methanol and distilled water (50:50, v/v), the flow rate was 1 mL min^{-1}, and the column temperature was 35 °C. Both trans-resveratrol and polydatin were analyzed at 306 nm [23].

2.7. Loading Efficiency

Loading efficiency of resveratrol was determined by isoelectric precipitation method [24]. The samples with WPI, SPI and SC were adjusted to pH 4.8–4.6 using 0.1 M NaOH or HCl. Loading efficiency of resveratrol was calculated according to following formulation:

$$\text{Loading efficiency}(\%) = \left(1 - \frac{C_s}{C_0}\right) \times 100 \qquad (3)$$

where, C_0 and C_s were resveratrol in samples and in the supernatant, centrifuged at $5000 \times g$ for 20 min, respectively.

2.8. Antioxidant Activity

ABTS assay was analyzed according to previous methods [25]. In brief, 7.4 mM ABTS and 2.6 mM $K_2S_2O_8$ were mixed in the dark for 12 h to produce $ABTS^{·+}$ solution, which was diluted and mixed with samples or buffer at a volume ratio of 19:1 and kept in the dark for 6 min. The absorbance was measured at 729 nm using a UV-1800 UV–Vis spectrophotometer (Shimadzu Co., Tokyo, Japan). The radical-scavenging activity was calculated as follows:

$$\text{Scavenging capacity (\%)} = \frac{A_c - A_s}{A_c} \times 100 \qquad (4)$$

where A_c and A_s are the absorbance of radical plus buffer and sample, respectively.

2.9. Sulfhydryl Analysis

Samples were mixed at a volume ratio of 1:2 with 0.1 M phosphate buffer at pH 8.0 without and with 8 M urea for free and total sulfhydryl determination, respectively. Then absorbance at 412 nm was measured, after 10 mM DTNB was added under vigorous stirring and incubated in the dark for 1 h. Both reagent and sample blanks were subtracted. Content of free and total sulfhydryl was calculated by using a molar extinction coefficient of 13,600 $M^{-1}cm^{-1}$ and expressed as nmol per mg protein [26].

2.10. Carbonyl Analysis

Protein solutions in the presence and absence of resveratrol during storage at 45 °C were mixed at a volume ratio of 1:2 with 10 mM DNPH in 2 M HCl. After 10% (w/v) trichloroacetic acid was added and centrifuged at $10,000 \times g$ for 5 min, precipitate was washed with 50% ethyl acetate and then dissolved in 6 M guanidine HCl in 20 mM phosphate buffer at pH 2.3. Absorbance at 370 nm was measured and carbonyl content was calculated using an extinction coefficient of 22,000 $M^{-1}cm^{-1}$ and expressed as nmol per mg protein [27].

2.11. Amino Acid Analysis

Amino acids except tryptophan were analyzed through acid hydrolysis of proteins by mixing 4 mL of samples with the same volume of 12 M HCl under blown nitrogen for 3 min, followed by hydrolysis at 120 °C for 22 h. Then a certain amount of NaOH was added to neutralize, and water was added to give a total volume of 25 mL. Tryptophan was determined by alkaline hydrolysis of proteins with 10 M NaOH and neutralized with a certain amount of HCl. The supernatant was centrifuged after filtering with filter paper. Amino acids were analyzed on the Agilent 1100 HPLC system equipped with an Agilent Hypersil ODS column (Angelon Co., Ltd., New York, NY, USA). Proline was detected at 262 nm, and the other amino acids were detected at 338 nm [6].

2.12. Statistical Analysis

All experiments were repeated three times. Data are presented as mean ± standard deviation. An analysis of variance (ANOVA) of the data was carried out and identified using the Duncan procedure. All statistical analyses were performed using the software package SPSS 20.0 (SPSS Inc., Chicago, IL, USA). A p value < 0.05 was considered significant.

3. Results

3.1. Particle Characterization

Pyrene is often used to investigate the association of macromolecules and the critical micelle concentration (CMC). Its intensity ratio I_1/I_3 decreased as the hydrophobicity of surrounding microenvironment increased [28]. The I_1/I_3 ratio of pyrene in water was

1.75 ($\pm$0.01). When the concentration of SC was 0.01%, the I_1/I_3 ratio was 1.67 (Figure 1), and its size distribution had three peaks around 1.5, 25 and 215 nm by intensity (Figure 2A). According to the submicelle model, each casein forms small submicelle units through hydrophobic interactions, and these subunits use calcium phosphate as the cement and further aggregates together to form SC micelles [29]. The relatively low concentration of SC solution is not sufficient to drive the formation of micelles, 0.01% SC mainly dissolved in the molecular level [30]. As the protein concentrations increased above 0.5%, the I_1/I_3 of SC gradually decreased to about 1.10 (Figure 1). The size distribution of SC at 0.1% showed a major peak around 230 nm and a minor peak around 25 nm, while only a peak at around 380 nm was observed at 1% (Figure 2A). These results indicate that SC aggregates to form micelles at 1% concentration [31]. As for WPI and SPI, the I_1/I_3 ratios of 0.01% protein were respectively 1.35 and 1.40 (Figure 1). Meanwhile, WPI had two peaks around 220 and 520 nm (Figure 2B), and SPI had two peaks around 110 and 380 nm (Figure 2C). The relatively low I_1/I_3 and large particle size suggest that WPI and SPI had already aggregated at 0.01%. The size peaks of WPI were not dependent on its concentrations (Figure 2B), while SPI became bigger with increasing concentrations, with two major peaks around 180 and 660 nm at 1% (Figure 2C). This is consistent with the results of the I_1/I_3 of pyrene. From Figure 1, the I_1/I_3 ratios of WPI decreased slightly from 1.35 at 0.01% to 1.18 at 0.1%, and then remained unchanged as the protein concentration further increased. The I_1/I_3 ratios decreased as the concentrations of proteins increased, reaching around 0.80 for 2% SPI (Figure 1). By comparative analysis, these results indicate that SPI particles have the most hydrophobic core, which is consistent with the highest content of hydrophobic amino acids (Tables 1 and 2 vs. Table 3). It has been reported that SPI had lower solubility compared with WPI and SC, while hydrophilic groups and/or water molecules were entrapped in the core of WPI particles [32,33].

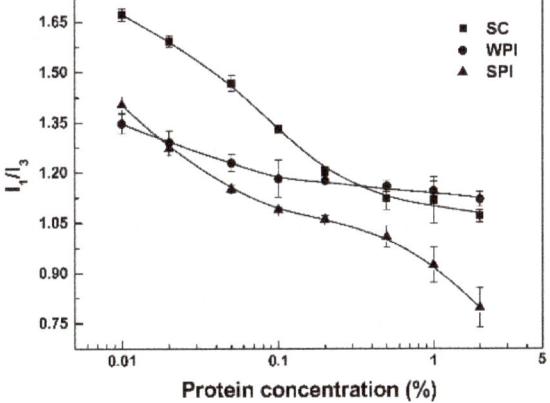

Figure 1. I_1/I_3 of whey protein isolate (WPI), sodium caseinate (SC) and soy protein isolate (SPI) solutions at various concentrations.

Since WPI, SC or SPI have an isoelectric point (pI) around pH 4.5~5, the ζ-potential values of their particles are negative at pH 7.0 (Figure 3). ζ-Potential absolute values of the protein particles ranked in order WPI > SC > SPI at the same concentration (Figure 3). This is consistent with their molar ratio of acidic (Asp and Glu) and basic (His, Lys and Arg) amino acids being around 2.38 for WPI, 2.03 for SC, and 1.98 for SPI, calculated from the data in Tables 1–3. Together with the most hydrophobic core of SPI particles in Figure 1, these results indicate that more negatively-charged groups were masked in SPI particles than WPI and SC particles. ζ-Potential absolute values of all complex particles decreased as the protein concentration increased (Figure 3), suggesting that negatively-charged groups were entrapped in the aggregated particle core.

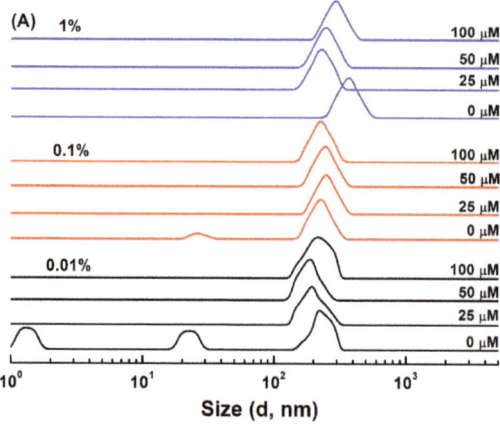
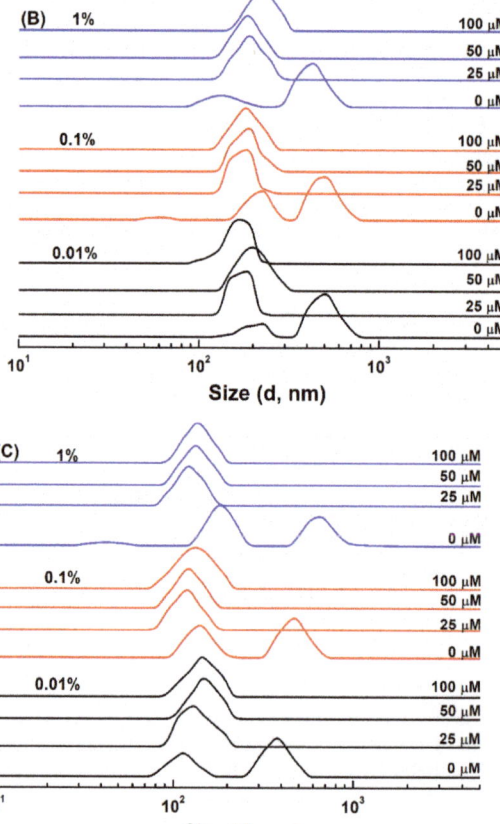

Figure 2. Size distribution of SC (**A**), WPI (**B**) and SPI (**C**) particles in the absence and presence of 25, 50 and 100 µM resveratrol. The protein concentrations were 0.01%, 0.1% and 1%.

Table 1. Amino acid composition of whey protein isolate (WPI) at 1% in the absence and presence of 100 μM resveratrol (RES) before and after storage for 30 days.

Amino Acid	Content of Amino Acid (μg/mL)			
	WPI (0)	WPI-RES (0)	WPI (30)	WPI-RES (30)
Cys	121 ± 4 [a]	119 ± 4 [a]	106 ± 4 [b]	104 ± 3 [b]
Trp	198 ± 7 [a]	195 ± 9 [a]	172 ± 8 [b]	179 ± 6 [b]
Tyr	298 ± 9 [a]	301 ± 7 [a]	277 ± 6 [b]	280 ± 6 [b]
Thr	433 ± 5 [a]	430 ± 7 [a]	417 ± 8 [b]	410 ± 8 [b]
Lys	942 ± 17 [a]	947 ± 12 [a]	933 ± 11 [b]	928 ± 9 [b]
Met	194 ± 5 [a]	192 ± 7 [a]	165 ± 5 [b]	163 ± 8 [b]
Phe	346 ± 2 [a]	351 ± 6 [a]	324 ± 8 [b]	320 ± 5 [b]
Asp	1195 ± 16 [a]	1203 ± 11 [a]	1196 ± 38 [a]	1158 ± 43 [a]
Arg	233 ± 11 [a]	235 ± 8 [a]	239 ± 8 [a]	235 ± 9 [a]
Glu	1790 ± 17 [a]	1784 ± 29 [a]	1777 ± 17 [a]	1783 ± 26 [a]
Ser	296 ± 7 [a]	292 ± 5 [a]	295 ± 8 [a]	299 ± 9 [a]
Gly	161 ± 2 [a]	158 ± 4 [a]	160 ± 3 [a]	158 ± 3 [a]
His	172 ± 2 [a]	169 ± 3 [a]	172 ± 5 [a]	173 ± 6 [a]
Val	506 ± 14 [a]	496 ± 29 [a]	508 ± 13 [a]	495 ± 10 [a]
Ala	474 ± 7 [a]	474 ± 5 [a]	475 ± 5 [a]	472 ± 9 [a]
Ile	586 ± 18 [a]	579 ± 16 [a]	573 ± 10 [a]	575 ± 15 [a]
Leu	991 ± 16 [a]	981 ± 22 [a]	980 ± 12 [a]	976 ± 20 [a]
Pro	399 ± 19 [a]	393 ± 7 [a]	398 ± 28 [a]	406 ± 10 [a]
Total	9335 ± 74 [a]	9299 ± 53 [a]	9167 ± 43 [b]	9114 ± 64 [b]

Note: Different lower-case letters in the same row represent significantly different mean values ($p < 0.05$).

Table 2. Amino acid composition of sodium caseinate (SC) in the absence and presence of resveratrol (RES) before and after storage for 30 days.

Amino Acid	Content of Amino Acid (μg/mL)			
	SC (0)	SC-RES (0)	SC (30)	SC-RES (30)
Cys	5 ± 0 [a]	5 ± 0 [a]	5 ± 0 [a]	4 ± 1 [a]
Trp	571 ± 9 [a]	566 ± 12 [a]	119 ± 14 [b]	74 ± 10 [c]
Tyr	451 ± 9 [a]	448 ± 12 [a]	409 ± 10 [b]	384 ± 6 [c]
Thr	344 ± 5 [a]	350 ± 8 [a]	330 ± 3 [b]	318 ± 2 [c]
Lys	744 ± 7 [a]	737 ± 11 [a]	690 ± 9 [b]	629 ± 7 [c]
Met	226 ± 8 [a]	222 ± 2 [a]	185 ± 2 [b]	184 ± 2 [b]
Phe	443 ± 8 [a]	439 ± 6 [a]	427 ± 10 [a]	433 ± 10 [a]
Asp	548 ± 14 [a]	556 ± 12 [a]	493 ± 18 [b]	460 ± 13 [c]
Arg	337 ± 10 [a]	345 ± 9 [a]	313 ± 12 [b]	294 ± 3 [c]
Glu	2053 ± 87 [a]	2014 ± 79 [a]	2048 ± 57 [a]	1998 ± 50 [b]
Ser	380 ± 7 [a]	378 ± 5 [a]	384 ± 8 [a]	375 ± 8 [b]
Gly	159 ± 4 [a]	153 ± 9 [a]	160 ± 2 [a]	152 ± 3 [b]
His	293 ± 3 [a]	290 ± 9 [a]	299 ± 7 [a]	288 ± 7 [a]
Val	600 ± 10 [a]	593 ± 19 [a]	588 ± 15 [a]	589 ± 11 [a]
Ala	260 ± 7 [a]	263 ± 7 [a]	252 ± 4 [a]	252 ± 4 [a]
Ile	487 ± 12 [a]	479 ± 12 [a]	473 ± 10 [a]	467 ± 16 [a]
Leu	788 ± 14 [a]	795 ± 22 [a]	783 ± 20 [a]	780 ± 17 [a]
Pro	727 ± 16 [a]	722 ± 10 [a]	720 ± 18 [a]	735 ± 11 [a]
Total	9416 ± 106 [a]	9355 ± 99 [a]	8678 ± 70 [b]	8416 ± 68 [c]

Note: Different lower-case letters in the same row represent significantly different mean values ($p < 0.05$).

Table 3. Amino acid composition of soy protein isolate (SPI) in the absence and presence of resveratrol (RES) before and after storage for 30 days.

Amino Acid	Content of Amino Acid (μg/mL)			
	SPI (0)	SPI-RES (0)	SPI (30)	SPI-RES (30)
Cys	20 ± 2 [a]	19 ± 4 [a]	13 ± 2 [b]	5 ± 2 [c]
Trp	160 ± 6 [a]	158 ± 8 [a]	93 ± 4 [b]	78 ± 3 [c]
Tyr	281 ± 8 [a]	276 ± 6 [a]	231 ± 4 [b]	200 ± 1 [c]
Thr	209 ± 3 [a]	211 ± 5 [a]	177 ± 8 [b]	170 ± 3 [c]
Lys	481 ± 7 [a]	476 ± 8 [a]	403 ± 10 [b]	382 ± 8 [c]
Met	84 ± 3 [a]	89 ± 8 [a]	79 ± 4 [b]	63 ± 6 [c]
Phe	418 ± 6 [a]	421 ± 9 [a]	358 ± 10 [b]	347 ± 7 [c]
Asp	735 ± 14 [a]	735 ± 13 [a]	685 ± 21 [b]	554 ± 10 [c]
Arg	585 ± 5 [a]	591 ± 6 [a]	584 ± 5 [a]	591 ± 8 [a]
Glu	1520 ± 67 [a]	1479 ± 67 [a]	1335 ± 79 [a]	1192 ± 106 [b]
Ser	371 ± 9 [a]	365 ± 8 [a]	325 ± 9 [b]	300 ± 4 [c]
Gly	352 ± 8 [a]	346 ± 9 [a]	328 ± 8 [b]	280 ± 9 [c]
His	213 ± 6 [a]	215 ± 3 [a]	195 ± 6 [b]	167 ± 3 [c]
Val	431 ± 8 [a]	442 ± 13 [a]	399 ± 6 [b]	352 ± 7 [c]
Ala	356 ± 4 [a]	350 ± 9 [a]	355 ± 11 [a]	349 ± 6 [a]
Ile	416 ± 11 [a]	409 ± 7 [a]	414 ± 3 [a]	402 ± 8 [a]
Leu	623 ± 12 [a]	627 ± 7 [a]	614 ± 10 [a]	606 ± 19 [a]
Pro	366 ± 6 [a]	357 ± 11 [a]	358 ± 8 [a]	347 ± 10 [a]
Total	7621 ± 90 [a]	7566 ± 89 [a]	6946 ± 98 [b]	6385 ± 121 [c]

Note: Different lower-case letters in the same row represent significantly different mean values ($p < 0.05$).

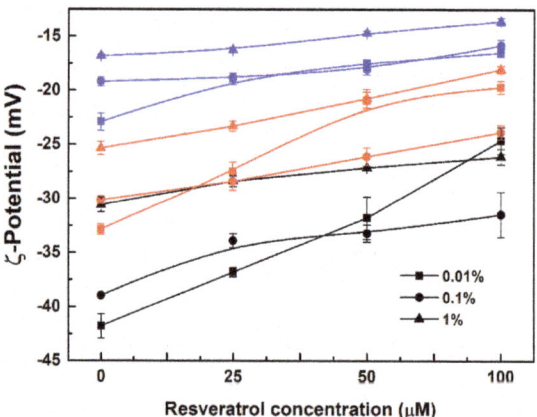

Figure 3. ζ–Potential of WPI (black), SC (red) and SPI (blue) particles in the absence and presence of 25, 50 and 100 μM resveratrol. The protein concentrations were 0.01%, 0.1% and 1%.

ζ–Potential absolute values of the protein particles decreased as the concentration of resveratrol increased, which was most pronounced at the protein concentration of 0.01% (Figure 3). The particles of SC, WPI and SPI became more homogeneous upon loading of resveratrol (Figure 2). These results are consistent with the formation of uniform particles of WPI with naringenin, a polyhydroxy flavonoid [34]. Meanwhile, the size distribution of all protein-resveratrol particles increased as the polyphenol concentration increased (Figure 2), which is consistent with the effect of hesperetin or hesperidin concentration on their individual particles with β-conglycinin, one of the major fractions of soy proteins, possibly due to polyphenols acting as bridging agents for protein molecules [35]. SC-resveratrol, WPI-resveratrol and SPI-resveratrol particles had a size distribution around 200–300 nm, 150–250 nm and 100–200 nm, respectively. At 25 μM resveratrol, the size distribution of

SC-resveratrol particles was close to the largest size distribution of SC particles (Figure 2A), while WPI-resveratrol and SPI-resveratrol particles had a size distribution close to the smallest ones of pure protein (Figure 2B,C). These results suggest that the addition of resveratrol favors the aggregation of SC but inhibits the formation of large WPI and SPI aggregates.

3.2. Resveratrol Loading

3.2.1. Microenvironment of Resveratrol

Pure resveratrol showed a crystalline structure, with sharp peaks at 6.58, 13.26, 16.39, 19.27, 22.40, 23.66, 25.28, 28.36 on a 2θ scale (Figure S2). Its characteristic peaks with less intensity were still observed in its physical mixtures with the proteins but disappeared in its protein particles (Figure S2), indicating that the polyphenol was amorphous when loaded in the protein particles [36]. From Figure 4, resveratrol in the absence of protein emits a relatively weak fluorescence, owing to its proton transfer tautomer fluorescence band [37]. The λ_{max} of resveratrol around 400 nm shifted to 392, 388 and 383 nm in the presence of 0.1% WPI, SC and SPI, respectively. At the same time, the fluorescence intensity at λ_{max} was 1.75, 3.06 and 4.09 times that of resveratrol alone. Similar changes were previously observed in the presence of β-lactoglobulin (β-LG) and bovine serum albumin (BSA), with respective fluorescence intensity at λ_{max} of 393 and 379 nm being 1.21 and 4.92 times that of resveratrol alone [38]. These results indicate that the microenvironment of resveratrol was more hydrophobic in protein particles, and the order of hydrophobicity was SPI > SC > WPI. The hydrophobicity of the resveratrol microenvironment (Figure 4) is consistent with the aggregation degree of pure protein at 1% but not that at 0.1% (SPI > WPI > SC, Figure 1). This is possibly attributed to the different impact of resveratrol loading on the aggregation of the three proteins. As discussed above, the added resveratrol as bridging agent favors the aggregation of SC (Figure 2).

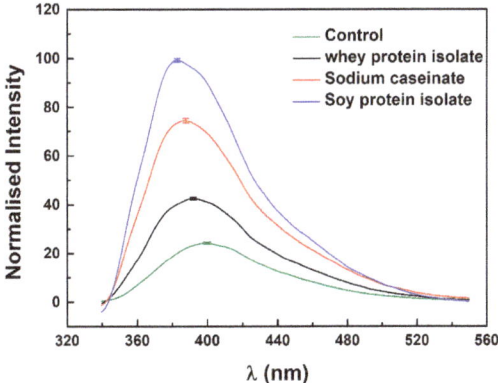

Figure 4. Fluorescence emission spectra of resveratrol in the absence (control) and presence of WPI, SC and SPI. Concentrations of resveratrol and proteins were 25 μM and 0.1%, respectively.

3.2.2. Loading Efficiency of Resveratrol

When the concentration of proteins was 0.01%, loading efficiencies of resveratrol were between 2% and 8% (Figure 5). The polyphenol loading efficiencies at 25 and 50 μM were greater in WPI and SPI particles than in SC particles. This may be due to WPI and SPI existing in the aggregate form at 0.01%, while SC exists in the molecular state (Figure 1). The loading efficiencies of resveratrol at 100 μM were 4% in all the protein particles (Figure 5), therefore resveratrol mainly exists in the free state in the presence of 0.01% proteins. As the concentration of WPI increased from 0.1% to 1%, the loading efficiencies of resveratrol increased by around 10%, of which the highest was 28%. The highest loading efficiencies of resveratrol in the presence of 0.1% SC and SPI were 31% and 27%, which further increased at 1% SC and SPI to around 80% and 76%, respectively. In the case of 0.1% and 1% proteins, the loading efficiencies of resveratrol ranked in order SC > SPI > WPI. As mentioned above, resveratrol is conducive to the micellization of SC (Figures 2 and 4). The complex of resveratrol and protein masked the charged group, and the absolute value of the ζ-potential of the system decreased in the order of WPI > SC > SPI (Figure 3). It is speculated that the loading of resveratrol in SC particles not only depends on the transfer of the hydrophobic environment, but also refers to the bridging of resveratrol to submicelles. These results supported the hypothesis that the resveratrol was mainly located in the hydrophobic core of SPI, while both entrapped in the hydrophobic core and partially bound to the surface of the SC micelles. For WPI, more resveratrol complexed with the protein. Meanwhile, the loading efficiencies of the remaining resveratrol in protein particles were similar before and after storage at 45 °C for 30 days (Figure 5 and Figure S3).

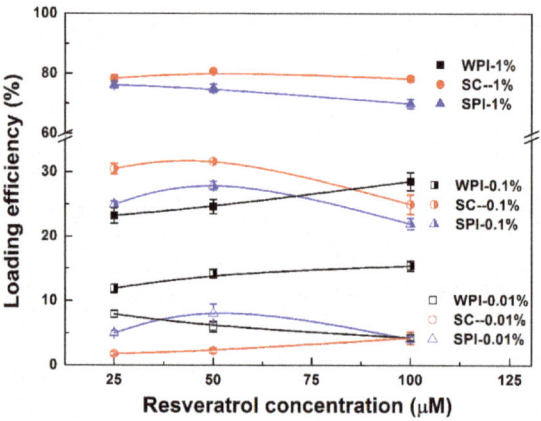

Figure 5. Loading efficiency of resveratrol in its complex particles with WPI (black), SC (red) and SPI (blue) at 0.01%, 0.1% and 1%.

3.2.3. Antioxidant Activity

It has been reported that casein contains more powerful antioxidant peptides than whey protein [39]. The presence of small peptides and C-terminal aromatic tyrosine residues contribute to the radical scavenging ability of SPI [40]. From Figure 6, the ABTS·+ scavenging capacity of proteins ranked in order SC > SPI > WPI under the same concentration. Resveratrol contains three phenolic hydroxyl groups and possesses antioxidant activity [41]. When the concentrations of resveratrol were 25, 50 and 100 μM, its ABTS·+ scavenging capacities were 11%, 20% and 44% (Figure 6), respectively. The scavenging capacities of WPI-resveratrol particles were similar to the sum of the individual capacity at the polyphenol concentrations of 25 and 50 μM (Figure 6A,B), suggesting an additive effect. At 100 μM, the scavenging capacities of WPI-resveratrol particles were less than the sum of the individual capacities (Figure 6C), suggesting partial screening of total antioxidant activity. As for SPI, resveratrol at 25 and 50 μM showed an additive effect with 0.01% and 0.1% protein but a masking effect with 1% protein (Figure 6A,B), and the masking effect was also observed at 100 μM resveratrol with all the investigated concentrations (Figure 6C). A masking effect was also observed in the case of SC, except for 25 and 50 μM resveratrol and 0.01% protein, which showed an additive effect (Figure 6). The masking effect is due to the protein–polyphenol interaction and the encapsulation of polyphenol in particles masking the phenolic hydroxyl groups [42,43]. It is worth noting that at 1% SPI and SC systems, the masked antioxidant activity was almost equal to that of resveratrol alone. This further confirms that resveratrol is mainly embedded in the hydrophobic core of SPI aggregations and SC micelles.

3.3. Protein Oxidation

3.3.1. Sulfhydryl Groups

It is well known that the ability of susceptible proteins for scavenging and generating reactive oxygen radical changes with the environment, which is related to the oxidation of the protein [44]. Protein oxidation is commonly accompanied by a decrease in the number of sulfhydryl (SH) groups [45]. The surface and total sulfhydryl contents of WPI were 12 and 17 nmol/mg, respectively, and for SPI, they were 3 nmol/mg and 5 nmol/mg, respectively. The lower sulfhydryl content of SPI in the initial state compared to WPI reflects that the initial oxidation state of SPI is greater than WPI, which may be related to the protein extraction process. As reported in the process of preparing SPI from defatted soy flour, lipoxygenase (LOX) was inevitably present in the system. The weakly alkaline extraction conditions caused LOX in the soybean flour to catalyze the oxidation of residual lipids [27]. The free and total sulfhydryl content of WPI and SPI decreased after storage (Figure 7), indicating that the accessible cysteine residues located at both surface and buried in the protein were attacked by free radicals [46]. The decrease in free and total sulfhydryl contents of SPI was greater in the presence than in the absence of resveratrol, while their contents of WPI were not affected by resveratrol (Figure 7). The interference of environmental factors on WPI and SPI sulfhydryl groups has been studied. After ultrasonic treatment, the disulfide bonds of SPI were destroyed, which significantly increased the free sulfhydryl content [47]. However, sonication did not change the thiol content of the whey protein concentrate. As reported, the oxidative susceptibility of free SH groups may depend on the constituent of mixture proteins. The intramolecular positions of the free thiol groups in β-lactoglobulin and α-lactalbumin may make WPI less sensitive [48].

3.3.2. Carbonyl Groups

Carbonyl groups (aldehydes and ketones) are produced on the side chains of the protein when they are oxidized [49]. The carbonyl contents of WPI, SC and SPI were 1.32, 1.63 and 2.26 nmol/mg (Figure 8), respectively. Due to the preparation process of SC, it contains about 6% of small ions in addition to the pure casein, mainly calcium, phosphate, magnesium and citric acid [15], leading to the worst oxidation stability. The extraction of SPI from soy flour may accelerate its carbonylation [50], since soy protein is extremely vulnerable to the attack of peroxyl radicals, and its degree of oxidation is related to the residual lipid content and LOX activity during the preparation process [27,51]. The carbonyl content of WPI increased during storage, reaching 2.7 nmol/mg after 30 days, which was invariable with the addition of resveratrol (Figure 8). The carbonyl content of SPI and SC increased as the concentration of resveratrol increased after 10 days. When the resveratrol concentrations were 0, 25, 50 and 100 μM, the carbonyl content of SPI increased from 3.95 to 5.11 nmol/mg, while the carbonyl content of SC increased from 2.77 to 3.29 nmol/mg after 30 days. The increase in carbonyl content may be related to the formation of peroxides in the system, which is generated by oxygen molecules attacking free radicals. The formation of peroxides on the α-carbon or other carbons of protein amino acid residues will result in an increase in the carbonyl content [27]. From Figure 8, it indicated that the peroxide content in the three protein solutions was in the order of SPI > SC > WPI, and the addition of resveratrol to SC and SPI solutions produced more peroxides. Together with the sulfhydryl contents in Figure 7, these results indicated that the SPI was more labile to oxidation than SC in the presence of resveratrol.

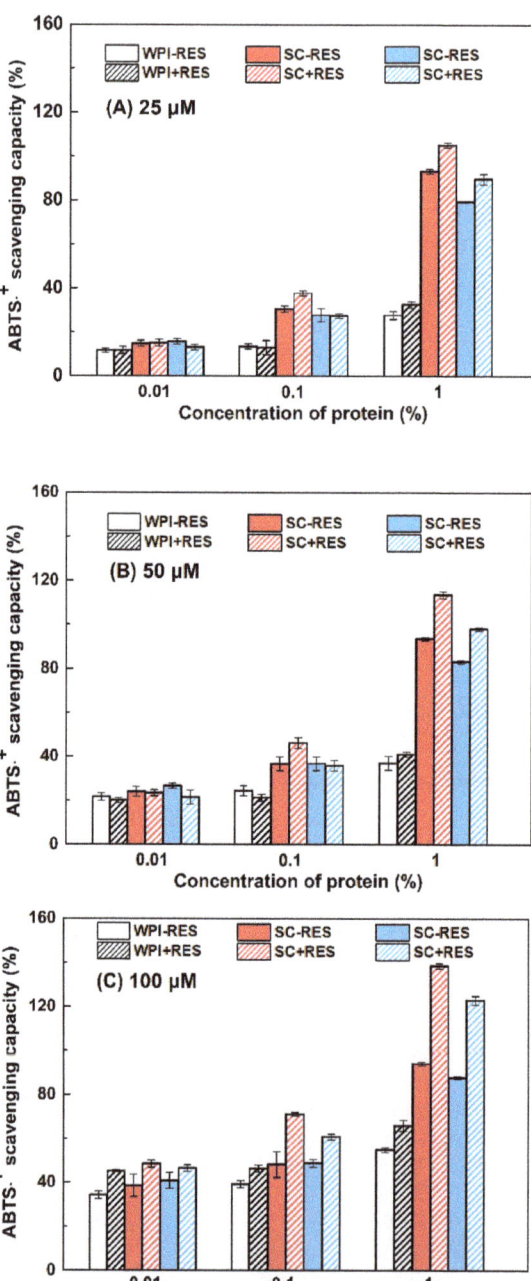

Figure 6. ABTS·+ scavenging capacity of resveratrol, WPI (black), SC (red), SPI (blue) and WPI-resveratrol, SC-resveratrol and SPI-resveratrol complex nanoparticles. The concentrations of proteins were 0.01%, 0.1% and 1%, while the concentrations of resveratrol were 25 (**A**), 50 (**B**) and 100 (**C**) μM.

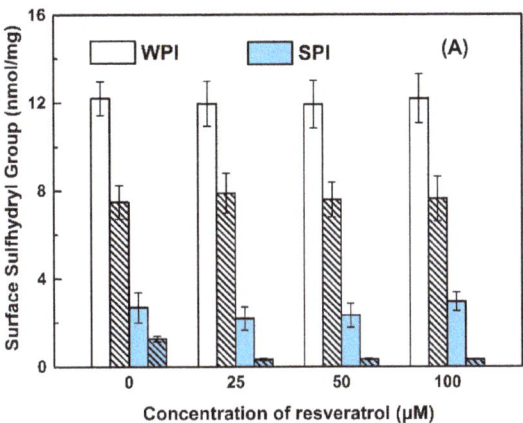

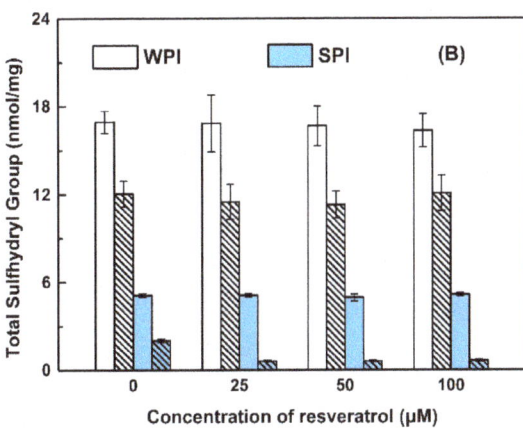

Figure 7. Surface (**A**) and total (**B**) sulfhydryl content of WPI-resveratrol and SPI-resveratrol complex particles before (no pattern) and after (sparse pattern) storage at 45 °C for 30 days. The concentration of proteins was 1%.

3.3.3. Amino Acid Composition

The oxidative attack of proteins modifies the side-chain groups of amino acid residues [52]. Tables 1–3 show the amino acid composition of WPI, SC and SPI in the absence and presence of resveratrol before and after storage for 30 days. The addition of resveratrol had no significant effect on the amino acid composition of the proteins before storage. The content of Cys ranked in order WPI > SPI > SC, and the surface and total sulfhydryl contents of SC were too low to be detected by the method of sulfhydryl analysis with DTNB (Figure 7). In the case of WPI alone, the content of Trp, Tyr, Thr, Lys, Met and Phe reduced after storage (Table 1), consistent with the indirect oxidation of WPI caused by the photodecomposition of folic acid [6]. Resveratrol had no effect on the change in the amino acid contents of WPI (Table 1). As for SC alone, the content of Trp, Tyr, Thr, Lys, Met, Asp and Arg reduced after storage and was more pronounced in the presence of resveratrol (Table 2). In addition, the content of Glu, Ser, Gly also reduced in the presence of resveratrol. The losses of Trp were about 11% for WPI and 79% and 87% for SC in the absence and presence of resveratrol, respectively. These results are consistent with a previous study that the tryptophan oxidation product, kynurenine, was higher in casein than β-LG upon photo-oxidation induced by riboflavin [4]. In the case of SPI alone, Asp, Ser, His, Gly, Thr, Tyr, Cys, Val, Met, Lys, Trp reduced after storage and was more pronounced in the presence of resveratrol (Table 3). In addition, the content of Glu also reduced in the presence of resveratrol. The reduction in the kinds and contents of total amino acids ranked in the order of SPI > SC > WPI (Tables 1–3).

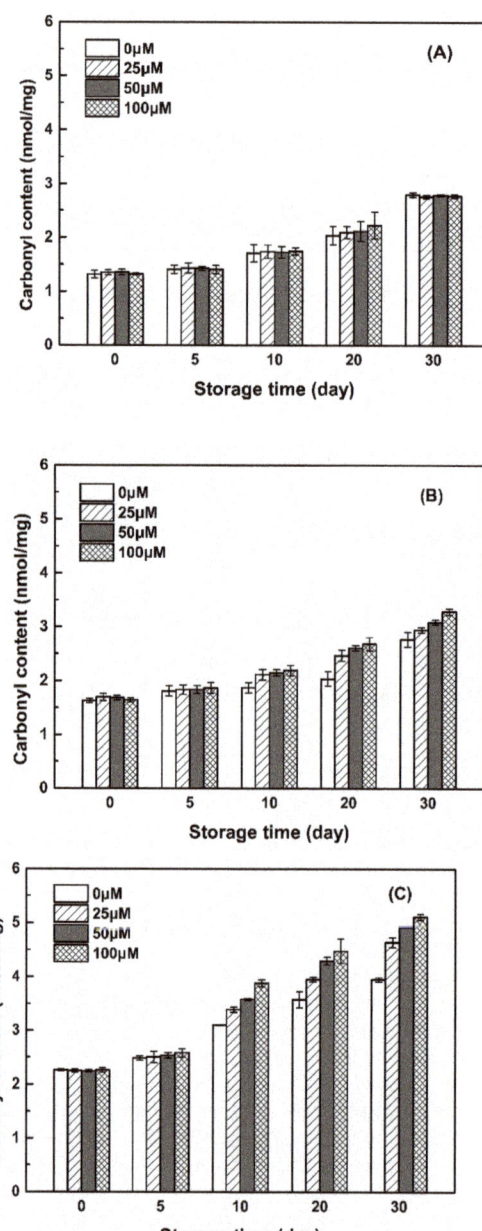

Figure 8. Carbonyl content of proteins in WPI-resveratrol (**A**), SC-resveratrol (**B**) and SPI-resveratrol (**C**) complex nanoparticles with various resveratrol concentrations during storage at 45 °C for 30 days. The concentration of proteins was 1%.

3.4. Storage Stability of Resveratrol

By visual observation, all protein-resveratrol solutions were transparent and colorless except that SPI-resveratrol solutions were turbid at 1% protein (Table S1 and Figure S1). No significant change was observed for WPI-resveratrol solutions at 45 °C after 30 days. However, SPI-resveratrol and SC-resveratrol solutions changed from colorless to light yellow after storage. It has been reported that the wine with resveratrol changed from colorless to light yellow, due to its sensitivity to atmospheric oxidation [53]. After storage at 45 °C for 30 days, the total color difference (ΔE) and chroma change (ΔC^*) of resveratrol alone increased, respectively, from 1.24 to 3.06 and from 1.17 to 2.94, as its concentration increased from 25 to 100 μM (Table 4). The ΔE and ΔC^* of WPI, SC, SPI, and WPI-resveratrol solutions were less than those of resveratrol alone. However, the ΔE and ΔC^* of SC-resveratrol and SPI-resveratrol solutions increased as the polyphenol concentration increased and were greater than the sum of correspondingly individual values at each concentration. A previous study also reported that WPI as emulsifier showed a better effect on inhibiting color changes of lutein-loaded emulsions relative to SC [54].

Table 4. Total color difference (ΔE) and chroma change (ΔC^*) of WPI-resveratrol, SC-resveratrol and SPI-resveratrol complex solutions before and after storage at 45 °C for 30 days. The concentration of proteins was 1%.

Protein	Concentration of Resveratrol (μM)			
	0	25	50	100
	ΔE			
		1.24 ± 0.19 [Aa]	1.85 ± 0.45 [Aa]	3.06 ± 0.27 [Bb]
WPI	0.74 ± 0.65 [Aa]	0.59 ± 0.31 [Aa]	0.58 ± 0.47 [Aa]	0.87 ± 0.92 [Aa]
SPI	1.16 ± 0.70 [Aa]	3.79 ± 0.83 [Bb]	6.95 ± 0.96 [Cb]	9.16 ± 1.54 [Dc]
SC	0.30 ± 0.20 [Aa]	2.82 ± 1.19 [Bb]	6.04 ± 1.49 [Cb]	8.59 ± 0.71 [Dc]
	ΔC^*			
		1.17 ± 0.15 [Aa]	1.73 ± 0.35 [Ba]	2.94 ± 0.23 [Cb]
WPI	0.29 ± 0.22 [Aa]	0.56 ± 0.31 [Aa]	0.42 ± 0.25 [Aa]	0.30 ± 0.12 [Aa]
SPI	0.58 ± 0.47 [Aa]	3.30 ± 0.46 [Bb]	5.94 ± 0.43 [Cb]	8.31 ± 1.39 [Dd]
SC	0.14 ± 0.09 [Aa]	2.63 ± 1.02 [Bb]	4.44 ± 1.23 [Cb]	6.19 ± 0.50 [Dc]

Note: Different lower-case letters in the same column represent significantly different mean values, different upper-case letters in the same row represent significant different mean values ($p < 0.05$).

Resveratrol alone degraded during storage at 45 °C and its content remained 68–74% after 30 days (Figure 9). The retention of resveratrol was improved by WPI, and the protective effect decreased slightly as the protein concentration increased. After 30 days of storage, the retention of resveratrol at 25 μM was around 88, 84, and 74% at 0.01, 0.1, and 1% WPI (Figure 9A), respectively, and the polyphenol retention was proportional to its initial concentration (Figure 9). In contrast, the loss of resveratrol was accelerated by SPI, the effect of which was more pronounced when the protein concentrations were 0.1% and 1% than 0.01% (Figure 9). SC also accelerated the degradation of resveratrol, the effect of which was less than that of SPI and decreased as the polyphenol concentration increased. The retention of resveratrol was consistent with the color change of its corresponding samples (Table 4).

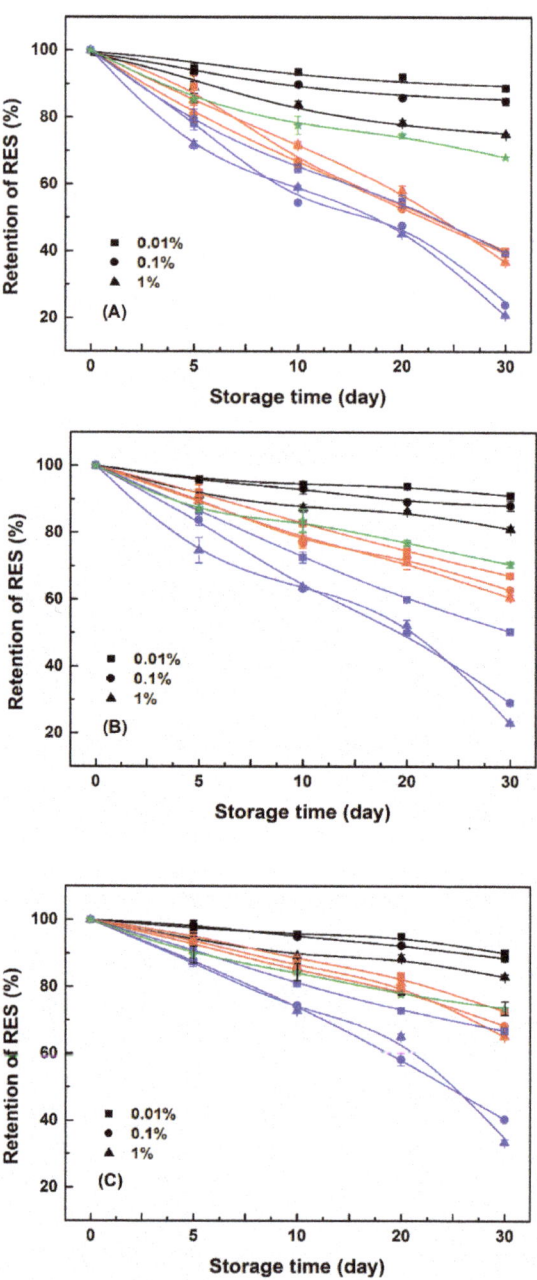

Figure 9. Retention of resveratrol alone (green) and in its WPI (black), SC (red) and SPI (blue) complex particles at various protein concentrations during storage at 45 °C. The concentrations of resveratrol were 25 (**A**), 50 (**B**) and 100 (**C**) μM.

4. Discussion

Resveratrol self-aggregates at a concentration higher than 40 μM, due to the hydrophobic stacking of aromatic phenol rings [55]. The aggregation of resveratrol reduces its contact with the external environment and affects its antioxidant activity with the highest value observed at a concentration of 30 µM [56]. Therefore, the retention of resveratrol increased from 68% to 74% as its concentration increased from 25 to 100 µM (Figure 9). α-Tocopherol (Log P ~ 8.84, https://go.drugbank.com/drugs/DB00163/ accessed on 8 July 2021), a hydrophobic vitamin E, was reported both bound in the molecular level and encapsulated as the aggregate in WPI particles, while naringenin (LogP ~ 2.84, https://go.drugbank.com/drugs/DB03467/ accessed on 8 July 2021), a polyhydroxy flavonoid, was bound in the molecular level [34]. Resveratrol (LogP ~ 3.4, https://go.drugbank.com/drugs/DB02709/ accessed on 8 July 2021) is more hydrophobic than naringenin but more hydrophilic than α-tocopherol. When calculated, based on the loading efficiency of resveratrol in Figure 5, the encapsulated amount of resveratrol in protein particles increased as the polyphenol concentration increased (Figure S3). It is thus possible that the aggregated resveratrol in protein particles increased as its concentration increased, which was supported by the transfer from the additive to the masking effect of total antioxidant activity (Figure 6). Therefore, the polyphenol retention increased with its concentration in protein particles (Figure 9).

For WPI, the solvent-accessible (bounded in the molecular level and in free state) resveratrol can scavenge and control the free radicals in the system within a certain range. Its oxidation was the least and not affected by resveratrol during storage for 30 days (Figures 7 and 8 and Table 1). At the same time, the stability of resveratrol was improved by WPI, with a retention of above 74% after 30 days (Figure 9). It is thus speculated that there is no reciprocal oxidation between WPI and resveratrol during storage. As the concentration of WPI increased, the loading efficiency of resveratrol increased (Figure 5), but the polyphenol stability decreased (Figure 9). These results suggest that the loaded microenvironment is not conducive to the polyphenol stability, compared to the free part in the WPI solution. The protective effect of WPI on resveratrol stability might not be attributed to the complex property of the protein.

For SPI, the encapsulated resveratrol located in the hydrophobic core could not exert its antioxidant capacity. Thus its oxidation was the most at the beginning and accelerated by resveratrol during storage after 10 days (Figures 7 and 8 and Table 3). At the same time, the stability of resveratrol decreased upon loading in SPI particles (Figure 9). These results suggest the occurrence of reciprocal oxidation between SPI and resveratrol. The co-oxidation has been reported for whey protein and Antarctic krill oil in oil-in-water emulsion [57]. The initial state of the SPI system contained more peroxides than SC and WPI (Figures 7 and 8), free radicals and hydroperoxides generated during protein oxidation may accelerate the degradation of resveratrol [58] (Figure 9). It has also been reported that ascorbic acid acted as a co-oxidant by generating superoxide anions in the presence of air and extracting hydrogen from the carrier [59]. Resveratrol is oxidized to generate H_2O_2 [60]. When the retention of resveratrol was between 59 and 73% after 10 days (Figure 9), the polyphenol may act as a co-oxidant to accelerate the oxidation of SPI (Figure 8).

However, most of the resveratrol in the SC system was encapsulated in the hydrophobic core of the protein, but also partially bounded with submicelles in the molecular level, which can play their antioxidant effect to a certain extent. The oxidation of SC was more pronounced than that of WPI but less than that of SPI at the beginning and during storage in the absence and presence of resveratrol (Figures 7 and 8 and Table 2). At the same time, the impact of SC on resveratrol stability basically changed from a protective to a harmful effect during storage (Figure 9). The antioxidant activity of SC was greater than that of WPI and SPI (Figure 6), and the loading efficiencies of resveratrol in SC particles were greater than those in SPI and WPI particles at protein concentrations of 0.1% and 1% (Figure 5). Therefore, the stability of resveratrol was initially improved by SC (Figure 9). A stable protein carrier can maintain the stability of polyphenols through scavenging free radicals and isolating the interference of external unfavorable factors [61]. Then, with the increasing oxidation of SC, the ability to scavenge free radicals was not enough to resist the auto-oxidation of SC. The system was out of balance and the protein changed from antioxidant to pro-oxidant to cause the co-oxidation with resveratrol (Figures 8 and 9).

According to the molecular mechanism of the protein–polyphenol interaction, the di-phenol part of polyphenol is easily oxidized by molecular oxygen and side-chain amino groups under certain conditions to form quinine, which can form a dimer in a side reaction and interact with the amino group of polypeptide or the irreversible reaction of the sulfhydryl side chain leads to the formation of protein cross-links. The closer the distance between the formed oxidation product and the α-carbon or other carbons of protein amino acid residues, the more easily the reaction occurs (Figure 4). Meanwhile, quinine can undergo condensation reactions to form high molecular weight, highly reactive brown tannins [17], which is verified in Table 4 and Figure S1. The formation of a covalent EGCG-protein complex involved the reaction of dimer quinone with protein nucleophilic side chains, such as lysine and cysteine residues, which is consistent with the results of amino acid composition in SC/SPI-resveratrol complex particles after storage (Tables 1–3). It has been assumed that the structure of SC and SPI gradually became flexible during storage and the exposed active groups benefited from the covalent interactions of protein-resveratrol complexations [62].

5. Conclusions

WPI improved the storage stability of resveratrol, but SPI accelerated the loss of resveratrol, while the impact of SC on resveratrol stability basically changed from a protective to a harmful effect. The stability of polyphenols increased as the polyphenol concentration increased but decreased as the protein concentration increased. The loading efficiency of resveratrol in protein particles and the initial antioxidant activity of proteins were not the dominant factors to affect the storage stability of resveratrol. The effect of proteins on the stability of resveratrol was mainly dependent on their oxidation sensitivity. The co-oxidation of resveratrol with SPI and SC occurred during storage. The oxidation degree of WPI was the least and not affected by resveratrol. The results obtained suggest that WPI might be a better material to design an effective carrier for the long-term protection of resveratrol than SPI and SC. To our knowledge, it is the first time that the important role of protein oxidability on the stability of polyphenols during storage has been reported and provides useful guidelines for the long-term protection of polyphenols by protein-based carriers.

Supplementary Materials: The following supporting information can be downloaded at: https://www.mdpi.com/article/10.3390/antiox11040647/s1, Figure S1: Appearance of WPI-resveratrol, SC-resveratrol and SPI-resveratrol complex nanoparticles before (A–C, respectively) and after (a–c, respectively) storage at 45 °C for 30 days. The concentration of proteins from left to right was 0.01%, 0.1% and 1%. The concentrations of resveratrol from left to right was 25, 50 and 100 μM; Figure S2: XRD patterns of resveratrol (black), proteins (blue), their physical mixtures (red) and resveratrol-loaded protein particles (green). The concentration of protein was 1%; Figure S3: Loading efficiency of resveratrol in its complex particles with WPI (black), SC (red) and SPI (blue) at 0.01%, 0.1% and 1% after storage at 45 °C; Table S1: Turbidity of WPI-resveratrol, SC-resveratrol and SPI-resveratrol complex nanoparticles at various concentrations of proteins and resveratrol.

Author Contributions: X.Y.: conceptualization, investigation, writing—original draft, formal analysis. H.C.: resources, methodology. W.: writing—review and editing. H.D.: methodology, writing—review and editing. W.H.: conceptualization. L.L.: conceptualization, resources, writing—review and editing, supervision. All authors have read and agreed to the published version of the manuscript.

Funding: This research was funded by the Postgraduate Research & Practice Innovation Program of Jiangsu Province (KYCX20-1863).

Institutional Review Board Statement: Not applicable.

Informed Consent Statement: Not applicable.

Data Availability Statement: The data are contained within the article and supplementary materials.

Conflicts of Interest: The authors declare no conflict of interest.

Abbreviations

WPI	whey protein isolate
SC	sodium caseinate
SPI	soy protein isolate
BSA	bovine serum albumin
RES	resveratrol
GRAS	generally recognized as safe
ABTS	2,2′-azino-bis-3- ethylbenzthiazoline-6-sulphonic acid
CMC	critical micelle concentration
β-LG	β-lactoglobulin
LOX	lipoxygenase
SH	sulfhydryl
ΔE	total color difference
ΔC*	chroma change

References

1. Zhang, J.; Field, C.J.; Vine, D.; Chen, L. Intestinal Uptake and Transport of Vitamin B12-loaded Soy Protein Nanoparticles. *Pharm. Res.* **2015**, *32*, 1288–1303. [CrossRef]
2. McClements, D.J. The future of food colloids: Next-generation nanoparticle delivery systems. *Curr. Opin. Colloid Interface Sci.* **2017**, *28*, 7–14. [CrossRef]
3. Livney, Y.D. Milk proteins as vehicles for bioactives. *Curr. Opin. Colloid Interface Sci.* **2010**, *15*, 73–83. [CrossRef]
4. Dalsgaard, T.K.; Otzen, D.; Nielsen, J.H.; Larsen, L.B. Changes in Structures of Milk Proteins upon Photo-oxidation. *J. Agric. Food Chem.* **2007**, *55*, 10968–10976. [CrossRef] [PubMed]
5. Hellwig, M. The Chemistry of Protein Oxidation in Food. *Angew. Chem. Int. Ed.* **2019**, *58*, 16742–16763. [CrossRef] [PubMed]
6. Fu, X.; Wusigale; Cheng, H.; Fang, Z.; Liang, L. Mechanism for improved protection of whey protein isolate against the photodecomposition of folic acid. *Food Hydrocoll.* **2018**, *79*, 439–449. [CrossRef]
7. Xu, D.; Wang, X.; Jiang, J.; Yuan, F.; Decker, E.A.; Gao, Y. Influence of pH, EDTA, α-tocopherol, and WPI oxidation on the degradation of β-carotene in WPI-stabilized oil-in-water emulsions. *LWT-Food Sci. Technol.* **2013**, *54*, 236–241. [CrossRef]
8. Salminen, H.; Heinonen, M. Plant Phenolics Affect Oxidation of Tryptophan. *J. Agric. Food Chem.* **2008**, *56*, 7472–7481. [CrossRef]
9. Hematyar, N.; Rustad, T.; Sampels, S.; Dalsgaard, T.K. Relationship between lipid and protein oxidation in fish. *Aquac. Res.* **2019**, *50*, 1393–1403. [CrossRef]
10. Haratifar, S.; Corredig, M. Interactions between tea catechins and casein micelles and their impact on renneting functionality. *Food Chem.* **2014**, *143*, 27–32. [CrossRef] [PubMed]
11. Chen, K.; Chen, X.; Liang, L.; Xu, X. Gallic Acid-Aided Cross-Linking of Myofibrillar Protein Fabricated Soluble Aggregates for Enhanced Thermal Stability and a Tunable Colloidal State. *J. Agric. Food Chem.* **2020**, *68*, 11535–11544. [CrossRef] [PubMed]
12. Joye, I.J.; Davidov-Pardo, G.; Ludescher, R.D.; McClements, D. Fluorescence quenching study of resveratrol binding to zein and gliadin: Towards a more rational approach to resveratrol encapsulation using water-insoluble proteins. *Food Chem.* **2015**, *185*, 261–267. [CrossRef] [PubMed]
13. Xiong, W.; Ren, C.; Li, J.; Li, B. Enhancing the photostability and bioaccessibility of resveratrol using ovalbumin–carboxymethylcellulose nanocomplexes and nanoparticles. *Food Funct.* **2018**, *9*, 3788–3797. [CrossRef] [PubMed]
14. Wu, Y.; Cheng, H.; Chen, Y.; Chen, L.; Fang, Z.; Liang, L. Formation of a multiligand complex of bovine serum albumin with retinol, resveratrol, and (−)-epigallocatechin-3-gallate for the protection of bioactive components. *J. Agric. Food Chem.* **2017**, *65*, 3019–3030. [CrossRef] [PubMed]
15. Ranadheera, C.S.; Liyanaarachchi, W.S.; Chandrapala, J.; Dissanayake, M.; Vasiljevic, T. Utilizing unique properties of caseins and the casein micelle for delivery of sensitive food ingredients and bioactives. *Trends Food Sci. Technol.* **2016**, *57*, 178–187. [CrossRef]
16. Cheng, H.; Dong, H.; Liang, L. A comparison of beta-casein complexes and micelles as vehicles for trans-/cis-resveratrol. *Food Chem.* **2020**, *330*, 127209. [CrossRef]
17. Ozdal, T.; Capanoglu, E.; Altay, F. A review on protein–phenolic interactions and associated changes. *Food Res. Int.* **2013**, *51*, 954–970. [CrossRef]
18. Morr, C.V.; Ha, E.Y.W. Whey protein concentrates and isolates: Processing and functional properties. *Crit. Rev. Food Sci. Nutr.* **1993**, *33*, 431–476. [CrossRef] [PubMed]
19. Yi, J.; Lam, T.I.; Yokoyama, W.; Cheng, L.W.; Zhong, F. Cellular Uptake of β-Carotene from Protein Stabilized Solid Lipid Nanoparticles Prepared by Homogenization–Evaporation Method. *J. Agric. Food Chem.* **2014**, *62*, 1096–1104. [CrossRef] [PubMed]
20. Peng, S.; Zhou, L.; Cai, Q.; Zou, L.; Liu, C.; Liu, W.; McClements, D.J. Utilization of biopolymers to stabilize curcumin nanoparticles prepared by the pH-shift method: Caseinate, whey protein, soy protein and gum Arabic. *Food Hydrocoll.* **2020**, *107*, 105963. [CrossRef]

21. Esmaili, M.; Ghaffari, S.M.; Moosavi-Movahedi, Z.; Atri, M.S.; Sharifizadeh, A.; Farhadi, M.; Yousefi, R.; Chobert, J.-M.; Haertlé, T.; Moosavi-Movahedi, A.A. Beta casein-micelle as a nano vehicle for solubility enhancement of curcumin; food industry application. *LWT-Food Sci. Technol.* **2011**, *44*, 2166–2172. [CrossRef]
22. McGuire, R.G. Reporting of Objective Color Measurements. *HortScience* **1992**, *27*, 1254–1255. [CrossRef]
23. Cheng, H.; Fang, Z.; Liu, T.; Gao, Y.; Liang, L. A study on β-lactoglobulin-triligand-pectin complex particle: Formation, characterization and protection. *Food Hydrocoll.* **2018**, *84*, 93–103. [CrossRef]
24. Zimet, P.; Rosenberg, D.; Livney, Y.D. Re-assembled casein micelles and casein nanoparticles as nano-vehicles for ω-3 polyunsaturated fatty acids. *Food Hydrocoll.* **2011**, *25*, 1270–1276. [CrossRef]
25. Re, R.; Pellegrini, N.; Proteggente, A.; Pannala, A.; Yang, M.; Rice-Evans, C. Antioxidant activity applying an improved ABTS radical cation decolorization assay. *Free Radic. Biol. Med.* **1999**, *26*, 1231–1237. [CrossRef]
26. Ellman, G.L. Tissue sulfhydryl groups. *Arch. Biochem. Biophys.* **1959**, *82*, 70–77. [CrossRef]
27. Huang, Y.; Hua, Y.; Qiu, A. Soybean protein aggregation induced by lipoxygenase catalyzed linoleic acid oxidation. *Food Res. Int.* **2006**, *39*, 240–249. [CrossRef]
28. Li, M.; Jiang, M.; Wu, C. Fluorescence and light-scattering studies on the formation of stable colloidal nanoparticles made of sodium sulfonated polystyrene ionomers. *J. Polym. Sci. Part B Polym. Phys.* **1997**, *35*, 1593–1599. [CrossRef]
29. Horne, D.S. Casein micelle structure: Models and muddles. *Curr. Opin. Colloid Interface Sci.* **2006**, *11*, 148–153. [CrossRef]
30. Almajano, M.; Delgado, M.E.; Gordon, M.H. Changes in the antioxidant properties of protein solutions in the presence of epigallocatechin gallate. *Food Chem.* **2007**, *101*, 126–130. [CrossRef]
31. Ghayour, N.; Hosseini, S.M.H.; Eskandari, M.H.; Esteghlal, S.; Nekoei, A.-R.; Gahruie, H.H.; Tatar, M.; Naghibalhossaini, F. Nanoencapsulation of quercetin and curcumin in casein-based delivery systems. *Food Hydrocoll.* **2019**, *87*, 394–403. [CrossRef]
32. Webb, M.; Naeem, H.; Schmidt, K. Food Protein Functionality in a Liquid System: A Comparison of Deamidated Wheat Protein with Dairy and Soy Proteins. *J. Food Sci.* **2002**, *67*, 2896–2902. [CrossRef]
33. Liu, W.; Chen, X.D.; Cheng, Z.; Selomulya, C. On enhancing the solubility of curcumin by microencapsulation in whey protein isolate via spray drying. *J. Food Eng.* **2016**, *169*, 189–195. [CrossRef]
34. Yin, X.; Fu, X.; Cheng, H.; Liang, L. α-Tocopherol and naringenin in whey protein isolate particles: Partition, antioxidant activity, stability and bioaccessibility. *Food Hydrocoll.* **2020**, *106*, 105895. [CrossRef]
35. Tian, Y.; Xu, G.; Cao, W.; Li, J.; Taha, A.; Hu, H.; Pan, S. Interaction between pH-shifted beta-conglycinin and flavonoids hesperetin/hesperidin: Characterization of nanocomplexes and binding mechanism. *LWT-Food Sci. Technol.* **2021**, *140*, 110698. [CrossRef]
36. Lu, Z.; Chen, R.; Fu, R.; Xiong, J.; Hu, Y. Cytotoxicity and inhibition of lipid peroxidation activity of resveratrol/cyclodextrin inclusion complexes. *J. Incl. Phenom. Macrocycl. Chem.* **2012**, *73*, 313–320. [CrossRef]
37. Wang, Y.; Wang, X. Binding, stability, and antioxidant activity of quercetin with soy protein isolate particles. *Food Chem.* **2015**, *188*, 24–29. [CrossRef]
38. Cheng, H.; Fang, Z.; Bakry, A.M.; Chen, Y.; Liang, L. Complexation of trans-and cis-resveratrol with bovine serum albumin, β-lactoglobulin or α-lactalbumin. *Food Hydrocoll.* **2018**, *81*, 242–252. [CrossRef]
39. Ahmed, A.S.; El-Bassiony, T.; Elmalt, L.M.; Ibrahim, H.R. Identification of potent antioxidant bioactive peptides from goat milk proteins. *Food Res. Int.* **2015**, *74*, 80–88. [CrossRef] [PubMed]
40. Beermann, C.; Euler, M.; Herzberg, J.; Stahl, B. Anti-oxidative capacity of enzymatically released peptides from soybean protein isolate. *Eur. Food Res. Technol.* **2009**, *229*, 637–644. [CrossRef]
41. Brewer, M. Natural antioxidants: Sources, compounds, mechanisms of action, and potential applications. *Compr. Rev. Food Sci. Food Saf.* **2011**, *10*, 221–247. [CrossRef]
42. Stojadinovic, M.; Radosavljevic, J.; Ognjenovic, J.; Vesic, J.; Prodic, I.; Stanic-Vucinic, D.; Velickovic, T.C. Binding affinity between dietary polyphenols and β-lactoglobulin negatively correlates with the protein susceptibility to digestion and total antioxidant activity of complexes formed. *Food Chem.* **2013**, *136*, 1263–1271. [CrossRef] [PubMed]
43. Wang, L.; Zhang, Y. Eugenol Nanoemulsion Stabilized with Zein and Sodium Caseinate by Self-Assembly. *J. Agric. Food Chem.* **2017**, *65*, 2990–2998. [CrossRef] [PubMed]
44. Amamcharla, J.K.; Metzger, L.E. Modification of the ferric reducing antioxidant power (FRAP) assay to determine the susceptibility of raw milk to oxidation. *Int. Dairy J.* **2014**, *34*, 177–179. [CrossRef]
45. Eaton, P. Protein thiol oxidation in health and disease: Techniques for measuring disulfides and related modifications in complex protein mixtures. *Free Radic. Biol. Med.* **2006**, *40*, 1889–1899. [CrossRef]
46. Feng, X.; Li, C.; Ullah, N.; Cao, J.; Lan, Y.; Ge, W.; Hackman, R.M.; Li, Z.; Chen, L. Susceptibility of whey protein isolate to oxidation and changes in physicochemical, structural, and digestibility characteristics. *J. Dairy Sci.* **2015**, *98*, 7602–7613. [CrossRef]
47. Hu, H.; Wu, J.; Li-Chan, E.C.; Zhu, L.; Zhang, F.; Xu, X.; Fan, G.; Wang, L.; Huang, X.; Pan, S. Effects of ultrasound on structural and physical properties of soy protein isolate (SPI) dispersions. *Food Hydrocoll.* **2013**, *30*, 647–655. [CrossRef]
48. Chandrapala, J.; Zisu, B.; Palmer, M.; Kentish, S.; Ashokkumar, M. Effects of ultrasound on the thermal and structural characteristics of proteins in reconstituted whey protein concentrate. *Ultrason. Sonochem.* **2011**, *18*, 951–957. [CrossRef]
49. Dalle-Donne, I.; Rossi, R.; Giustarini, D.; Milzani, A.; Colombo, R. Protein carbonyl groups as biomarkers of oxidative stress. *Clin. Chim. Acta* **2003**, *329*, 23–38. [CrossRef]

50. Duque-Estrada, P.; Kyriakopoulou, K.; de Groot, W.; van der Goot, A.J.; Berton-Carabin, C.C. Oxidative stability of soy proteins: From ground soybeans to structured products. *Food Chem.* **2020**, *318*, 126499. [CrossRef]
51. Wu, W.; Zhang, C.; Kong, X.; Hua, Y. Oxidative modification of soy protein by peroxyl radicals. *Food Chem.* **2009**, *116*, 295–301. [CrossRef]
52. Li, C.; Xiong, Y.L.; Chen, J. Oxidation-induced unfolding facilitates myosin cross-linking in myofibrillar protein by microbial transglutaminase. *J. Agric. Food Chem.* **2012**, *60*, 8020–8027. [CrossRef] [PubMed]
53. Fan, E.; Zhang, K.; Yao, C.; Yan, C.; Bai, Y.; Jiang, S. Determination of trans-Resveratrol in China Great Wall "Fazenda"Red Wine by Use of Micellar Electrokinetic Chromatography. *Chromatographia* **2005**, *62*, 289–294. [CrossRef]
54. Weigel, F.; Weiss, J.; Decker, E.A.; McClements, D.J. Lutein-enriched emulsion-based delivery systems: Influence of emulsifiers and antioxidants on physical and chemical stability. *Food Chem.* **2018**, *242*, 395–403. [CrossRef]
55. Liang, L.; Tajmir-Riahi, H.A.; Subirade, M. Interaction of β-lactoglobulin with resveratrol and its biological implications. *Biomacromolecules* **2007**, *9*, 50–56. [CrossRef]
56. López-Nicolás, J.M.; Pérez-Gilabert, M.; García-Carmona, F. Effect of Protonation and Aggregation State of (E)-Resveratrol on Its Hydroperoxidation by Lipoxygenase. *J. Agric. Food Chem.* **2009**, *57*, 4630–4635. [CrossRef]
57. Wang, Y.; Liu, Y.; Ma, L.; Yang, L.; Cong, P.; Lan, H.; Xue, C.; Xu, J. Co-oxidation of Antarctic krill oil with whey protein and myofibrillar protein in oil-in-water emulsions. *J. Food Sci.* **2020**, *85*, 3797–3805. [CrossRef]
58. Hellwig, M. Analysis of Protein Oxidation in Food and Feed Products. *J. Agric. Food Chem.* **2020**, *68*, 12870–12885. [CrossRef]
59. Zoldners, J.; Kiseleva, T.; Kaiminsh, I. Influence of ascorbic acid on the stability of chitosan solutions. *Carbohydr. Polym.* **2005**, *60*, 215–218. [CrossRef]
60. Yang, N.C.; Lee, C.H.; Song, T.Y. Evaluation of resveratrol oxidation in vitro and the crucial role of bicarbonate ions. *Biosci. Biotechnol. Biochem.* **2010**, *74*, 63–68. [CrossRef]
61. Shpigelman, A.; Israeli, G.; Livney, Y.D. Thermally-induced protein–polyphenol co-assemblies: Beta lactoglobulin-based nanocomplexes as protective nanovehicles for EGCG. *Food Hydrocoll.* **2010**, *24*, 735–743. [CrossRef]
62. Wei, Z.; Yang, W.; Fan, R.; Yuan, F.; Gao, Y. Evaluation of structural and functional properties of protein–EGCG complexes and their ability of stabilizing a model β-carotene emulsion. *Food Hydrocoll.* **2015**, *45*, 337–350. [CrossRef]

Article

Encapsulation of Phenolic Compounds from a Grape Cane Pilot-Plant Extract in Hydroxypropyl Beta-Cyclodextrin and Maltodextrin by Spray Drying

Danilo Escobar-Avello [1,2], Javier Avendaño-Godoy [3], Jorge Santos [4,5], Julián Lozano-Castellón [1,6], Claudia Mardones [7], Dietrich von Baer [7], Javiana Luengo [3], Rosa M. Lamuela-Raventós [1,6], Anna Vallverdú-Queralt [1,6,*] and Carolina Gómez-Gaete [2,3,*]

Citation: Escobar-Avello, D.; Avendaño-Godoy, J.; Santos, J.; Lozano-Castellón, J.; Mardones, C.; von Baer, D.; Luengo, J.; Lamuela-Raventós, R.M.; Vallverdú-Queralt, A.; Gómez-Gaete, C. Encapsulation of Phenolic Compounds from a Grape Cane Pilot-Plant Extract in Hydroxypropyl Beta-Cyclodextrin and Maltodextrin by Spray Drying. *Antioxidants* **2021**, *10*, 1130. https://doi.org/10.3390/antiox10071130

Academic Editors: Li Liang, Hao Cheng and Daniel Franco Ruiz

Received: 10 June 2021
Accepted: 13 July 2021
Published: 15 July 2021

Publisher's Note: MDPI stays neutral with regard to jurisdictional claims in published maps and institutional affiliations.

Copyright: © 2021 by the authors. Licensee MDPI, Basel, Switzerland. This article is an open access article distributed under the terms and conditions of the Creative Commons Attribution (CC BY) license (https:// creativecommons.org/licenses/by/ 4.0/).

1. Department of Nutrition, Food Science and Gastronomy XaRTA, Faculty of Pharmacy and Food Sciences, Institute of Nutrition and Food Safety (INSA-UB), University of Barcelona, 08028 Barcelona, Spain; daniescobar01@ub.edu (D.E.-A.); julian.lozano@ub.edu (J.L.-C.); lamuela@ub.edu (R.M.L.-R.)
2. Unidad de Desarrollo Tecnológico, Universidad de Concepción, 4191996 Coronel, Chile
3. Departamento de Farmacia, Facultad de Farmacia, Universidad de Concepción, 4191996 Concepción, Chile; jaavendano@udec.cl (J.A.-G.); jluengo@udec.cl (J.L.)
4. DEMad, Instituto Politécnico de Viseu, 3504-510 Viseu, Portugal; jsantosu@estgv.ipv.pt
5. LEPABE—Faculty of Engineering, University of Porto, 4200-465 Porto, Portugal
6. Consorcio CIBER, M.P. Fisiopatología de la Obesidad y la Nutrición (CIBERObn), Instituto de Salud Carlos III (ISCIII), 28029 Madrid, Spain
7. Departamento de Análisis Instrumental, Facultad de Farmacia, Universidad de Concepción, 4070386 Concepción, Chile; cmardone@udec.cl (C.M.); dvonbaer@udec.cl (D.v.B.)
* Correspondence: avallverdu@ub.edu (A.V.-Q.); cargomez@udec.cl (C.G.-G.); Tel.: +34-934020834 (A.V.-Q.); +56-41-2204226 (C.G.-G.)

Abstract: Grape canes, the main byproducts of the viticulture industry, contain high-value bioactive phenolic compounds, whose application is limited by their instability and poorly solubility in water. Encapsulation in cyclodextrins allows these drawbacks to be overcome. In this work, a grape cane pilot-plant extract (GC$_{PPE}$) was encapsulated in hydroxypropyl beta-cyclodextrin (HP-β-CD) by a spray-drying technique and the formation of an inclusion complex was confirmed by microscopy and infrared spectroscopy. The phenolic profile of the complex was analyzed by LC-ESI-LTQ-Orbitrap-MS and the encapsulation efficiency of the phenolic compounds was determined. A total of 42 compounds were identified, including stilbenes, flavonoids, and phenolic acids, and a complex of (*epi*)catechin with β-CD was detected, confirming the interaction between polyphenols and cyclodextrin. The encapsulation efficiency for the total extract was 80.5 ± 1.1%, with restrytisol showing the highest value (97.0 ± 0.6%) and (*E*)-resveratrol (32.7 ± 2.8%) the lowest value. The antioxidant capacity of the inclusion complex, determined by ORAC-FL, was 5300 ± 472 µmol TE/g DW, which was similar to the value obtained for the unencapsulated extract. This formulation might be used to improve the stability, solubility, and bioavailability of phenolic compounds of the GC$_{PPE}$ for water-soluble food and pharmaceutical applications.

Keywords: microencapsulation; cyclodextrin; vine shoots; food waste; *Vitis vinifera* L.; polyphenols; stilbenoids; mass spectrometry; Fourier Transform Infrared Spectroscopy; Scanning Electron Microscopy

1. Introduction

The generation of food and agricultural waste is a growing problem, with negative impacts on the economy, environment, and human health. Therefore, the integral valorization of these wastes by conversion into bioenergy or recovery of chemical compounds for biobased products is a technological challenge for achieving a circular economy. Alternative ways of disposing of food waste include the valorization of byproducts as a source of phenolic compounds used to fortify high-consumption foods or to formulate new functional foods [1].

Grape canes of *V. vinifera* L. produced during pruning are the main byproduct of viticulture, and millions of tons are generated worldwide every year [2,3]. With the aim of developing a circular economy by an integrated biorefinery strategy, grape canes are being investigated as a high-value resource due to their attractive chemical composition and potential industrial applications. These include SO_2 substitution in wine to improve wine quality [4], usage as a cosmetic ingredient [5] and as a filler in food packaging [4,6], and for the recovery of hemicellulosic oligosaccharides, lignin, and cellulosic substrates [7]. Above all, however, grape canes have been studied for their nutraceutical applications, as they contain high-added-value phenolic compounds with wide-ranging biological properties.

Grape canes contain a complex mixture of phenolic compounds, including phenolic acids (hydroxybenzoic acid and hydroxycinnamic acids), flavonoids (mainly proanthocyanidins, flavonols, flavanonol and flavanones), and stilbenes (monomers, dimers, and oligomers). The most abundant phenolic compounds are proanthocyanidins and stilbene oligomers [8,9]. The phenolic composition of *V. vinifera* plants is genetically determined and strongly influenced by environmental conditions such as water stress [10]. Other determining factors, which particularly affect the stilbenes content, are the cultivar [11], the geographic region of cultivation [12], and the time, temperature, and humidity of storage after pruning [13]. Moreover, the yields of phenolic compounds such as resveratrol and ε-viniferin are highly dependent on variables of the extraction methods such as grape cane particle size, the type of solvent, temperature, duration, and the effects of light [14]. Furthermore, the phenolic profile of grape canes can also be altered by the extraction and scale-up process, where oxidation, degradation, or polymerization of proanthocyanidins have been observed, as well as the formation of phenolic aldehydes [2]. Phenolic compounds are known to be unstable and sensitive to high temperature, light, pH, and oxidative and degradative enzymes, which affects the phenolic profile of extracts. It is therefore important to find a strategy to protect phenolic compounds, and preserve their biological activities and properties. Enhancing their bioaccessibility and bioavailability and promoting their transport for absorption by the human body are also of great interest. Accordingly, new approaches, such as encapsulation with cyclodextrins, have been developed to overcome these drawbacks. [15].

Cyclodextrins are highly biocompatible and have been approved by the Food and Drug Administration (FDA) as safe for humans. Several research studies have described the complexation of phenolic compounds with cyclodextrins [15], which provides protection from environmental conditions and improves bioactive shelf-life. Cyclodextrins, which have a truncated cone-shaped structure, possess a hydrophobic interior and a hydrophilic outer surface. Complexation in cyclodextrins improves the water solubility of phenolic compounds, which is otherwise relatively poor. The stability of the complex is maintained via hydrophobic forces, van der Waals interactions and hydrogen bonding [15]. Therefore, encapsulation of the bioactive molecule by cyclodextrin alters the physicochemical properties of both agents. Nevertheless, the effect of cyclodextrins on the profile of phenolic compounds from grape canes after encapsulation needs further study.

In this context, we encapsulated a previously characterized grape cane pilot-plant extract [2] in HP-β-CD by a spray-drying technique, using maltodextrin (MD) as a coating material. The physicochemical properties and parameters of the encapsulation process were determined, and the complex formation was verified by scanning electron microscopy (SEM) and Fourier Transform Infrared Spectroscopy with diamond attenuated total reflectance (FTIR-ATR). In addition, the phenolic profile of the inclusion complex was investigated using LC-ESI-LTQ-Orbitrap-MS, and encapsulation efficiency and antioxidant capacity were determined. The microencapsulated extract is envisaged as a functional ingredient of food, cosmetics, biomaterials, and other biobased products.

2. Materials and Methods

2.1. Chemicals and Reagents

Gallic, 4-hydroxybenzoic, and ellagic acids, catechin, epicatechin, (E)-resveratrol, (E)-ε-viniferin, (E)-piceatannol, eriodictyol, taxifolin, quercetin, quercetin-3-O-glucoside, and quercetin-3-O-glucuronide were purchased from Sigma-Aldrich (St. Louis, MO, USA). Gallic acid and kaempferol-3-O-glucoside were acquired from Extrasynthèse (Genay, Auvergne-Rhône-Alpes, France). Isohopeaphenol and hopeaphenol were kindly given by the research group of Prof. Dr. Peter Winterhalter (Institute of Food Chemistry, Technical University Braunschweig, Lower Saxony, Germany). Light exposure was avoided when manipulating the standards.

HPLC-grade acetonitrile, formic acid, ethanol, and water were purchased from Merck (Darmstadt, Hesse, Germany). Ultrapure water was generated by a Milli-Q water purification system Millipore (Bedford, Massachusetts, USA). Potable ethanol (96%) from molasses employed for pilot-plant scale extraction was purchased from Oxiquim S.A. (Coronel, Concepción, Chile).

Hydroxypropyl beta-cyclodextrin (HP-β-CD, Kleptose®, HP oral grade) was purchased from Roquette Frères (Lestrem, Lillers, France). Maltodextrin (MD) (dextrose equivalent 16.5–19.5) was obtained from Merck KGaA, (Darmstadt, Hesse, Germany).

2.2. Pilot-Plant Scale Extraction

Grape cane extraction was performed on a pilot-plant scale (750 L) at 80 °C for 100 min, following our previously reported extraction process [2].

After winter pruning, we collected a total of 500 kg of wet sample grape canes (*V. vinifera* L. cv. Pinot noir) from plants in an organic vineyard at Viña De Neira, located in Ránquil, Itata Valley, Biobio Region, Chile. All samples were cut into 30–50 cm long pieces and stored for three months at room temperature (19 °C $\pm$ 5) and 30–70% relative humidity, according to previous reports by Riquelme et al. [16] and Patent [13]. We used 67 kg of dry grape canes from a total sample of 500 kg of wet samples for the pilot-plant extraction, before which grape canes were crushed in a Retsch grinder (model SM) at 300–2000 rpm until particle size was less than 1 cm. After extraction, the ethanol used as a solvent for the extraction was removed and recovered by distillation (absolute pressure 0.05 bar). The liquid extract was collected in a dark container and protected from light.

2.3. Preparation of Microcapsules by Spray-Drying

The GC$_{PPE}$ was in a mixed ethanol/water solution (30:70 v/v). HP-β-CD was used to prepare the microcapsules in a proportion of 2.2% w/v with the extract. MD in a proportion of 10% w/v was used as the coating material. HP-β-CD was slowly added to a beaker containing 200 mL of GC$_{PPE}$ to avoid its agglomeration. The mixture was continuously stirred at room temperature and protected from the light for 24 h, during which the encapsulation took place. Then, MD was added slowly, and the mixture was stirred for a few minutes [17]. The microencapsulated and unencapsulated GC$_{PPE}$ were dried using a Büchi Mini Spray Dryer B-290 (Büchi, Flawil, Switzerland). Inlet temperature was maintained at 130 °C, while the outlet air temperature was 71 °C. Air inlet and airflow were 35–40 m^3/h and 473 L/h, respectively. The spray dryer was equipped with a nozzle tip diameter of 0.7 mm and a peristaltic pump operated at 6–7 mL/min. The dried powder was collected and stored in an amber airtight container at 4 °C until analysis.

2.4. Determination of the Physical Properties of the Microencapsulated Powders

2.4.1. Moisture Content and Total Solids

The moisture content of the samples was calculated from the weight loss after heating the sample to 105 °C for 6 h [18].

2.4.2. Process Yield (PY%)

The yield of the powder process was calculated considering the number of solids introduced into the spray drying system and the powder obtained at the end of the technological process [19]. The results were determined according to Equation (1):

$$PY\ (\%) = \frac{\text{Powder after spray dry}}{\text{Solids introduced in the feeding}} \times 100 \tag{1}$$

2.4.3. Bulk Density

To determine the bulk density, about 10 g of the spray-dried sample was weighed, placed in a 25 mL graduated test tube, and the occupied volume was recorded [18].

2.4.4. Angle of Repose

The angle of repose was determined using a fixed funnel by the following Equation (2) as in Dadi et al. [18]:

$$\text{Angle of repose}\ (°) = \tan^{-1}(H/R) \tag{2}$$

where H is the height of the pile and R is its radius at the base.

2.4.5. Size Distribution

The particle size distribution was determined in a particle size analyzer by laser diffraction with a Mastersizer 3000 (Malvern Instruments, Worcestershire, UK). Samples were dispersed in MilliQ water (900 mL) under constant stirring (2300 RPM) using a Hydro EV dispersion unit to achieve a homogeneous suspension. The particle size distribution in the powder (*span*) was calculated using Equation (3):

$$Span = (d_{90} - d_{10})/d_{50} \tag{3}$$

where d_{90}, d_{10}, and d_{50} are the equivalent volume diameters at 90%, 10%, and 50% cumulative volume, respectively [20].

2.5. Scanning Electron Microscopy

The microcapsules and GC_{PPE} were analyzed by scanning electron microscopy (SEM) using the JSM 6380 LV system (JEOL Techniques Ltd., Tokyo, Japan). The microscope was operated at 20 kV accelerating voltage. The samples were coated with a gold layer of about 150 Å in thickness, using an Edwards S 150 sputter coater (Agar Scientific, Standsted, UK) [21].

2.6. Fourier Transform Infrared Analysis

The FTIR absorption spectra of the individual samples of GC_{PPE}, HP-β-CD, the inclusion complex (IC) (GC_{PPE}+HP-β-CD+MD), and the physical mixture were analyzed separately. The physical mixture (PM) (GC_{PPE}/HP-β-CD/MD) was prepared by accurately weighing HP-β-CD (100 mg), GC_{PPE} (100 mg), and MD (20 mg), which were ground in a mortar until the mixture was homogeneous. The resulting physical mixture was immediately analyzed by FTIR-ATR.

The FTIR spectra were recorded using a Bruker Alpha T FTIR (Bruker, Germany) spectrophotometer equipped with a diamond attenuated total reflectance (ATR) unit. The 32 scans were acquired over a spectral range of 4000–500 cm^{-1} with a resolution of 4 cm^{-1} [22,23]. All spectra were acquired and processed using the OPUS 7.0 software.

2.7. LC-ESI-LTQ-Orbitrap-MS Analyses

Liquid chromatography (LC) analysis was conducted using an Accela chromatograph (Thermo Scientific, Hemel Hempstead, UK) equipped with a quaternary pump, photodiode array detector, and thermostated autosampler. Chromatographic separation was performed in an Atlantis T3 column 2.1 × 100 mm, 3μm (Waters, Milford, MA, USA). Gradient elution of analytes was performed utilizing H$_2$O/0.1% HCOOH (solvent A) and CH$_3$CN (solvent B) at a continuous flow rate of 0.350 mL/min, and an injection volume of 5 µL. The following gradient was applied: 0 min, 2% B; 0–2 min, 8% B; 2–12 min, 20% B; 12–13 min, 30% B; 13–14 min, 100% B; 14–17 min, 100% B; 17–18 min, 2% B and the column was equilibrated to the initial conditions for 5 min [24].

The LC equipment was coupled to an LTQ-Orbitrap Velos mass spectrometer (Thermo Scientific, Hemel Hempstead, UK) employed for accurate mass measurements and equipped with an electrospray ionization (ESI) source operating in negative mode. The working parameters were as follows: source voltage, 4 kV; sheath gas, 20 a.u. (arbitrary units); auxiliary gas, 10 a.u.; sweep gas, 2 a.u.; and capillary temperature, 275 °C. Default values were used for most of the other acquisition parameters (FT Automatic gain control target $5 \cdot 10^5$ for MS mode and $5 \cdot 10^4$ for MS^n mode). Samples were analyzed in FTMS mode with a resolving power of 30,000 (FWHM at m/z 400) and data-dependent MS/MS events were acquired with a resolving power of 15,000. The most intense ions detected in the FTMS spectrum were chosen for the data-dependent scan. The parent ions were fragmented by high-energy C-trap dissociation by normalized collision energy of 35 V and an activation time of 10 ms. The mass range in FTMS mode was from m/z 100 to 1500. Instrument control and data recovery were conducted using Xcalibur 3.0 software (Thermo Fisher Scientific). The tentative identification of analytes was performed by comparing MS/MS spectra with fragments found in databases and the literature when no standard compound was available [25].

Individual compounds were semi-quantified using pure standards or the most similar compounds. Some analytes, such as glycosylated forms, dimers, or oligomers, were semi-quantified using the aglycone form of the monomer [2].

2.8. Encapsulation Efficiency

Encapsulation efficiency was calculated by considering the nonencapsulated compounds present in the GC$_{PPE}$ before the encapsulation process and the encapsulated compounds after the spray drying process. The identification and quantification of phenolic compounds from GC$_{PPE}$ were specified in a previous article [2] (see detail Supplementary Materials Table S1). Extraction of polyphenols from the microcapsules was performed according to the procedure of Robert et al. [26] with minor modifications. Before the analysis, samples were dissolved in highly pure deionized water containing 0.1% v/v formic acid (1 mg/mL). The GC$_{PPE}$ was centrifuged at 4000 rpm for 5 min at 4 °C. The supernatant was recovered, and the extraction procedure was repeated twice. The supernatants were combined and evaporated under nitrogen flow, and the residue was reconstituted in 0.1% aqueous formic acid (5 mL). The samples were filtered through 0.20 µm PTFE membrane filters (Waters Corporation, Milford, CT, USA) into an amber vial. Subsequently, both samples were analyzed using LC -ESI-LTQ-Orbitrap- MS to determine the individual degree of encapsulation according to the method described above (Section 2.7). We estimated the encapsulation efficiency (EE), according to Radünz et al. [27], based on Equation (4):

$$EE\ (\%) = \frac{\text{Phenolic compound of GCPPE}\ -\ \text{Phenolic compound of the capsule}}{\text{Phenolic compound of GCPPE}} \times 100 \qquad (4)$$

2.9. Antioxidant Capacity Assay

The assay of oxygen radical absorbance capacity using fluorescein (ORAC-FL) was carried out according to the method reported by Ou et al. [28]. The calibration curves were prepared with Trolox, and results reported as µmol Trolox equivalents (TE) by grams of dried weight (DW) (TE/g DW). All assays were performed in triplicate and protected from light.

2.10. Statistical Analysis

All samples were run at least three times, and results are expressed as means ± standard deviations. A p-value < 0.05 was considered statistically significant using Student's *t*-test with 95% confidence. Statistical analyses were determined using GraphPad Prism 8.0.1 (GraphPad Software, San Diego, CA, USA).

3. Results and Discussion

3.1. Physical Characterization of the Microencapsulated Powder

Table 1 shows the physicochemical properties and process parameters for the GC_{PPE} and inclusion complex (IC) (GC_{PPE}+HP-β-CD+MD). The process yield obtained by spray-drying for IC (GC_{PPE}+HP-β-CD+MD) was 83.8 ± 2.6%, which was two-fold higher than for GC_{PPE} alone (38.4 ± 1.2%). Similarly, the total solids increased 2.6-fold for the microencapsulated formulation. The high yield of powdered microparticles can be attributed to the rapid formation of the drying crust, which prevents the powder from adhering to the drying chamber [29]. Our result constitutes an improvement on the yield (64.5 ± 1.5%) reported by Davidov-Pardo et al. [30], who microencapsulated a grape seed extract using MD. The high values obtained in our work are promising for the development of industrial-scale applications.

Table 1. Physical characteristics and process parameters for the grape cane phenolic extract (GC_{PPE}) and inclusion complex (GC_{PPE}+HP-β-CD+MD).

	Process Parameters			Size Distribution		Property of Powders	
	Moisture (%)	Solids (g)	PY (%)	D50 (μm)	Span	Bulk Density (g/mL)	Angle of Repose(°)
GC_{PPE}	6.3 ± 1.5 [a]	11.0 ± 0.1 [a]	38.4 ± 1.2 [a]	17.5 ± 0.4 [a]	6.15 ± 0.1 [a]	0.10 ± 0.01 [a]	34.8 ± 0.5 [a]
IC (GC_{PPE}+HP-β-CD+MD)	7.2 ± 0.3 [a]	28.9 ± 0.1 [b]	83.8 ± 2.6 [b]	10.9 ± 0.9 [b]	6.14 ± 0.1 [a]	0.19 ± 0.01 [b]	36.9 ± 1.3 [a]

Results are expressed as means ± standard deviations, and values with different superscripts letters in a column indicate significant differences at $p < 0.05$.

The particle size distribution and median particle diameter were smaller in the IC (GC_{PPE}+HP-β-CD+MD) than in the GC_{PPE} alone; the particle sizes were 10.9 μm and 17.5 μm, respectively. According to the literature, the diameter of spray-dried particles depends on the properties of the material, the drying conditions, the atomization method used, and the concentration and viscosity of the encapsulated material [31]. The *span* values of the IC (GC_{PPE}+HP-β-CD+MD) and GC_{PPE} were very similar, 6.14 and 6.15, respectively, and were higher than those reported for an aqueous grape skin extract microencapsulated with Arabic gum, polydextrose, and partially hydrolyzed guar gum, which ranged from 1.91 to 5.99 [32]. A lower *span* value is a desirable result, as it indicates a more homogeneous particle size distribution [32].

The bulk density was 0.10 ± 0.01 g/mL and 0.19 ± 0.01 g/mL for the GC_{PPE} and IC (GC_{PPE}+HP-β-CD+MD), respectively, being lower than the values reported for encapsulated rosemary essential oil (0.25–0.34 g/mL) [31] or soy milk (0.21–0.22 g/mL) [20]. The slightly higher bulk density of the IC (GC_{PPE}+HP-β-CD+MD) vs. the GC_{PPE} indicates an improved powder flow, as a more densely packed powder reflects weaker forces between the particles [20]. Density is an important factor for the packaging, transportation, and marketing of a microencapsulated product. A dry product with high density can be stored in a smaller container compared to a less dense product [31]. The flowability of the samples was also determined by the angle of repose, which was 34.8° ± 0.5 for the GC_{PPE} and 36.9° ± 1.3 for the IC (GC_{PPE}+HP-β-CD+MD), showing no statistical difference between them. A similar value was obtained in a study on the microencapsulation of bioactive products from a *Moringa stenopetala* leaf extract using MD, where the angle of repose was 37.26° ± 1.01 [18].

3.2. Surface Morphology: SEM Analysis

SEM can be used to determine the surface morphology of materials and is recognized as an auxiliary method for monitoring the formation of inclusion complexes. The structure and size of the GC_{PPE} and IC (GC_{PPE}+HP-β-CD+MD) in the solid state obtained from the spray-drying process were analyzed through microscopy. Microcapsules should preferably have a slightly spherical form and a uniform and smooth cover with minimum fractures and signs of collapse [33]. The SEM micrographs showed that the GC_{PPE} was composed of a mixture of non-spherical particles with irregular surfaces and other larger spherical microparticles (Figure 1A–C). In contrast, the IC (GC_{PPE}+HP-β-CD+MD) was spherical, and without visible pores on a smooth surface; microparticles of a variable size but with similar morphology were observed together (Figure 1D–F). These significant morphological changes are probably due to a loss of crystallinity of the guest molecule after its inclusion in the cyclodextrin [34]. The SEM results provided evidence for the formation of the IC (GC_{PPE}+HP-β-CD+MD), which was subsequently supported by mass spectrometry and FTIR analysis.

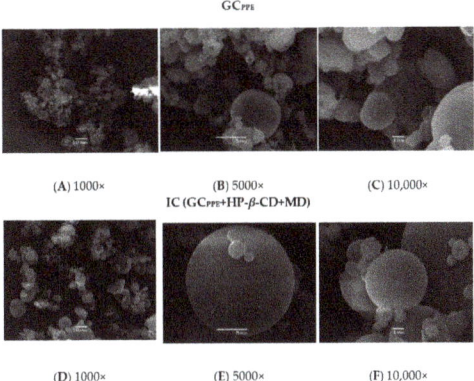

Figure 1. Scanning electron microscopy micrographs of the GC_{PPE} (**A–C**) and IC (GC_{PPE}+HP-β-CD+MD) (**D–F**) at different magnifications.

3.3. FTIR Analysis of Spray-Dried Powders

FTIR-ATR is a useful method to detect the formation of inclusion complexes, which are revealed by changes in the FTIR spectra, such as the reduction, disappearance, or shift of absorption bands, due to weak intermolecular interactions [35]. The FTIR spectra of the GC_{PPE}, HP-β-CD, IC (GC_{PPE}+HP-β-CD+MD), and physical mixture (PM) (GC_{PPE}/HP-β-CD/MD) are shown in Figure 2.

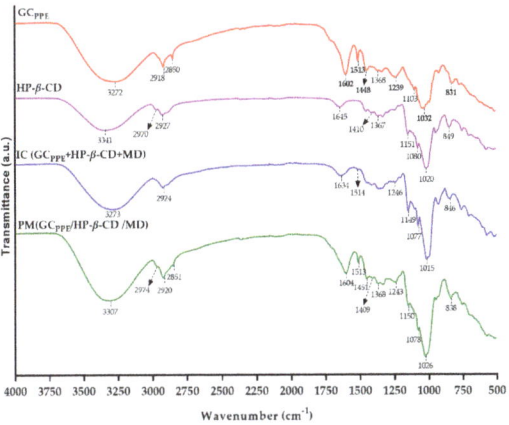

Figure 2. FTIR-ATR spectra of the GC_{PPE} (**red**), HP-β-CD (**purple**), IC (GC_{PPE}+HP-β-CD+MD) (**blue**), and PM (GC_{PPE}/HP-β-CD/MD) (**green**).

The FTIR spectrum of the GC$_{PPE}$ (Figure 2, red) showed specific bands associated with phenolic compounds (in bold). The peaks at 1602 cm^{-1} and 1448 cm^{-1} were owing to the C=C stretching vibration of the phenolic aromatic ring and the C–H bending vibrations of the CH$_2$ groups. The peak at 1239 cm^{-1} was assigned to C=O stretching of gallic or ellagic acid components due to the presence of hydrolyzable tannins [36]. The band at 1513 cm^{-1} was due to the C–C benzene skeletal vibrations of stilbenoids. The strong band at 1032 cm^{-1} is attributed to C–O stretching in phenolic compounds [22]. Moreover, the band at 831 cm^{-1} was due to the C-H out-of-plane bending vibrations of aromatic compounds. On the other hand, the signals at 2918 cm^{-1} and 2850 cm^{-1} were related to the CH$_2$ asymmetric and symmetric stretch vibration in aliphatic hydrocarbons [37]. Additionally, a peak at 1103 cm^{-1} was observed owing to the C–H in-plane bending vibration [36].

The FTIR spectrum of HP-β-CD (Figure 2, purple) showed bands at 3341 cm^{-1} due to O–H stretching vibrations and 2927 cm^{-1} due to C–H stretching vibrations. The peaks at 1645 cm^{-1} correspond to the bending of H–O–H, at 1151 cm^{-1} to C–O vibration, and at 1020 cm^{-1} to the C–O–C symmetric stretching vibration. The peak at 849 cm^{-1} was due to an α-type glycosidic bond [38]. The band at 1080 cm^{-1} was ascribed to C–C stretching vibrations, and the peak at 1410 cm^{-1} to C–C–H and O–C–H bending [39]. The presence of the hydroxypropyl group was recognized by a peak at 2970 cm^{-1} corresponding to the antisymmetric vibration of methyl groups. Additionally, a peak was observed at 1367 cm^{-1}, which was ascribed to the bending vibration of the methyl group [40].

The spectrum of the IC (GC$_{PPE}$+HP-β-CD+MD) (Figure 2, blue) shows that some of the characteristic peaks of the GC$_{PPE}$ and HP-β-CD have shifted, decreased, or disappeared. The bands at 1645 cm^{-1} and 1020 cm^{-1} of HP-β-CD have shifted to 1634 cm^{-1} and 1015 cm^{-1}, respectively, in the IC. Two peaks of the GC$_{PPE}$ at 1239 cm^{-1} and 1513 cm^{-1}, ascribed to the presence of hydrolysable tannins and stilbenes, respectively, have shifted to 1246 cm^{-1} and 1514 cm^{-1}, showing a sharp reduction in intensity due to the complexation, whereas the peak at 2850 cm^{-1} disappeared. Furthermore, the bands of the GC$_{PPE}$ at 1602 cm^{-1} and 1448 cm^{-1}, related to the phenolic aromatic ring, completely disappeared in the IC (GC$_{PPE}$+HP-β-CD+MD). These findings may indicate that the phenolic rings became embedded in the HP-β-CD cavities.

The FTIR spectrum of the PM (Figure 2, green) showed a simple overlap between the individual components of GC$_{PPE}$, HP-β-CD, and MD. Insignificant variations in intensity were detected, indicating a mixture of the three components without interactions between them.

3.4. Phenolic Profile of the IC (GC$_{PPE}$+HP-β-CD+MD) by LC-ESI-LTQ-Orbitrap-MS

We performed a targeted analysis of phenolic compounds in the IC (GC$_{PPE}$+HP-β-CD+MD) using LC-ESI-LTQ-Orbitrap-MS. Table 2 shows 42 identified compounds with their accurate mass, theoretical mass, retention times (min), molecular formula, error (ppm) between the found mass and the accurate mass of each compound, and the MS/MS fragment ions with their respective intensities used for identification. The identification was further supported by comparisons with mass spectra databases and the literature. In addition, 13 phenolic compounds were identified by comparing the retention times and their masses with pure standards. We have reported the fragmentation patterns of most of these compounds in previous studies using an analytical [8] and pilot-scale extraction [2].

Table 2. Identification of phenolic compounds in the IC (GC_{PPE}+HP-β-CD+MD), an adduct of β-CD and a complex between β-CD with (*epi*)catechin using LC-ESI-LTQ-Orbitrap-MS in negative mode.

Compounds	t_R (min)	Accurate Mass $[M-H]^-$	Theo. Mass	Error (ppm)	MS/MS Ions (Intensity)	Molecular Formula
Gallic acid *	4.33	169.0138	169.0142	−2.41	125.0236(100)	$C_7H_6O_5$
Monogalloyl-glucose	5.79	331.0662	331.0671	−2.65	169.0134(100), 125.0238(5)	$C_{13}H_{16}O_{10}$
[β-CD+HCOO]$^-$	7.21	1179.3679	1179.3680	−0.04	675.8472(100), 797.1872(80), 332.8267(80), 734.6646(80)	$C_{43}H_{72}O_{37}$
Protocatechuic acid-O-hexoside (1)	7.35	315.0718	315.0722	−1.07	153.0186(100), 109.0288(10)	$C_{13}H_{16}O_9$
Protocatechuic acid	7.55	153.0190	153.0193	−2.16	109.0288(100)	$C_7H_6O_4$
Protocatechuic acid-O-hexoside (2)	8.46	315.0717	315.0722	−1.30	153.0189(100), 109.0290(10)	$C_{13}H_{16}O_9$
Syringic acid hexoside	8.66	359.0933	359.0925	2.21	197.0445(100)	$C_{15}H_{20}O_{10}$
Protocatechuic aldehyde	9.32	137.0241	137.0244	−2.02	93.0338(100), 109.0285(70)	$C_7H_6O_3$
Procyanidin dimer (1)	10.34	577.1348	577.1351	−0.69	425.0864(100), 407.0760(70), 451.1019(45), 289.0706(35)	$C_{30}H_{26}O_{12}$
Procyanidin dimer (2)	10.77	577.1346	577.1351	−0.35	425.0861(100), 407.0757(60), 451.1017(35), 289.0705(25)	$C_{30}H_{26}O_{12}$
Catechin *	11.29	289.0722	289.0718	1.43	245.0817(100), 205.0504(40), 179.0348(20)	$C_{15}H_{14}O_6$
Complex [β-CD+(*epi*)catechin]$^-$	13.00	1423.5708	1423.5718	−0.71	1245.4839(100), 1303.5255(85), 1101.4424(60), 1365.5260(60), 1.083.4313(60)	$C_{57}H_{84}O_{41}$
Epicatechin *	13.29	289.0715	289.0718	−1.00	245.0812(100), 205.0499(40), 179.0343(15)	$C_{15}H_{14}O_6$
Restrytisol (A or B)	15.08	471.1441	471.1449	−1.86	255.0653(100), 377.1017(65), 349.1068(45)	$C_{28}H_{24}O_7$
Epicatechin gallate *	16.86	441.0821	441.0827	−1.30	289.0705(100), 169.0136(30)	$C_{22}H_{18}O_{10}$
Taxifolin *	17.00	303.0508	303.0510	−0.78	285.0397(100), 177.0186(15), 125.0238(10)	$C_{15}H_{12}O_7$
Astilbin (1)	17.36	449.1084	449.1089	−1.12	303.0495(100), 285.0392(85), 151.0028(30)	$C_{21}H_{22}O_{11}$
Taxifolin isomer	17.43	303.0503	303.0510	−2.32	285.0390(100), 177.0184(15), 125.0237(10)	$C_{15}H_{12}O_7$
Stilbenoid heterodimer (caraphenol B/C)	17.78	469.1292	469.1293	−0.06	451.1189(100), 363.0875(35), 375.0872(30), 281.0452(2)	$C_{28}H_{22}O_7$
(*E*)-Piceatannol *	17.95	243.0665	243.0663	0.79	225.0551(100), 201.0552(75), 159.0447(20)	$C_{14}H_{12}O_4$
Kaempferol-3-O-glucoside *	18.39	447.0931	447.0933	−0.49	284.0315(100), 285.0392(75), 327.0497(15), 255.0286(10)	$C_{21}H_{20}O_{11}$
Ethyl protocatechuate	18.49	181.0504	181.0506	−1.02	153.0187(100), 152.0110(15), 109.0289(5)	$C_9H_{10}O_4$
Dihydrokaempferol-O-rhamnoside	18.77	433.1139	433.1140	−0.38	269.0446(100), 287.0550(40), 259.0603(15)	$C_{21}H_{22}O_{10}$
Eriodictyol-O-glucoside	19.01	449.1089	449.1089	−0.04	287.0552(100), 151.0031(5)	$C_{21}H_{22}O_{11}$
Stilbenoid dimer	19.16	469.1292	469.1293	−0.07	363.0857(100), 375.0857(20), 451.1168(5)	$C_{28}H_{22}O_7$
Undefined (tetrahydroxyisoflavanone)	19.18	287.0561	287.0561	−0.18	259.0602(100), 243.0652(20), 201.0547(5)	$C_{15}H_{12}O_6$
Viniferin diglycoside	19.25	777.2397	777.2400	−0.36	615.1854(100), 453.1330(80)	$C_{40}H_{42}O_{16}$
Myricetin	19.39	317.0301	317.0303	−0.63	178.9981(100), 151.1032(45), 192.0058(10)	$C_{15}H_{10}O_8$
Pallidol	19.50	453.1349	453.1344	2.24	359.0922(100), 265.0499(10)	$C_{28}H_{22}O_6$
(*E*)-resveratrol *	20.23	227.0715	227.0714	0.77	185.0607(100), 183.0816(35), 159.0814(30)	$C_{14}H_{12}O_3$
Stilbenoid dimer (resveratrol dimer)	20.56	453.1351	453.1344	1.57	359.0916(100), 289.0861(5)	$C_{28}H_{22}O_6$

Table 2. Cont.

Compounds	t_R (min)	Accurate Mass $[M - H]^-$	Theo. Mass	Error (ppm)	MS/MS Ions (Intensity)	Molecular Formula
Stilbenoid dimer (maackin)	20.62	485.1242	485.1242	0.01	375.0865(100), 467.1125(15), 363.0863(10)	$C_{28}H_{22}O_8$
Resveratrol-O-hexoside	20.79	615.1868	615.1872	−0.62	453.1334(100)	$C_{34}H_{32}O_{11}$
Eriodictyol *	20.97	287.0558	287.0561	−0.92	151.0031(100), 135.0446(5)	$C_{15}H_{12}O_6$
Stilbenoid tetramer (viniferol E)	21.11	923.2680	923.2674	0.68	905.2576(100), 707.1898(55), 801.2318(50), 881.2573(20), 783.2209(10)	$C_{56}H_{44}O_{13}$
Stilbenoid hexamer (viniphenol A)	21.12	679.1969$[M - 2H]^{2-}$	679.1974$[M - 2H]^{2-}$	−1.12	585.1543(100), 491.1126(10), 359.0914(5), 905.2584(2.5), 453.1333(2)	$C_{84}H_{64}O_{18}$
Quercetin *	21.16	301.0354	301.0354	0.06	178.9987(100), 151.0038(80)	$C_{15}H_{10}O_7$
Hopeaphenol *	21.34	905.2607	905.2627	−2.24	811.2152(100), 717.1741(65), 451.1173(10)	$C_{56}H_{42}O_{12}$
Isohopeaphenol *	21.47	905.2588	905.2568	2.15	811.2178(100), 717.1752(40)	$C_{56}H_{42}O_{12}$
(E)-ε-viniferin *	21.59	453.1348	453.1344	1.03	359.0928(100), 347.0923(50), 435.1238(25)	$C_{28}H_{22}O_6$
(E)-ω-viniferin	21.72	453.1346	453.1344	0.62	359.0927(100), 347.0925(50), 435.1234(30), 411.1233(20)	$C_{28}H_{22}O_6$
Stilbenoid tetramer (vitisin A/B/C/D)	21.90	905.2599	905.2627	−3.05	799.2157(100), 887.2482(70), 811.2158(50), 359.0913(35), 545.1599(15)	$C_{56}H_{42}O_{12}$

* Compounds identified via matching with authentic standards. t_R, retention times. Isomers are presented in brackets. Stilbenoid hexamer appeared as a doubly charged ion.

Of the 42 compounds identified in the present work, 22 had been previously detected in both the analytical and pilot-plant extractions [2,8]: gallic acid (m/z 169.0138, −2.41 ppm), monogalloyl-glucose (m/z 331.0662, −2.65 ppm), protocatechuic acid-O-hexoside (1) (m/z 315.0718, −1.07 ppm), protocatechuic acid (m/z 153.0190, −2.16 ppm), protocatechuic acid-O-hexoside (2) (m/z 315.0717, −1.30 ppm), syringic acid hexoside (m/z 359.0933, 2.21 ppm), catechin (m/z 289.0722, 1.43 ppm), epicatechin (m/z 289.0715, −1.00 ppm), restrytisol (A or B) (m/z 471.1441, −1.86 ppm), taxifolin (m/z 303.0508, −0.78 ppm), astilbin (m/z 449.1084, −1.12 ppm), stilbenoid heterodimer (caraphenol B/C) (m/z 469.1292, −0.06 ppm), eriodictyol-O-glucoside (m/z 449.1089, −0.04 ppm), stilbenoid dimer (m/z 469.1292, −0.07 ppm), pallidol (m/z 453.1349, 2.24 ppm), (E)-resveratrol (m/z 227.0715, 0.77 ppm), stilbenoid dimer (resveratrol dimer) (m/z 453.1351, 1.57 ppm), resvertarol-O-hexoside (m/z 615.1868, −0.62 ppm), eriodictyol (m/z 287.0558, −0.92 ppm), hopeaphenol (m/z 905.2607, −2.24 ppm), isohopeaphenol (m/z 905.2588, 2.15 ppm) and (E)-ε-viniferin (m/z 453.1348, 1.03 ppm). These findings indicate that these compounds are stable through each step of the production process, including microencapsulation.

On the other hand, seven compounds identified in the microencapsulated extract had previously been detected only in the analytical extraction [8]: procyanidin dimer (1) (m/z 577.1348, −0.69 ppm), procyanidin dimer (2) (m/z 577.1346, −0.35 ppm), epicatechin gallate (m/z 441.0821, −1.30 ppm), (E)-piceatannol (m/z 243.0665, 0.79 ppm), viniferin diglycoside (m/z 777.2397, −0.36 ppm), (E)-ω-viniferin (m/z 453.1346, 0.62 ppm), and stilbenoid tetramer (vitisin A/B/C/D) (m/z 905.2599, −3.05 ppm). Thus, although these compounds were not detected in the pilot-scale extraction, they were recovered after the microencapsulation process. Similar to our results, the incorporation of β-CD enabled an effective and selective recovery of flavan-3-ols [41] and stilbenes [42], resulting in a cleaner analytical extract phenolic profile. Additionally, three compounds previously detected in the pilot-scale extraction [2]: protocatechuic aldehyde (m/z 137.0241, −2.02 ppm), kaempferol-3-O-glucoside (m/z 447.0931, −0.49 ppm), and ethyl protocatechuate (m/z 181.0504, −1.02 ppm), were also recovered and identified in the microencapsulated extract.

Finally, ten compounds were identified only in the IC (GC$_{PPE}$+HP-β-CD+MD), and not in the analytical or pilot extracts: five flavonoids, three stilbenes, an adduct of β-CD, and a complex of β-CD with (epi)catechin.

Flavonoids. The taxifolin isomer (m/z 303.0503, −2.32 ppm) showed ions at m/z 285.0390, owing to the initial loss of a water molecule and ions at m/z 177.0184 and 125.0237 due to cleavage of the C ring, respectively. Quercetin (m/z 301.0354, 0.06 ppm) was identified and confirmed by comparison with a pure standard. Myricetin (m/z 317.0301, −0.63 ppm), tentatively identified by its fragmentation pattern, produced ions at m/z 178.9981 ($^{1,2}A^-$) and 151.0032 ($^{1,3}A^-$) due to retro-Diels–Alder fragmentation [43], and at m/z 192.0058 due to the loss of the B ring. Although they were not identified in our previous studies, these compounds have been recovered, detected, and quantified by other authors using microwave-assisted, subcritical water, and conventional extraction techniques [44].

Dihydrokaempferol-O-rhamnoside (engeletin) (m/z 433.1139, −0.38 ppm), a compound previously reported in grape stems [45], was tentatively identified and showed fragment ions at m/z 269.0446, 287.0550, and 259.0603. An undefined tetrahydroxyisoflavanone (m/z 287.0561, −0.18 ppm) gave product ions at m/z 259.0602, 243.0652, and 201.0547, and was provisionally identified as 2,6,7,4′-tetrahydroxyisoflavanone, based on the exact mass and fragmentation pattern. However, as dihydrokaempferol and eriodictyol chalcone have similar structures, the identity of this compound could not be accurately defined using our spectrometric approach.

Stilbenes. A stilbenoid dimer (maackin, Figure 3A) (m/z 485.1242, 0.01 ppm) showed product ions at m/z 467.1125, 375.0865, and 363.0863, which were generated by the loss of a water molecule (18 Da), resorcinol (110 Da), and 2-hydroxy-4-methylenecyclohexa-2,5-dienone (122 Da), respectively [46]. This compound has a structure consisting of two piceatannol units, and the most likely assignment is maackin A, which was identified previously in *V. vinifera* stalks [47].

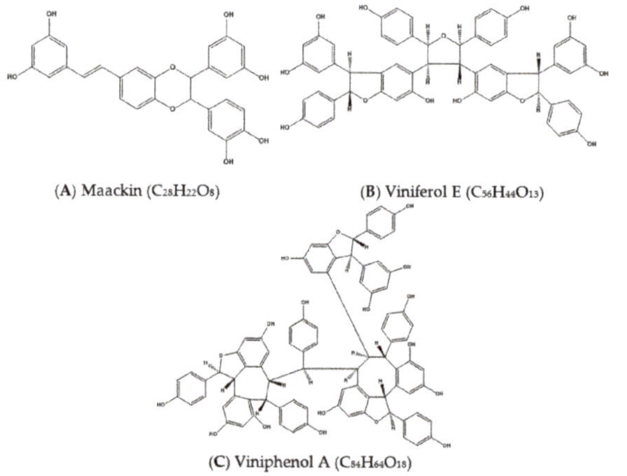

Figure 3. Representative stilbenes tentatively identified in the microencapsulated extract. (**A**) Maackin ($C_{28}H_{22}O_8$); (**B**) Viniferol E ($C_{56}H_{44}O_{13}$); (**C**) Viniphenol A ($C_{84}H_{64}O_{18}$).

A stilbenoid tetramer (Figure 3B) (m/z 923.2680, 0.68 ppm) was tentatively identified as viniferol E and yielded product ions at m/z 905.2576, 881.2573, 801.2318, 783.2209 and 707.1898. The product ions at m/z 905.2576 and 881.2573 were due to a loss of H_2O (18 Da) and C_2H_2O (42 Da), respectively, and at m/z 801.2318 probably to the loss of the group $C_7H_6O_2$ (122 Da). The ion at m/z 801.2318 was further fragmented to ions at m/z 783.2209 and m/z 707.1898 by the loss of a water molecule (18 Da) and a phenol group (94 Da), respectively. Viniferol E was previously detected and quantified from grapevine canes by subcritical water extraction [48].

A stilbenoid hexamer, viniphenol A (Figure 3C) (m/z 679.1969 [M − 2H]$^{2-}$, −1.12 ppm), was detected as a doubly charged ion with product ions at m/z 905.2584, 585.1543, 491.1126, 453.1333 and 359.0914. The product ion at m/z 905.2584 shows the presence of a stilbenoid tetramer molecule, probably formed by the loss of a stilbenoid dimer (454 Da) from the deprotonated hexamer. The high-intensity product ion at m/z 585.1543 could be attributed to the loss of a phenol group (94 Da) from the stilbenoid trimer (m/z 679). The product ion at m/z 585.1543 underwent fragmentation to ions at m/z 491.1126 and m/z 359.0914, which could be attributed to the loss of a phenol group (94 Da) and a stilbenoid dimer (226 Da), respectively. Finally, a low intensity stilbenoid dimer fragment was observed at m/z 453.1333. Viniphenol A was previously isolated from *V. vinifera* stalks by centrifugal partition chromatography, while its structure was proposed based on the analysis of spectroscopic data and molecular modeling under NMR conditions [47].

According to the supplier, the HP-β-CD used for encapsulation has a maximum β-CD impurity of 1.5%. The presence of [β-CD+ HCOO]$^-$ (m/z 1179.3679, −0.04 ppm) and a complex of (*epi*)catechin with β-CD [β-CD +(*epi*)catechin]$^-$ (m/z 1423.5708, −0.71 ppm) were detected and identified by comparison with the mass spectra reported by Żyżelewicz et al. [49]. The detection of this complex by mass spectrometry confirmed the interaction between polyphenols and cyclodextrins, in agreement with the SEM and FTIR analysis.

3.5. Encapsulation Efficiency

GC$_{PPE}$ is a complex mixture of various compounds that have different physical and chemical properties and abilities to form interactions and bind within the HP-β-CD cavity. As shown earlier, the phenolic compounds in the microencapsulated extract are phenolic acids and derivatives, flavonoids, and stilbenes. The individual EE (%) for twenty of these compounds was calculated and presented in Table 3. The other compounds were identified but not quantified due to their low amounts (see detail Supplementary Materials Table S1).

Table 3. Encapsulation efficiency (%).

Compounds	EE (%)
Phenolic Acids and Derivatives	
Protocatechuic acid-O-hexoside (1)	54.3 ± 2.4
Protocatechuic acid	66.8 ± 9.8
Ethyl protocatechuate	69.7 ± 5.5
Protocatechuic aldehyde	75.6 ± 4.2
Gallic acid	83.4 ± 0.6
Caftaric acid	87.1 ± 2.2
Ellagic acid pentoside	87.6 ± 0.5
Hydroxybenzaldehyde	95.8 ± 0.6
Weighted Average	**81.5 ± 0.7**
Flavonoids	
Eriodictyol	74.2 ± 4.7
Quercetin-O-glucoside	81.8 ± 7.0
Quercetin-3-O-glucuronide	88.0 ± 3.1
Astilbin (1)	82.7 ± 2.3
Astilbin (2)	92.0 ± 1.1
Weighted Average	**85.2 ± 2.4**
Stilbenes	
(E)-resveratrol	32.7 ± 2.8
Stilbenoid tetramer (hopeaphenol/isohopeaphenol)	65.8 ± 8.6
Pallidol	74.6 ± 3.8
(E)-ε-viniferin	76.8 ± 3.5
Stilbene dimer (resveratrol dimer)	94.1 ± 2.1
Stilbenoid heterodimer (caraphenol B/C)	96.8 ± 0.4
Restrytisol (A or B)	97.0 ± 0.6
Weighted Average	**78.6 ± 1.9**
Total Weighted Average	**80.5 ± 1.1**

Results are given as means ± standard deviations. For each phenolic class, we calculated the weighted average of the concentration of the respective metabolites. The total weighted average was calculated by weighting all quantified compounds. All encapsulation efficiencies were calculated according to Equation (4) (Section 2.8).

The average EE for all the analyzed compounds was 80.5 ± 1.1%. Several authors have studied the encapsulation of phenolic compounds from wine byproducts using different encapsulation materials. Davidov-Pardo et al. [30] reported a similar polyphenol EE of 82% for a commercial grape seed extract microencapsulated by spray drying using MD as the wall material. Moschona and Liakopoulou-Kyriakides [50] found a low EE of 55% to 79% for grape marc and lees phenolic extracts from white and red wine encapsulated with alginate and chitosan. Lavelli and Sri Harsha [51] also reported a low EE of 68% in a study where alginate hydrogel was used as an agent to encapsulate phenolic compounds from grape skin. Another study using grape skin extracts, encapsulated in water-in-oil-in-water (W/O/W) double emulsions, observed an improved EE of 87.74 ± 3.12% for anthocyanins [52]. The EE depends on a variety of factors, such as the technique used, the solubility and size of the guest molecule relative to the cavity of the host, the concentrations of the host and guest molecules, the binding constant within the guest and host, etc. [53]. As we did not find a similar study in the literature that employed the same raw material, encapsulant, and analytical methods to determine the concentration of guest molecules, we also analyzed the individual encapsulation of twenty compounds present in the IC (GC$_{PPE}$+HP-β-CD+MD).

The average EE for the phenolic acids and derivatives was 81.5 ± 0.7%, the lowest value being obtained for protocatechuic acid-O-hexoside 1 (54.3 ± 2.4%) and the highest for hydroxybenzaldehyde (95.8 ± 0.6%). The EE for protocatechuic acid was 66.8 ± 9.8%, higher than the value reported by Taofiq et al. [54], who determined an EE of 50.3% for protocatechuic acid encapsulated by the atomization/coagulation technique, using sodium alginate in combination with calcium chloride ($CaCl_2$) to promote alginate gelation. The EE for hydroxybenzaldehyde obtained here is higher than the value (46.50%) reported for *p*-hydroxybenzaldehyde in a β-CD inclusion complex [55].

The EE for gallic acid, 83.4 ± 0.6%, was slightly higher than the value reported by da Rosa et al. [56], who determined an EE of 80.0 ± 1.4% for microencapsulated gallic acid using β-CD and the lyophilization method. However, Olga et al. [53] obtained a higher EE of 89.22% in a complex with HP-β-CD, a result that was reduced to 77.34% when the gallic acid was co-encapsulated with *trans*-ferulic acid. The authors suggested a possible antagonistic relationship between the two phenols in the HP-β-CD cavity.

The EE of 87.6 ± 0.5% observed for ellagic acid pentoside was considerably higher than the 55.2% reported for ellagic acid (aglycone) encapsulated with polyvinyl alcohol [57]. The EE for caftaric acid—a hydroxycinnamic acid—was very high (87.1 ± 2.2%) in comparison with the values reported for other hydroxycinnamic acids, such as the essential oil-encapsulated chitosan-*p*-coumaric acid (42 ± 1%) [58] and chlorogenic acid (77.5%) in β-CD nanosponges [59], respectively. Finally, to our knowledge, this is the first time that the EE for ethyl protocatechuate (69.7 ± 5.5%) and protocatechuic aldehyde (75.6 ± 4.2%) has been reported.

The average EE for flavonoids was 85.2 ± 2.4%. The values for astilbin isomer (2) and astilbin (1) were particularly high, 92.0 ± 1.1% and 82.7 ± 2.3%, respectively. Zheng and Zhang [60] reported a lower EE of 80.1% for astilbin encapsulated with zein–caseinate nanoparticles by the antisolvent method.

Quercetin-3-*O*-glucuronide (88.0 ± 3.1%) and quercetin-*O*-glucoside (81.8 ± 7.0) also had high EE values. Tchabo et al. [61] obtained a lower EE (63.90–66.45%) for quercetin-3-*O*-glucoside in a spray-dried mulberry leaf extract prepared with MD, yet the EE was higher (91.71–93.95%) when sodium carboxymethyl cellulose was used. To the best of our knowledge, no EE values have been previously reported for quercetin-3-*O*-glucuronide.

The EE for eriodictyol was 74.2 ± 4.7%, similar to the almost 70% reported for naringenin (another flavanone) in polymer PLGA nanoparticles prepared by an emulsion-diffusion-evaporation method [62].

The stilbene group had the lowest mean EE (78.6 ± 1.9%), mainly because of the low value obtained for (*E*)-resveratrol (32.7 ± 2.8%). Nevertheless, previous studies have shown that the cyclodextrin encapsulation of resveratrol increases its solubility, stability, and bioactivity (antioxidant and anticarcinogenic properties) [15]. Furthermore, the HP-β-CD complex exhibits a strong H-bonding interaction with molecules such as oxyresveratrol [63]. In our study, we found a high EE (97.0 ± 0.6%) for restrytisol (A or B), an oxidized resveratrol dimer, which may be related to the van der Waals force interaction and hydrogen bonding between the guest compound and the HP-β-CD.

A good EE was obtained for stilbene dimers such as pallidol (74.6 ± 3.8%), (*E*)-ε-viniferin (76.8 ± 3.5%), and an undefined resveratrol dimer (94.1 ± 2.1%). Previously, ε-viniferin, a resveratrol dimer, was encapsulated in phospholipid-based multi-lamellar liposomes called spherulites or onions. This formulation gave a lower EE (58 ± 3%) than in our study, but increased the water solubility of this stilbene more than five-fold and provided protection against its UV-induced isomerization [64]. Finally, the low EE (65.8 ± 8.6%) obtained for the stilbenoid tetramer is probably due to its large size and polar surface area.

3.6. Antioxidant Capacity

Several authors suggest that the antioxidant capacity of phenolic compounds is improved by encapsulation in cyclodextrins [15]. The ORAC value has been applied to standardize the antioxidant activity of herbal extracts and foods, and is widely used as an accurate indicator of antioxidant activity in vivo [65].

The antioxidant capacity of the IC (GC_{PPE}+HP-β-CD+MD) by ORAC-FL was 5300 ± 472 µmol TE/g DW, similar to the 4612 ± 155 µmol TE/g DW reported in our previous GC$_{PPE}$ study [2]. The GC$_{PPE}$ microencapsulated with HP-β-CD retains its antioxidant capacity and its formulation may improve stability, solubility, and bioavailability for applications in the food, cosmetic and pharmaceutical industries.

4. Conclusions

Phenolic compounds from a grape cane pilot-plant extract were successfully encapsulated in an inclusion complex (GC$_{PPE}$+HP-β-CD+MD). The microencapsulated extract was rich in stilbenes, especially oligomers, flavonoids, and phenolic acids. A complex of (*epi*)catechin and β-CD was detected by mass spectrometry, which confirmed the interaction between polyphenols and cyclodextrin. The formation of the inclusion complex was also supported by FTIR-ATR and SEM analyses. HP-β-CD provided a high EE for phenolic compounds, with a mean of 80.5 ± 1.1%, the highest values being obtained for restrytisol (97.0 ± 0.6%), stilbenoid heterodimer (1) (96.8 ± 0.4%) and hydroxybenzaldehyde (95.8 ± 0.6%), and the lowest for (*E*)-resveratrol (32.7 ± 2.8%). The antioxidant capacity of the inclusion complex was similar to the unencapsulated extract. Considering the protection afforded the phenolic compounds by the inclusion complex, it is expected that the formulation may improve their stability, solubility, and bioavailability in water-soluble applications for the food and pharmaceutical industries.

5. Patents

The preparation of grape canes before extraction in the pilot plant followed the procedure described in the Patent [13]. The formulation of the microparticles of phenolic compounds from the extract of grape canes of *V. vinifera* and the application of HP-β-CD to form the IC and the use of MD as a coating agent are described in the Patent [17].

Supplementary Materials: The following are available online at https://www.mdpi.com/article/10.3390/antiox10071130/s1, Table S1: Identification and quantification of phenolic compounds from GC$_{PPE}$ using LC-ESI-LTQ-Orbitrap-MS in negative mode.

Author Contributions: Conceptualization, D.E.-A., C.G.-G., C.M., J.L., D.v.B. and A.V.-Q.; methodology, D.E.-A., J.A.-G., J.S., J.L.-C., C.G.-G., J. L and A.V.-Q.; validation, A.V.-Q. and C.G.-G.; formal analysis, D.E.-A., J.A.-G., J.L.-C. and J.S.; investigation, D.E.-A., J.A.-G. and J.S.; resources, A.V.-Q., C.G.-G., R.M.L.-R., C.M., J.L. and D.v.B.; data curation, D.E.-A., J.A.-G. and J.S.; writing—original draft preparation, D.E.-A.; writing—review and editing, D.E.-A., A.V.-Q., C.G.-G., C.M., J.A.-G., D.v.B., R.M.L.-R. and J.S.; visualization, D.E.-A.; supervision, A.V.-Q., C.G.-G., R.M.L.-R. and C.M.; project administration, D.E.-A., A.V.-Q., C.G.-G., C.M. and R.M.L.-R.; funding acquisition, D.E.-A., C.M., A.V.-Q. and R.M.L.-R. All authors have read and agreed to the published version of the manuscript.

Funding: This research was supported by the Agencia Nacional de Investigación y Desarrollo (ANID)/PCI-REDES170051; ANID PIA/APOYO CCTE AFB170007; CORFO 14 IDL2- 30156, from Chile; CICYT [AGL2016-75329-R], CIBEROBN from the Instituto de Salud Carlos III, ISCIII from the Ministerio de Ciencia, Innovación y Universidades, (AEI/FEDER, UE) and Generalitat de Catalunya (GC) [2017SGR 196]. Danilo Escobar-Avello is grateful to ANID/Scholarship Program/DOCTORADO BECAS CHILE/2017—72180476. Javier Avendaño-Godoy is grateful to ANID/Scholarship Program/DOCTORADO NACIONAL/2020-21202096. Anna Vallverdú-Queralt thanks to the Ministry of Science, Innovation and Universities for the Ramon y Cajal contract (RYC-2016-19355).

Institutional Review Board Statement: Not applicable.

Informed Consent Statement: Not applicable.

Data Availability Statement: Data is contained within the article and Supplementary Materials.

Acknowledgments: The authors would like to thank the CCiT-UB for the mass spectrometry equipment and to Yamil Neira from Viña de Neira, who provided the grape cane samples.

Conflicts of Interest: Rosa M. Lamuela-Raventós reports receiving lecture fees from Cerveceros de España and Wine in moderation. She also received lecture fees and travel support from Adventia. The other authors declare no conflicts of interest. The funders had no role in the design of the study;

in the collection, analyses, or interpretation of data; in the writing of the manuscript, or in the decision to publish the results.

References

1. Melini, V.; Melini, F.; Luziatelli, F.; Ruzzi, M. Functional ingredients from agri-food waste: Effect of inclusion thereof on phenolic compound content and bioaccessibility in bakery products. *Antioxidants* **2020**, *9*, 1216. [CrossRef]
2. Escobar-Avello, D.; Mardones, C.; Saéz, V.; Riquelme, S.; von Baer, D.; Lamuela-Raventós, R.M.; Vallverdú-Queralt, A. Pilot-plant scale extraction of phenolic compounds from grape canes: Comprehensive characterization by LC-ESI-LTQ-Orbitrap-MS. *Food Res. Int.* **2021**, *143*, 110265. [CrossRef] [PubMed]
3. Sun, X.; Wei, X.; Zhang, J.; Ge, Q.; Liang, Y.; Ju, Y.; Zhang, A.; Ma, T.; Fang, Y. Biomass estimation and physicochemical characterization of winter vine prunings in the Chinese and global grape and wine industries. *Waste Manag.* **2020**, *104*, 119–129. [CrossRef]
4. Troilo, M.; Difonzo, G.; Paradiso, V.M.; Summo, C.; Caponio, F. Bioactive Compounds from Vine Shoots, Grape Stalks, and Wine Lees: Their Potential Use in Agro-Food Chains. *Foods* **2021**, *10*, 342. [CrossRef] [PubMed]
5. Malinowska, M.A.; Billet, K.; Drouet, S.; Munsch, T.; Unlubayir, M.; Tungmunnithum, D.; Giglioli-Guivarc'H, N.; Hano, C.; Lanoue, A. Grape cane extracts as multifunctional rejuvenating cosmetic ingredient: Evaluation of sirtuin activity, tyrosinase inhibition and bioavailability potential. *Molecules* **2020**, *25*, 2203. [CrossRef]
6. Díaz-Galindo, E.P.; Nesic, A.; Cabrera-Barjas, G.; Mardones, C.; Von Baer, D.; Bautista-Baños, S.; Garcia, O.D. Physical-chemical evaluation of active food packaging material based on thermoplastic starch loaded with grape cane extract. *Molecules* **2020**, *25*, 1306. [CrossRef]
7. Dávila, I.; Gullón, P.; Andrés, M.A.; Labidi, J. Coproduction of lignin and glucose from vine shoots by eco-friendly strategies: Toward the development of an integrated biorefinery. *Bioresour. Technol.* **2017**, *244*, 328–337. [CrossRef]
8. Escobar-Avello, D.; Lozano-Castellón, J.; Mardones, C.; Pérez, A.J.; Saéz, V.; Riquelme, S.; Von Baer, D.; Vallverdú-Queralt, A. Phenolic profile of grape canes: Novel compounds identified by LC-ESI-LTQ-orbitrap-MS. *Molecules* **2019**, *24*, 3763. [CrossRef]
9. Sáez, V.; Pastene, E.; Vergara, C.; Mardones, C.; Hermosín-gutiérrez, I.; Gómez-Alonso, S.; Gómez, M.V.; Theoduloz, C.; Riquelme, S.; Baer, D. Von Oligostilbenoids in *Vitis vinifera* L. Pinot Noir grape cane extract: Isolation, characterization, in vitro antioxidant capacity and anti-proliferative effect on cancer cells. *Food Chem.* **2018**, *265*, 101–110. [CrossRef] [PubMed]
10. Pinasseau, L.; Vallverdú-Queralt, A.; Verbaere, A.; Roques, M.; Meudec, E.; Le Cunff, L.; Péros, J.-P.; Ageorges, A.; Sommerer, N.; Boulet, J.-C.; et al. Cultivar diversity of grape skin polyphenol composition and changes in response to drought investigated by LC-MS based metabolomics. *Front. Plant Sci.* **2017**, *8*. [CrossRef]
11. Lambert, C.; Richard, T.; Renouf, E.; Bisson, J.; Waffo-Téguo, P.; Bordenave, L.; Ollat, N.; Mérillon, J.M.; Cluzet, S. Comparative analyses of stilbenoids in canes of major *Vitis vinifera* L. cultivars. *J. Agric. Food Chem.* **2013**, *61*, 11392–11399. [CrossRef]
12. Vergara, C.; Von Baer, D.; Mardones, C.; Wilkens, A.; Wernekinck, K.; Damm, A.; Macke, S.; Gorena, T.; Winterhalter, P. Stilbene levels in grape cane of different cultivars in southern Chile: Determination by HPLC-DAD-MS/MS method. *J. Agric. Food Chem.* **2012**, *60*, 929–933. [CrossRef] [PubMed]
13. Mardones, C.; Von Baer, D.; Vergara, C.; Escobar-Avello, D.; Fuentealba, C.; Riquelme, S. Un procedimiento para aumentar el contenido de estilbenos, esencialmente resveratrol, en sarmientos provenientes de las podas de Vitis vinífera. 2014. CL2014003417A1. Available online: https://patentscope.wipo.int/search/es/detail.jsf?docId=CL174158667 (accessed on 13 July 2021).
14. Zwingelstein, M.; Draye, M.; Besombes, J.L.; Piot, C.; Chatel, G. Trans-Resveratrol and trans-ε-Viniferin in Grape Canes and Stocks Originating from Savoie Mont Blanc Vineyard Region: Pre-extraction Parameters for Improved Recovery. *ACS Sustain. Chem. Eng.* **2019**, *7*, 8310–8316. [CrossRef]
15. Pinho, E.; Grootveld, M.; Soares, G.; Henriques, M. Cyclodextrins as encapsulation agents for plant bioactive compounds. *Carbohydr. Polym.* **2014**, *101*, 121–135. [CrossRef]
16. Riquelme, S.; Sáez, V.; Escobar, D.; Vergara, C.; Fuentealba, C.; Bustamante, L.; von Baer, D.; Jara, P.; Lamperti, L.; Mardones, C. Bench-scale extraction of stilbenoids and other phenolics from stored grape canes (*Vitis vinifera*): Optimization process, chemical characterization, and potential protection against oxidative damage. *J. Chil. Chem. Soc.* **2019**, *64*, 4414–4420. [CrossRef]
17. Mardones, C.; Escobar-Avello, D.; Luengo-Contreras, J.; Avendaño-Godoy, J.; Ortega-Medina, E. Formulación de micropartículas de extracto de sarmientos de Vitis vinífera útil en las industrias nutracéutica y cosmética, además de su proceso de elaboración y sus usos. 2017. CL2017002902A1. Available online: https://patentscope.wipo.int/search/es/detail.jsf?docId=CL250856147 (accessed on 13 July 2021).
18. Dadi, D.W.; Emire, S.A.; Hagos, A.D.; Eun, J.B. Physical and Functional Properties, Digestibility, and Storage Stability of Spray- and Freeze-Dried Microencapsulated Bioactive Products from Moringa stenopetala Leaves Extract. *Ind. Crops Prod.* **2020**, *156*, 112891. [CrossRef]
19. Leyva-Jiménez, F.J.; Lozano-Sánchez, J.; de la Luz Cádiz-Gurrea, M.; Fernández-Ochoa, Á.; Arráez-Román, D.; Segura-Carretero, A. Spray-drying microencapsulation of bioactive compounds from lemon verbena green extract. *Foods* **2020**, *9*, 1547. [CrossRef] [PubMed]
20. Jinapong, N.; Suphantharika, M.; Jamnong, P. Production of instant soymilk powders by ultrafiltration, spray drying and fluidized bed agglomeration. *J. Food Eng.* **2008**, *84*, 194–205. [CrossRef]

21. Gómez-Gaete, C.; Retamal, M.; Chávez, C.; Bustos, P.; Godoy, R.; Torres-Vergara, P. Development, characterization and in vitro evaluation of biodegradable rhein-loaded microparticles for treatment of osteoarthritis. *Eur. J. Pharm. Sci.* **2017**, *96*, 390–397. [CrossRef]
22. García, D.E.; Delgado, N.; Aranda, F.L.; Toledo, M.A.; Cabrera-Barjas, G.; Sintjago, E.M.; Escobar-Avello, D.; Paczkowski, S. Synthesis of maleilated polyflavonoids and lignin as functional bio-based building-blocks. *Ind. Crops Prod.* **2018**, *123*, 154–163. [CrossRef]
23. Santos, J.; Delgado, N.; Fuentes, J.; Fuentealba, C.; Vega-Lara, J.; García, D.E. Exterior grade plywood adhesives based on pine bark polyphenols and hexamine. *Ind. Crops Prod.* **2018**, *122*, 340–348. [CrossRef]
24. Escobar-Avello, D.; Olmo-Cunillera, A.; Lozano-Castellón, J.; Marhuenda-Muñoz, M.; Vallverdú-Queralt, A. A Targeted Approach by High Resolution Mass Spectrometry to Reveal New Compounds in Raisins. *Molecules* **2020**, *25*, 1281. [CrossRef]
25. Sasot, G.; Martínez-Huélamo, M.; Vallverdú-Queralt, A.; Mercader-Martí, M.; Estruch, R.; Lamuela-Raventós, R.M. Identification of phenolic metabolites in human urine after the intake of a functional food made from grape extract by a high resolution LTQ-Orbitrap-MS approach. *Food Res. Int.* **2017**, *100*, 435–444. [CrossRef]
26. Robert, P.; Gorena, T.; Romero, N.; Sepulveda, E.; Chavez, J.; Saenz, C. Encapsulation of polyphenols and anthocyanins from pomegranate (*Punica granatum*) by spray drying. *Int. J. Food Sci. Technol.* **2010**, *45*, 1386–1394. [CrossRef]
27. Radünz, M.; Hackbart, H.C.D.S.; Bona, N.P.; Pedra, N.S.; Hoffmann, J.F.; Stefanello, F.M.; Da Rosa Zavareze, E. Glucosinolates and phenolic compounds rich broccoli extract: Encapsulation by electrospraying and antitumor activity against glial tumor cells. *Colloids Surf. B Biointerfaces* **2020**, *192*. [CrossRef]
28. Ou, B.; Chang, T.; Huang, D.; Prior, R.L. Determination of total antioxidant capacity by oxygen radical absorbance capacity (ORAC) using fluorescein as the fluorescence probe: First action 2012.23. *J. AOAC Int.* **2013**, *96*, 1372–1376. [CrossRef] [PubMed]
29. Urzúa, C.; González, E.; Dueik, V.; Bouchon, P.; Giménez, B.; Robert, P. Olive leaves extract encapsulated by spray-drying in vacuum fried starch–gluten doughs. *Food Bioprod. Process.* **2017**, *106*, 171–180. [CrossRef]
30. Davidov-Pardo, G.; Arozarena, I.; Marín-Arroyo, M.R. Optimization of a Wall Material Formulation to Microencapsulate a Grape Seed Extract Using a Mixture Design of Experiments. *Food Bioprocess Technol.* **2013**, 941–951. [CrossRef]
31. Fernandes, R.V.D.B.; Borges, S.V.; Botrel, D.A. Gum arabic/starch/maltodextrin/inulin as wall materials on the microencapsulation of rosemary essential oil. *Carbohydr. Polym.* **2014**, *101*, 524–532. [CrossRef] [PubMed]
32. Siede Kuck, L.; Pelayo Zapata Noreña, C. Microencapsulation of grape (*Vitis labrusca* var. Bordo) skin phenolic extract using gum Arabic, polydextrose, and partially hydrolyzed guar gum as encapsulating agents. *Food Chem.* **2015**. [CrossRef]
33. Tolun, A.; Altintas, Z.; Artik, N. Microencapsulation of grape polyphenols using maltodextrin and gum arabic as two alternative coating materials: Development and characterization. *J. Biotechnol.* **2016**, *239*, 23–33. [CrossRef]
34. Li, W.; Ran, L.; Liu, F.; Hou, R.; Zhao, W.; Li, Y.; Wang, C.; Dong, J. Preparation and Characterisation of Polyphenol-HP-β-Cyclodextrin Inclusion Complex that Protects Lamb Tripe Protein against Oxidation. *Molecules* **2019**, *24*, 4487. [CrossRef] [PubMed]
35. Franco, P.; De Marco, I. Preparation of non-steroidal anti-inflammatory drug/β-cyclodextrin inclusion complexes by supercritical antisolvent process. *J. CO2 Util.* **2021**, *44*. [CrossRef]
36. Grasel, F.D.S.; Ferrão, M.F.; Wolf, C.R. Development of methodology for identification the nature of the polyphenolic extracts by FTIR associated with multivariate analysis. *Spectrochimica Acta Part A Mol. Biomol. Spectrosc.* **2016**, *153*, 94–101. [CrossRef]
37. Santos, J.; Pereira, J.; Ferreira, N.; Paiva, N.; Ferra, J.; Magalhães, F.D.; Martins, J.M.; Dulyanska, Y.; Carvalho, L.H. Valorisation of non-timber by-products from maritime pine (*Pinus pinaster*, Ait) for particleboard production. *Ind. Crop. Prod.* **2021**, *168*, 113581. [CrossRef]
38. Rodríguez-López, M.I.; Mercader-Ros, M.T.; López-Miranda, S.; Pellicer, J.A.; Pérez-Garrido, A.; Pérez-Sánchez, H.; Núñez-Delicado, E.; Gabaldón, J.A. Thorough characterization and stability of HP-β-cyclodextrin thymol inclusion complexes prepared by microwave technology: A required approach to a successful application in food industry. *J. Sci. Food Agric.* **2019**, *99*, 1322–1333. [CrossRef] [PubMed]
39. Reddy, C.K.; Jung, E.S.; Son, S.Y.; Lee, C.H. Inclusion complexation of catechins-rich green tea extract by β-cyclodextrin: Preparation, physicochemical, thermal, and antioxidant properties. *LWT* **2020**, *131*, 109723. [CrossRef]
40. Yuan, C.; Liu, B.; Liu, H. Characterization of hydroxypropyl-β-cyclodextrins with different substitution patterns via FTIR, GC-MS, and TG-DTA. *Carbohydr. Polym.* **2015**, *118*, 36–40. [CrossRef]
41. Ratnasooriya, C.C.; Rupasinghe, H.P.V. Extraction of phenolic compounds from grapes and their pomace using b-cyclodextrin. *Food Chem.* **2012**, *134*, 625–631. [CrossRef]
42. Chemat, F.; Vian, M.A.; Cravotto, G. Green extraction of natural products: Concept and principles. *Int. J. Mol. Sci.* **2012**, *13*, 8615–8627. [CrossRef]
43. Lin, Y.; Wu, B.; Li, Z.; Hong, T.; Chen, M.; Tan, Y.; Jiang, J.; Huang, C. Metabolite identification of myricetin in rats using HPLC coupled with ESI-MS. *Chromatographia* **2012**, *75*, 655–660. [CrossRef]
44. Moreira, M.M.; Barroso, M.F.; Porto, J.V.; Ramalhosa, M.J.; Švarc-Gajić, J.; Estevinho, L.; Morais, S.; Delerue-Matos, C. Potential of Portuguese vine shoot wastes as natural resources of bioactive compounds. *Sci. Total Environ.* **2018**, *634*, 831–842. [CrossRef] [PubMed]
45. Goufo, P.; Singh, R.K.; Cortez, I. A reference list of phenolic compounds (Including stilbenes) in grapevine (*Vitis vinifera* L.) roots, woods, canes, stems, and leaves. *Antioxidants* **2020**, *9*, 398. [CrossRef]

46. Moss, R.; Mao, Q.; Taylor, D.; Saucier, C. Investigation of monomeric and oligomeric wine stilbenoids in red wines by ultra-high-performance liquid chromatography/electrospray ionization quadrupole time-of-flight mass spectrometry. *Rapid Commun. Mass Spectrom.* **2013**, *27*, 1815–1827. [CrossRef]
47. Papastamoulis, Y.; Richard, T.; Nassra, M.; Badoc, A.; Krisa, S.; Harakat, D.; Monti, J.P.; Mérillon, J.M.; Waffo-Teguo, P. Viniphenol A, a complex resveratrol hexamer from *Vitis vinifera* stalks: Structural elucidation and protective effects against amyloid-β-induced toxicity in PC12 cells. *J. Nat. Prod.* **2014**, *77*, 213–217. [CrossRef]
48. Gabaston, J.; Leborgne, C.; Valls, J.; Renouf, E.; Richard, T.; Waffo-Teguo, P.; Mérillon, J.M. Subcritical water extraction of stilbenes from grapevine by-products: A new green chemistry approach. *Ind. Crops Prod.* **2018**, *126*, 272–279. [CrossRef]
49. Żyżelewicz, D.; Oracz, J.; Kaczmarska, M.; Budryn, G.; Grzelczyk, J. Preparation and characterization of inclusion complex of (+)-catechin with β-cyclodextrin. *Food Res. Int.* **2018**, *113*, 263–268. [CrossRef]
50. Moschona, A.; Liakopoulou-Kyriakides, M. Encapsulation of biological active phenolic compounds extracted from wine wastes in alginate-chitosan microbeads. *J. Microencapsul.* **2018**, *35*, 229–240. [CrossRef]
51. Lavelli, V.; Sri Harsha, P.S.C. Microencapsulation of grape skin phenolics for pH controlled release of antiglycation agents. *Food Res. Int.* **2019**, *119*, 822–828. [CrossRef]
52. Xu, W.; Yang, Y.; Xue, S.J.; Shi, J.; Lim, L.T.; Forney, C.; Xu, G.; Bamba, B.S.B. Effect of in vitro digestion on water-in-oil-in-water emulsions containing anthocyanins from grape skin powder. *Molecules* **2018**, *23*, 2808. [CrossRef] [PubMed]
53. Olga, G.; Styliani, C.; Ioannis, R.G. Coencapsulation of Ferulic and Gallic acid in hp-b-cyclodextrin. *Food Chem.* **2015**, *185*, 33–40. [CrossRef]
54. Taofiq, O.; Heleno, S.A.; Calhelha, R.C.; Fernandes, I.P.; Alves, M.J.; Barros, L.; González-Paramás, A.M.; Ferreira, I.C.F.R.; Barreiro, M.F. Phenolic acids, cinnamic acid, and ergosterol as cosmeceutical ingredients: Stabilization by microencapsulation to ensure sustained bioactivity. *Microchem. J.* **2019**, *147*, 469–477. [CrossRef]
55. Zeng, Z.; Fang, Y.; Ji, H. Side chain influencing the interaction between β-cyclodextrin and vanillin. *Flavour Fragr. J.* **2012**, *27*, 378–385. [CrossRef]
56. Da Rosa, C.G.; Borges, C.D.; Zambiazi, R.C.; Nunes, M.R.; Benvenutti, E.V.; da Luz, S.R.; D'Avila, R.F.; Rutz, J.K. Microencapsulation of gallic acid in chitosan, β-cyclodextrin and xanthan. *Ind. Crops Prod.* **2013**, *46*, 138–146. [CrossRef]
57. Bala, I.; Bhardwaj, V.; Hariharan, S.; Kharade, S.V.; Roy, N.; Kumar, M.N.V.R. Sustained release nanoparticulate formulation containing antioxidant-ellagic acid as potential prophylaxis system for oral administration. *J. Drug Target.* **2006**, *14*, 27–34. [CrossRef]
58. Silva Damasceno, E.T.; Almeida, R.R.; de Carvalho, S.Y.B.; de Carvalho, G.S.G.; Mano, V.; Pereira, A.C.; de Lima Guimarães, L.G. Lippia origanoides Kunth. essential oil loaded in nanogel based on the chitosan and p-coumaric acid: Encapsulation efficiency and antioxidant activity. *Ind. Crops Prod.* **2018**, *125*, 85–94. [CrossRef]
59. Ramírez-Ambrosi, M.; Caldera, F.; Trotta, F.; Berrueta, L.; Gallo, B. Encapsulation of apple polyphenols in β-CD nanosponges. *J. Incl. Phenom. Macrocycl. Chem.* **2014**, *80*, 85–92. [CrossRef]
60. Zheng, D.; Zhang, Q.F. Bioavailability enhancement of astilbin in rats through zein− caseinate nanoparticles. *J. Agric. Food Chem.* **2019**, *67*, 5746–5753. [CrossRef] [PubMed]
61. Tchabo, W.; Ma, Y.; Kaptso, G.K.; Kwaw, E.; Cheno, R.W.; Xiao, L.; Osae, R.; Wu, M.; Farooq, M. Process Analysis of Mulberry (Morus alba) Leaf Extract Encapsulation: Effects of Spray Drying Conditions on Bioactive Encapsulated Powder Quality. *Food Bioprocess Technol.* **2019**, *12*, 122–146. [CrossRef]
62. Maity, S.; Chakraborti, A.S. Formulation, physico-chemical characterization and antidiabetic potential of naringenin-loaded poly D, L lactide-co-glycolide (N-PLGA) nanoparticles. *Eur. Polym. J.* **2020**, *134*, 109818. [CrossRef]
63. He, J.; Zheng, Z.P.; Zhu, Q.; Guo, F.; Chen, J. Encapsulation mechanism of oxyresveratrol by β-cyclodextrin and hydroxypropyl-β-cyclodextrin and computational analysis. *Molecules* **2017**, *22*, 1801. [CrossRef] [PubMed]
64. Courtois, A.; Garcia, M.; Krisa, S.; Atgié, C.; Sauvant, P.; Richard, T.; Faure, C. Encapsulation of ε-viniferin in onion-type multi-lamellar liposomes increases its solubility and its photo-stability and decreases its cytotoxicity on Caco-2 intestinal cells. *Food Funct.* **2019**, *10*, 2573–2582. [CrossRef] [PubMed]
65. Jaffe, R.; Mani, J. *Polyphenolics Evoke Healing Responses: Clinical Evidence and Role of Predictive Biomarkers*, 2nd ed.; Elsevier Inc.: Amsterdam, The Netherlands, 2018; ISBN 9780128130063.

Article

Thermo-Responsive Gel Containing Hydroxytyrosol-Chitosan Nanoparticles (Hyt@tgel) Counteracts the Increase of Osteoarthritis Biomarkers in Human Chondrocytes

Anna Valentino [1,†], Raffaele Conte [2,†], Ilenia De Luca [1], Francesca Di Cristo [3], Gianfranco Peluso [1,4], Michela Bosetti [5,*] and Anna Calarco [1,*]

1. Research Institute on Terrestrial Ecosystems (IRET)—CNR, Via Pietro Castellino 111, 80131 Naples, Italy; anna.valentino@iret.cnr.it (A.V.); ilenia.deluca@iret.cnr.it (I.D.L.); gianfranco.peluso@unicamillus.org (G.P.)
2. AMES Group Polydiagnostic Center, Via Padre Carmine Fico, 24, 80013 Casalnuovo di Napoli, Italy; raffaele.conte86@tiscali.it
3. Elleva Pharma s.r.l., Via P. Castellino 111, 80131 Naples, Italy; francesca.dicristo@ellevapharma.com
4. UniCamillus, International Medical University, 00131 Rome, Italy
5. Dipartimento di Scienze del Farmaco, Università del Piemonte Orientale "A. Avogadro", Largo Donegani, 2, 28100 Novara, Italy
* Correspondence: michela.bosetti@uniupo.it (M.B.); anna.calarco@cnr.it (A.C.)
† These authors contributed equally to this work.

Citation: Valentino, A.; Conte, R.; De Luca, I.; Di Cristo, F.; Peluso, G.; Bosetti, M.; Calarco, A. Thermo-Responsive Gel Containing Hydroxytyrosol-Chitosan Nanoparticles (Hyt@tgel) Counteracts the Increase of Osteoarthritis Biomarkers in Human Chondrocytes. *Antioxidants* 2022, 11, 1210. https://doi.org/10.3390/antiox11061210

Academic Editors: Li Liang and Hao Cheng

Received: 20 May 2022
Accepted: 19 June 2022
Published: 20 June 2022

Publisher's Note: MDPI stays neutral with regard to jurisdictional claims in published maps and institutional affiliations.

Copyright: © 2022 by the authors. Licensee MDPI, Basel, Switzerland. This article is an open access article distributed under the terms and conditions of the Creative Commons Attribution (CC BY) license (https://creativecommons.org/licenses/by/4.0/).

Abstract: Although osteoarthritis (OA) is a chronic inflammatory degenerative disease affecting millions of people worldwide, the current therapies are limited to palliative care and do not eliminate the necessity of surgical intervention in the most severe cases. Several dietary and nutraceutical factors, such as hydroxytyrosol (Hyt), have demonstrated beneficial effects in the prevention or treatment of OA both in vitro and in animal models. However, the therapeutic application of Hyt is limited due to its poor bioavailability following oral administration. In the present study, a localized drug delivery platform containing a combination of Hyt-loading chitosan nanoparticles (Hyt-NPs) and in situ forming hydrogel have been developed to obtain the benefits of both hydrogels and nanoparticles. This thermosensitive formulation, based on Pluronic F-127 (F-127), hyaluronic acid (HA) and Hyt-NPs (called Hyt@tgel) presents the unique ability to be injected in a minimally invasive way into a target region as a freely flowing solution at room temperature forming a gel at body temperature. The Hyt@tgel system showed reduced oxidative and inflammatory effects in the chondrocyte cellular model as well as a reduction in senescent cells after induction with H_2O_2. In addition, Hyt@tgel influenced chondrocytes gene expression under pathological state maintaining their metabolic activity and limiting the expression of critical OA-related genes in human chondrocytes treated with stressors promoting OA-like features. Hence, it can be concluded that the formulated hydrogel injection could be proposed for the efficient and sustained Hyt delivery for OA treatment. The next step would be the extraction of "added-value" bioactive polyphenols from by-products of the olive industry, in order to develop a green delivery system able not only to enhance the human wellbeing but also to promote a sustainable environment.

Keywords: hydroxytyrosol-chitosan nanoparticles; injectable hydrogel; anti-inflammatory; anti-oxidative; osteoarthritis

1. Introduction

Osteoarthritis (OA) is a chronic inflammatory degenerative disease affecting millions of people worldwide. OA leads to cartilage deterioration, inflammation of the synovial membrane, and subchondral bone sclerosis due to abnormal bone remodeling caused by an overproduction of enzymes degrading the extracellular matrix [1–3]. This disease considerably reduces the quality of life for patients and is associated with pain, transient morning stiffness, and crepitus felt in a joint on moving it [4–6]. A large body of evidence supports the involvement of inflammation and reactive oxygen species (ROS) production by chondrocytes in OA cartilage [7–9].

To date, numerous pharmacological and non-pharmacological therapies have been developed for the management of OA. However, the current therapies are limited to palliative care and do not exclude the necessity of surgical intervention [10,11]. Under alternative or adjuvant therapeutic schemes, regular dietary intake of natural functional foods containing polyphenols and other phytochemicals such as fruits, vegetables, whole grains, legumes, and olive oil [12–17] has been associated to a protective role in chronic disease prevention such as OA because of their anti-inflammatory and antioxidant properties [18].

Among all, Hydroxytyrosol (Hyt), a polyphenol found mainly in olive oil and raw olives, exerts strong antioxidant activity (as a potent radical scavenger and metals chelator) and acts as an anti-inflammatory as well as antithrombotic, antitumor, antimicrobial, and neuroprotective agent [19–22].

Fucelli et al. demonstrated the ability of Hyt to reduce inflammatory markers, such as Cyclooxygenase-2 (COX2) and Tumor Necrosis Factor alfa (TNF-α) and reduces oxidative stress on a mouse model of systemic inflammation [23]. Pre-treatment of Balb/c mice with Hyt (40 and 80 mg/Kg b.w.) prevented all lipopolysaccharide-induced effects and decreased oxidative stress. In another study, Cetrullo et al. demonstrated that Hyt inhibits the inflammatory response in vascular endothelial cells, macrophages, and monocytes [24]. Furthermore, Hyt reduces oxidative stress and damage, exerts pro-survival and anti-apoptotic actions, and favorably influences the expression of critical OA-related genes in human chondrocytes treated with stressors promoting OA-like features [25].

However, the amount of Hyt obtained through its natural sources' consumption is considerably lower than the recommended daily intake able to exert its claimed health-promoting properties [26]. Furthermore, the local therapeutic concentration of Hyt following oral administration is limited due to its poor bioavailability and enzyme degradation. Encapsulation of Hyt could be a functional alternative strategy to preserve its the biological activity and to ensure controlled release of the latter increasing the residence time inside the joint. Chitosan biopolymer has been extensively used as a matrix for the encapsulation of a wide range of natural products due to its beneficial properties including biodegradability, biocompatibility, and low cost [27]. Moreover, the ionic gelation method allows to obtain drug-loaded chitosan nanoparticles with a controlled size and satisfactory encapsulation capacity, protecting polyphenols from enzymatic oxidation or degradation. However, polymeric particles present some significant limitations such as initial burst release, escape from the joint's cavity, and in vivo rejection.

To overcome the above-mentioned drawbacks, in the present work, a localized drug delivery platform containing a combination of Hyt-loading chitosan nanoparticles (Hyt-NPs) and in situ forming hydrogel have been developed to derive the benefits of both hydrogels and nanoparticles. This thermo-sensitive formulation, based on Pluronic F-127 (F-127), hyaluronic acid (HA), and Hyt-NPs (called Hyt@tgel) presents the unique ability to be injected in a minimally invasive way into a target region as a freely flowing solution. When the temperature rises near the body temperature of 37 °C, hydrogel has in situ sol-to-gel transition accommodating the shape to the geometry of the treated area.

HA, a naturally polysaccharide, represents one of the largest components of the extracellular matrix of articular cartilage and plays an important endogenous role in the protection of articular cartilage decreasing the gene expression of inflammatory cytokines. Moreover, HA degrade ECM enzymes with stimulating in vitro chondrocytes proliferation, and chondrogenesis by directing mesenchymal stromal cells (MSCs) differentiation and increasing type 2 collagen production [28–30]. Although HA represents a conventional treatment in knee OA management, several lines of clinical evidence have questioned the effectiveness of such therapies due to HA prompt in vivo degradation mediated by hyaluronidases and oxidative stress [31,32]. To prolong HA residence time and confer optimized product functionality, Pluronic F-127 (F-127) consisting of hydrophilic poly (ethylene oxide) (PEO) and hydrophobic poly (propylene oxide) (PPO) (PEO-PPO-PEO) was added. Moreover, reported by Young-seok Jung and colleagues [33], the addition of high-molecular-weight HA (Mw: ~1000 kDa) increases the mechanical strength of thermos-responsive hydrogel hindering the interactions between water and poloxamer molecules due to HA-assisted inter-micellar packing. Starting from the results obtained in the work of Young-seok Jung, nanocomposite hydrogel formulations (Hyt@tgels) were optimized to ensure gelation around 37 °C, as well as allowing Hyt release exerting antioxidant and anti-inflammatory activity on an in vitro induced inflammatory environment mimicking OA.

Based on the above, the developed platform may serve as both a Hyt delivery system and as a tissue engineering scaffold to stimulate the regeneration of a lesioned tissue and to prevent chondrocytes senescence providing an alternative and potentially more effective loco-regional approach to manage OA.

2. Materials and Methods

2.1. Materials

Hyt (>98% purity), chitosan medium molecular weight (50,000–190,000 Da, 75–85% deacetylated, viscosity < 200 mPa.s, 1% in acetic acid), lactic acid (DL-Lactic acid, powder), sodium tripolyphosphate (TPP, technical grade), Pluronic F-127, and Fluorescein isothiocyanate (FITC), 3,3′,5,5′-Tetramethylbenzidine (TMB), Thiobarbitoric acid (TBA), Dichloro-dihydro-fluorescein diacetate (DCFH-DA) and Ultrapure HA in the form of sodium hyaluronate medium molecular weight were purchased from Sigma Aldrich (Milan, Italy) and used as received. All other reagents used in the experiment were of analytical grade and when not indicated were purchased from Sigma Aldrich (Milan, Italy).

2.2. Preparation and Physico-Chemical Characterization of Hyt-Loading Nanoparticles (Hyt-NPs)

A series of three Hyt-loading nanoparticles (Hyt-NPs) with varying chitosan concentration (0.1%, 0.5%, and 1% w/w) were obtained, with slight modifications, according to the well-known ionotropic gelation method [34]. Briefly, chitosan solution in 1% (v/v) lactic acid was prepared and stirred overnight at room temperature. TPP (5 mg/mL) and Hyt (10 mg) were dissolved in double distilled water to achieve different CS:TPP mass ratios. All solutions were filtered using 0.45 µm pore size membrane filters. TPP/Hyt solution was added dropwise into the chitosan solution under magnetic stirring (750 rpm) until a translucent Hyt-NPs suspension was formed. The suspension was stirred for 1 h at 1000 rpm at 25 °C to allow complete interaction. Then, the solution containing nanoparticles was ultrasonicated for 5 min at 40 kHz. Finally, Hyt-NPs were collected by cooling centrifugation (Frontiers 5718R, OHAUS, Milan, Italy) at 15,000 rpm for 45 min at 4 °C and washed with deionized water. FITC-loaded NPs were obtained by adding hydrophilic fluorescent probe into the TPP aqueous phase instead of Hyt, and the NPs prepared as described previously. Blank NPs were produced as negative control.

Particle Size (hydrodynamic diameter), polydispersity index (PDI), and zeta potential measurements were carried out on freshly prepared samples as reported in Conte et al. [35]. All samples were diluted in deionized water and measured at 25 °C using a Malvern Zetasizer (Malvern Instruments Ltd., Malvern, UK). The reported data are an average value of three measurements of the same sample. The particle size was confirmed by NanoSight NS300 Nanoparticles Tracking Analysis (NTA, Malvern Instruments, Amesbury, United Kingdom, UK). The fresh nanoparticle dispersions were centrifuged at 14,000 rpm for 30 min (5718R, OHAUS, Nänikon, Switzerland). The amount of drug entrapped in NPs was determined in triplicate indirectly by analyzing the amount of free Hyt in supernatant. The free Hyt in supernatant was quantified as described in Section 2.4.4 paragraph.

The encapsulation efficiencies of a series of Hyt-loaded nanoparticles were determined based on the following equation:

$$\text{Encapsulation Efficiency (EE \%)} = \frac{\text{Total amount of Hyt loaded} - \text{Free Hyt in supernatant}}{\text{Total amount of Hyt loaded}} \times 100$$

2.3. Hyt-Loaded Hydrogel (Hyt@tgel) Preparation

Injectable hydrogels were prepared according to the cold method as reported by [33]. Briefly, HA (100 mg) and Pluronic F-127 concentrations of 12–25% (w/v, 5 mL) were mixed in double distilled water at a temperature below 4 °C to form hydrogels. The polymer solution was left for at least 24 h to ensure the complete dissolution. Meanwhile, lyophilized Hyt-NPs (1, 5, and 10 mg) were added to the polymer dispersion and stirred for 1 h at 4 °C.

2.4. Hyt@tgel Characterization

2.4.1. Gelation Time and Syringeability

The sol-gel phase transition (gelation time) of Hyt@tgels was determined by modified test tube inversion method [36]. An aliquot (1.0 mL) of each sample was prepared in a glass tube and then placed in a low temperature digital water bath. The solution was heated at the rate of 0.5 °C/min and after each minute the glass vial rotated 90° to check the gelling of the sample. Tsol-gel was determined as the temperature at which the gel did not exhibit gravitational flow during a period of 2 min when the tube was reversed. Averages and standard deviations of each sample were determined in triplicate.

Syringeability was assayed after injection of Hyt@tgel through a syringe with a 30-gauge needle. The solutions which were easily passed from the syringe were termed as pass and the solutions which were difficult to pass were termed as fail.

2.4.2. Mechanical Strength Test

The mechanical strength of Hyt@tgels was measured in relation to its viscosity with a Brookfield viscometer (RVDV-II + P, Brookfield, WI, USA) set at 200 rpm router speed with increasing temperatures (20–65 °C, Equilibrium time: 1 °C/2 min (>35 °C) or 5 °C/10 min (<35 °C)).

2.4.3. Short-Term Stability Studies

The physico-chemical stability of Hyt@tgels was determined upon 14-day storage at different temperatures (4.0 ± 0.5 °C and 25.0 ± 0.5 °C). At predetermined times, aliquots were centrifuged (12,000× g, 4 h, 20 °C) to separate nanoparticles from the hydrogel. All samples were analyzed for particle size, PDI, and % drug entrapment efficiency, and the results were compared with the initial values. The Hyt stability during storage was confirmed by HPLC analysis as described in Section 2.4.4 paragraph.

2.4.4. In Vitro Hyt Release

The cumulative Hyt release from the hydrogel formulations was determined using the dialysis bag method in phosphate buffer saline (PBS, pH 7.4). The Hyt@tgels formulations (1 mL) were sealed in pre-swollen cellulose membrane dialysis bags (3.5–5.0 kDa cut-off, Spectrum) and immersed into 5 mL of PBS buffer (pH 7.4) in a water bath at 37 °C shaken at 100 rpm for 5 days. At set time intervals, 5 mL of the release media was collected for Hyt analysis and replaced with the same volume of fresh PBS to maintain the sink conditions. Hyt released in the PBS media from the hydrogels was measured with liquid chromatography–tandem mass spectrometry (LC-MS/MS) as reported by [37]. The LC-MS/MS system consisted of a Shimadzu NexeraXR UHPLC (Shimadzu Italy, Milan, Italy) coupled to an LCMS 8060 turbo spray ionization triple-quadrupole mass spectrometer (LCMS 8060, Shimadzu Italy, Milan, Italy). The whole system was controlled by Lab Solution software. Separation of analytes was achieved using a 2.6 μm Kinetex polar C18 column (Phenomenex, Torrance, CA, USA). The mobile phase included Buffer A (0.1% formic acid in water) and Buffer B (0.1% formic acid in acetonitrile) in isocratic flow. The total run time was 7 min for each injection. The mass spectrometer was operated in the turbo-spray mode with negative ion detection. The detection and quantification of Hyt was accomplished by multiple reaction monitoring (MRM) with the transitions m/z 153.05 → 123.0 (quantifier); 153.05 → 93.0 (qualifier). The instrumental parameters tuned to maximize the MRM signals were nebulizing gas flow 3 L/min, heating gas flow 10 L/min, interface temperature 370 °C, DL temperature 250 °C, heat block temperature 450 °C, and drying gas flow 10 L/min.

2.5. In Vitro Cell Studies

2.5.1. Cell Culture and Treatment

Human chondrocyte cells line C20A4 was obtained from American Type Culture Collection (ATCC, Manassas, VA, USA). It was maintained at 37 °C in a humidified atmosphere containing 5% CO_2 in Dulbecco's modified Eagle's Medium/Nutrient Mixture F-12 (DMEM/F12) supplemented with 10% fetal bovine serum (FBS), 1% L-glutamine, 50 U/mL penicillin, 50 mg/mL streptomycin, 50 µg/mL ascorbic acid, and 50 µM α-tocopherol (Euroclone, Milan, Italy). Cells were tested for contamination, including *Mycoplasma*, and used within 2–4 months. All experiments were performed with an 80% confluent monolayer. The protective effects of Hyt were studied with the acute toxicity model by pre-treating cells with Hyt@tgel for 24 and 96 h followed by 24 h H_2O_2 (230 µM) [38] treatment in the absence of hydroxytyrosol. A shorter exposure (4 h) was used to investigate effects on mRNA expression.

2.5.2. Intracellular Oxidative Stress

DCFH-DA assay was used to measure the production of intracellular reactive oxygen species (ROS) in C20A4 cells according to the manufacturer's protocol. Following treatment, cells were labeled with DCFH-DA (25 µM) for 1 h in the dark. The fluorescence was measured every 5 min for 1 h, with an excitation wavelength of 485 nm and an emission wavelength of 535 nm using a microplate reader (Cytation 3).

The malondialdehyde (MDA) concentration, as a lipid peroxidation index, was determined using the thiobarbituric acid reactive substances (TBARS) assay, according to the manufacturer's protocol. The basal concentration of MDA was established adding about 600 µL of TBARS solution to 50 µg of total protein dissolved in 300 µL of Milli-Q water. The mix was incubated for 40 min at 100 °C prior to centrifugation at 14,000 rpm for 2 min. The supernatant was analyzed with a microplate reader at a wavelength of 532 nm [39].

Total SOD-like activity was assessed with the SOD Assay Kit-WST according to the manufacturer's protocol. The activity was expressed as units per mg of protein, where one unit of enzyme inhibits reduction of cytochrome C by 50% in a coupled system formed by xanthine and xanthine oxidase.

2.5.3. Enzyme-Linked Immunosorbent Assay (ELISA)

Secreted IL-6, IL-8, and TNF-α protein levels were measured in supernatants of chondrocytes treated as reported in paragraph 2.5.1. Briefly, 100 µL of samples and standards were added into the wells already pre-coated with antibody specific for IL-6, IL-8, or TNF-α, and incubated for 2 h at 37 °C. Unbound substances were removed and 100 µL of biotin-conjugated antibody specific for IL-6, IL-8, or TNF-α was added to the well. After washing, 100 µL of avidin conjugated Horseradish Peroxidase (HRP) was added to the wells and incubated for 1 h at 37 °C, followed by addition of 90 µL of TMB substrate solution, and then incubation for 15–30 min at 37 °C. Stop solution was added to each well, the plate was gently tapped for thorough mixing, and the color intensity measured at 450 nm using a Cytation 3 Microplate Reader (ASHI, Milan, Italy).

2.5.4. Quantitative Senescence-Associated Beta-Galactosidase Assay

4-methylumbelliferyl-β-D-galactopyranoside (4-MUG) was used as substrate of β-galactosidase for the quantitative SA-β-gal assay [40]. 4-MUG does not fluoresce until cleaved by the enzyme to generate the fluorophore 4-methylumbelliferone. The assay was carried out on lysates obtained from cells that were grown as reported above. The production of the fluorophore was monitored at an emission/excitation wavelength of 365/460 nm.

2.5.5. RNA Isolation, Reverse Transcription, and Quantitative Real-Time PCR (qRT-PCR)

Total RNA was extracted from cell cultures using TriFast (EuroClone, Milan, Italy), according to the manufacturer's protocol, and mRNA levels quantified by RT-PCR amplification as reported by Calarco el al. [41]. For retro-transcription, total RNA (0.5 µg) was treated as described in EuroClone standard protocol and amplified by qPCR. Specific primers for SRY-Box Transcription Factor 9 (SOX9), Collagen Type II Alpha 1 Chain (COL2A1), Aggrecan (ACAN), Cartilage Oligomeric Matrix Protein (COMP), Interleukin-6 (IL-6), Interleukin-8 (IL-8), tumor necrosis factor (TNF)-α, Matrix Metallopeptidase 3 and 13 (MMP-3 and 13), and β-Actin (ACTB) were used and listed in Table 1. qRT-PCR was run on a 7900 HT fast real-time PCR System (Applied Biosystem, Milan, Italy). The reactions were performed according to the manufacturer's instructions using SYBR Green PCR Master mix (Euroclone, Italy). Data were normalized using the housekeeping gene (ACTB). All reactions were run in triplicate and the results expressed as mean ± SD. The $2^{-\Delta\Delta Ct}$ method was used to determine the relative quantification.

Table 1. Primers used for qRT-PCR.

Gene	Accession Number	Forward (5'-3')	Reverse (5'-3')
COL2A1	NM_001844.5	CTGGTGTGAAGGGTGAGAGT	AGTCCGTCCTCTTTCACCAG
ACAN	NM_001135.4	TCCCCAACAGATGCTTCCAT	GTACTTGTTCCAGCCCTCCT
SOX9	NM_000346.4	CCGCTCACAGTACGACTACA	GTGAAGGTGGAGTAGAGGCC
COMP	NM_000095.4	CCTTCAATGGCGTGGACTTC	TGACCACGTAGAAGCTGGAG
MMP-3	NM_002422.5	CCTCTGATGGCCCAGAATTGA	GAAATTGGCCACTCCCTGGGT
MMP-13	NM_002427.4	GTCCAGGAGATGAAGACCCC	CTCGGAGACTGGTAATGGCA
SOD2	NM_000636.4	CTGGACAAACCTCAGCCCTA	TGATGGCTTCCAGCAACTC
IL-6	NM_000600.5	CGCCTTCGGTCCAGTTGCC	GCCAGTGCCTCTTTGCTGCTTT
IL-8	NM_000584.4	CTCTTGGCAGCCTTCCTGATTTC	TTTTCCTTGGGGTCCAGACAGAG
TNF-α	NM_000594.4	AACATCCAACCTTCCCAAACGC	TGGTCTCCAGATTCCAGATGTCAGG
ACTB	NM_001101.5	ACTCTTCCAGCCTTCCTTCC	CGTACAGGTCTTTGCGGATG

2.6. Statistical Analysis

Statistical comparisons between the different experimental groups and controls were made using GraphPad Prism 6 software (GraphPad Software Inc., San Diego, CA, USA). Each experiment was performed at least three times and all quantitative data are expressed as mean ± standard deviation (SD).

3. Results and Discussion

3.1. Preparation and Physicochemical Characterization of Hyt-Loaded Nanoparticles (Hyt NPs)

OA, the most common musculoskeletal disease in the elderly population, involves the inflammatory immune response at both local (joint site) and systemic levels leading to severe articular joint pain and reduced joint mobility. To date, local anti-inflammatory treatment is usually insufficient because of their short intra-articular half-lives, while systemic administration is associated with more adverse events [42–44]. Chitosan-based nanoparticles have been extensively used as ideal drug carriers for wide range of biomedical applications due to their good compatibility and degradability [45,46].

In this work, Hyt-loaded nanoparticles (Hyt NPs) were successfully produced by the ionic gelation method, using tripolyphosphate (TPP) as the crosslink. Although the ionic gelation process represents a simple and robust route to obtain chitosan NPs in aqueous medium and under mild conditions, the optimal process parameters were determined to achieve NPs with high drug loading and narrow polydispersity index (PDI). Indeed, the ratio of chitosan/TPP, the chitosan concentration, and the concentration of the encapsulated drug could interfere with the NP size and size distribution during NP formation.

Results of polymer ratio on particle size and size distribution, polydispersity index (PDI), zeta potential (ZP), and encapsulation efficiency (EE%) of nine batches of Hyt NPs studied are summarized in Table 2. Particle size of the prepared formulations was in a nanometric range varying between 510.14 ± 13.21 nm (CS:TPP 1:1) and 137.56 ± 3.13 nm (CS:TPP 10:1) demonstrating that the size of the nanoparticles depends greatly on the ratio of CS to TPP. This behavior is achieved by the interaction of the phosphate charged groups of TPP with the –NH_3^+ groups within the CS structure. Indeed, as the amount of TPP increases the particle size decreases because of the increment in the cross-linking of CS macromolecules mediated by TPP, leading to a minimum particle size at 10:1 CS:TPP ratio.

Table 2. Effect of chitosan concentration and chitosan/TPP ratio on the size (hydrodynamic diameter), polydispersity index (PDI), zeta potential (ZP), and encapsulation efficiency (EE) of Hyt-loading nanoparticles (Hyt NPs). The Hyt concentration was kept constant at 10 mg.

Chitosan (mg/mL)	CS:TPP Mass Ratio	Size (nm ± SD)	PDI (mV ± SD)	EE (% ± SD)
0.1	1:1	510.14 ± 13.21	0.28 ± 0.03	18.31 ± 1.23
0.1	5:1	365.23 ± 9.01	0.15 ± 0.04	36.71 ± 1.26
0.1	10:1	298.73 ± 9.06	0.26 ± 0.02	39.27 ± 2.14
0.5	1:1	330.33 ± 7.52	0.17 ± 0.01	28.96 ± 1.34
0.5	5:1	279.61 ± 0.72	0.23 ± 0.04	41.25 ± 2.41
0.5	10:1	219.63 ± 0.46	0.21 ± 0.03	53.37 ± 2.62
1	1:1	348.16 ± 12.64	0.14 ± 0.02	29.81 ± 1.29
1	5:1	224.43 ± 0.46	0.25 ± 0.01	48.57 ± 1.85
1	10:1	137.56 ± 3.13	0.16 ± 0.03	74.18 ± 3.16

Note: In the same column, value with the same subscript letter (a–c) were not significantly different ($p > 0.05$). Data were mean of three replications ± standard deviation (SD).

Moreover, all formulations present a narrow size distribution and high positive surface charge indicating their better stability to aggregation due to the repulsive forces exerted by the positive surface charge. As reported in Table 2, the EE of Hyt NPs enhanced with an increase in CS:TPP ratio ranging between 18.31 ± 1.23% of 1:1 and 74.18 ± 3.16% of 10:1. This could likely be attributed to the number of crosslinking units associated with different TPP concentrations [47]. Moreover, the reduction in nanoparticle size obtained with 10:1 CS:TPP ratio resulted in increment of space for drug encapsulation.

According to the above results, the chitosan and TPP ratio of 10:1 was chosen for further study as the obtained Hyt NPs showed the highest EE with acceptable particle size and distribution.

Figure 1 shows representative images of: size (1A), zeta potential (1B) distribution, screenshot of nanoparticles tracking analysis video (NTA, 1C), and measurements (1D) of Hyt-NPs synthetized in the optimal condition.

3.2. In Vitro Hydrogel Formulation (Hyt@tgel) and Hyt Release

To obtain a sustained and localized drug delivery of Hyt at body temperature, different amounts of Hyt NPs were dispersed into injectable hydrogels composed of 20 wt% of Pluronic F127 and 1 wt% Hyaluronic acid (Hyt@tgel). According to Young-seok et al. [33] the Hyt@tgel formulation was optimized to reduce the Pluronic F-127 concentration needed to obtain gelation at body temperature. Moreover, the hydrophobic interaction between acetyl groups on HA and methyl groups on Pluronic could enhance the mechanical strength of the resulting hydrogel at temperatures above the critical gelation temperature (CGT). As shown in Figure 2A, the addition of different concentrations of Hyt-NPs (1, 5, and 10 mg) did not significantly affect the Hyt@tgel gelation temperature, suggesting that hydrogel structure organization was maintained after nanoparticles dispersion. These results are in agreement with previous studies at the same Pluronic concentrations [48,49]. The gelation time of the Hyt@tgel at 35 °C was slightly increased by Hyt NPs incorporation (Figure 2B). In particular, the presence of high nanoparticle concentrations increases the gelation time by 0.5 min (10 mg, Hyt@tgel$_{10}$) and 0.2 min (5 mg, Hyt@tgel$_5$) with respect to Hyt@tgel alone (10.6 min). Long in vivo gelation time, in fact, can cause nanoparticle loss by diffusion into the surrounding tissue. On the contrary, gelation that occurs too quickly could lead to clogging of the injection needle resulting in incomplete administration. Based on suitable gelation time and temperature, further analyses were conducted only on the Hyt@tgel$_{10}$ sample. As shown in Figure 2C, Hyt@tgel$_{10}$ demonstrated easy injectability through hypodermic needles at room temperature, while when the temperature increases at 35 °C, the extrusion needs the application of an extra force due to the increase in the viscosity. When the gel concentration reached the critical gelation concentration, the Hyt@tgel$_{10}$ passed from an aqueous solution to a gel as the temperature was increased from 4 to 35 °C as demonstrated by the inversion test tube (Figure 2D). Hyt@tgel$_{10}$ exhibited a viscous flowable form at low temperature becoming a semi-solid gel after incubation at temperature higher than 30 °C. This behavior was confirmed by the measure of viscosity as a function of temperatures (Figure 2E).

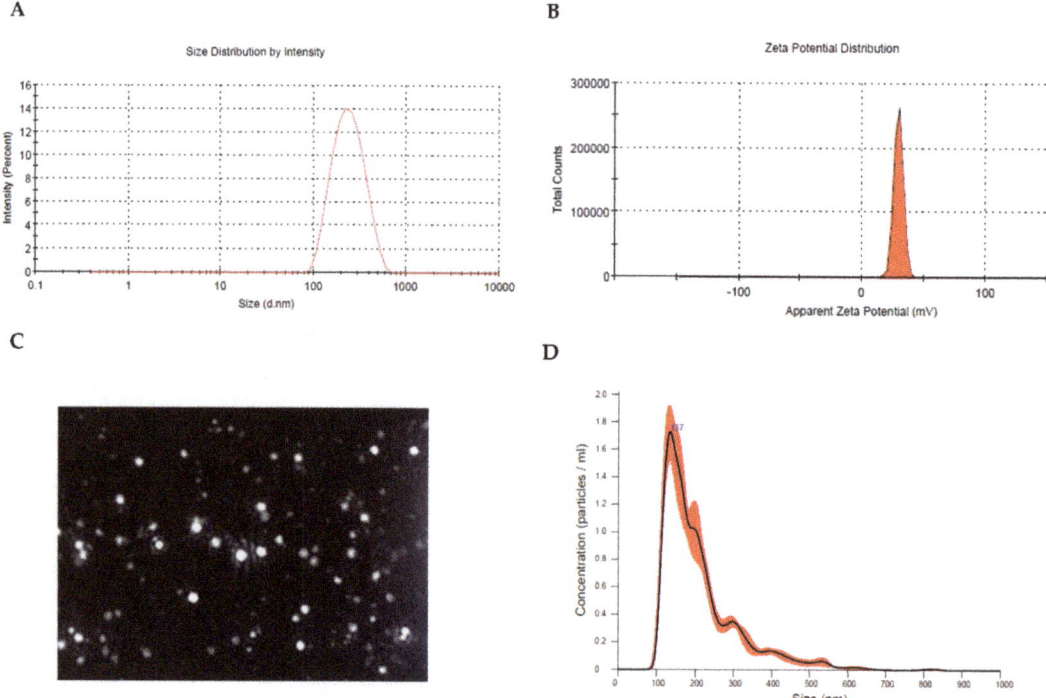

Figure 1. Physicochemical properties of Hyt-NPs. (**A**) Size distribution, (**B**) zeta potential profile, (**C**) screenshot of representative NTA video, and (**D**) NTA measurements for Hyt-NPs in suspension. Histograms are the average of 3 measurements.

There is a substantial body of evidence that encapsulation enhances the bioactivity of compounds improving their stability in aqueous medium and increasing upon the delivery at the target site. Chen et al. demonstrated the ability of chitosan microspheres dispersed in a thermally responsive chitosan hydrogel to load anti-inflammatory drugs. After injection into the knee joints of OA rabbits, drugs were released for more than 7 days in a controlled manner [50]. According to Wang et al., curcumin-loaded HA/chitosan nanoparticles exhibited a good sustained-release property leading to inflammation and cartilage apoptosis inhibition acting on the NF-κB pathway [51].

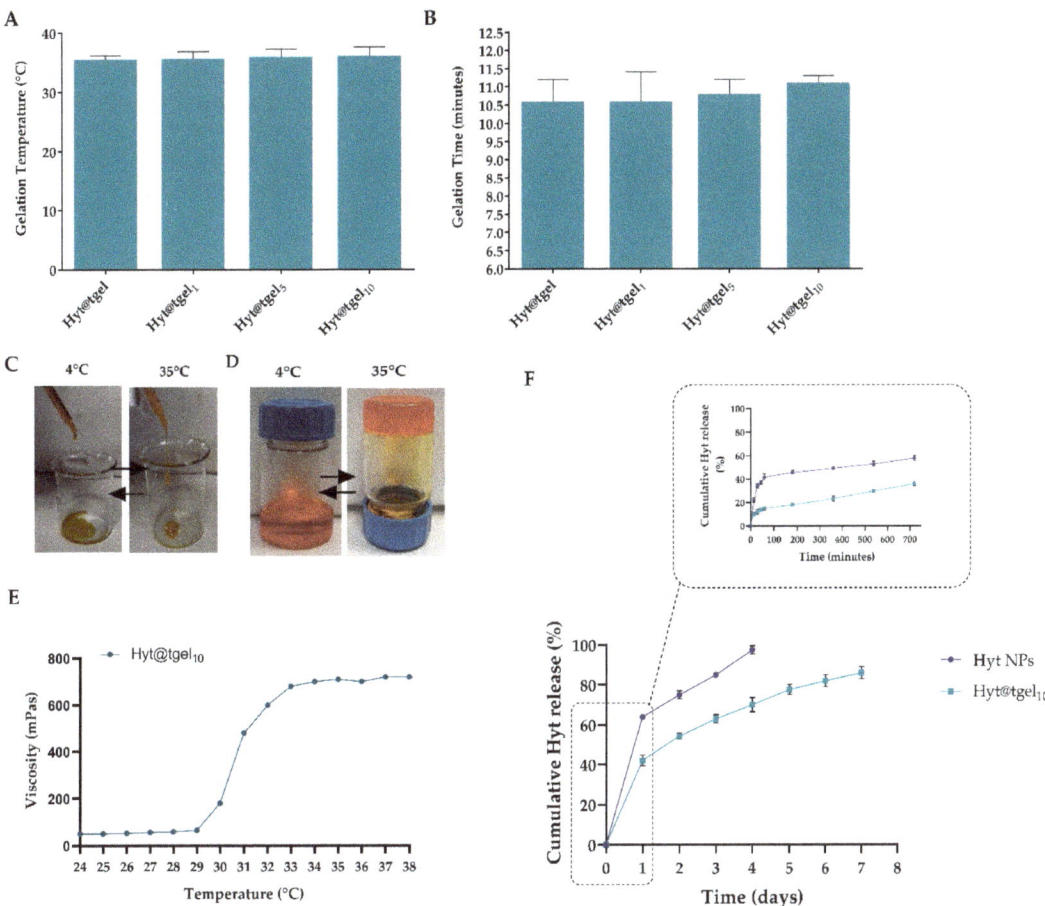

Figure 2. Characterizations of Hyt@tgel. Gelation temperature (**A**) and gelation time at 37 °C (**B**) of different hydrogel compositions (Hyt@tgel, Hyt@tgel$_1$, Hyt@tgel$_5$, and Hyt@tgel$_{10}$). Representative photographs of the Hyt@tgel$_{10}$ syringeability (**C**) and inverted test tube (**D**) obtained at 4 and 35 °C. Phenol red was added to facilitate hydrogel monitoring. (**E**) Solution viscosity measurement of the Hyt@tgel$_{10}$ as a function of temperature. (**F**) Cumulative Hyt release from Hyt NPs and Hyt@tgel$_{10}$ in phosphate buffer saline (PBS) after 24 h (and eight days). Six different experiments were conducted, and the results expressed as the mean of the values obtained (mean ± SD).

Chitosan nanoparticles have been recognized as a useful drug delivery tool in OA for their ability to prolong the drug retention time. To evaluate the sustained release properties of Hyt-NPs, an in vitro drug release study of Hyt from Hyt-NPs and from HYt@tgel$_{10}$ was carried out using dialysis membrane against phosphate buffer saline (PBS). As shown in Figure 2F, the in vitro release of Hyt by chitosan NPs exhibited a fast drug release rate with 41% of Hyt released within the first hour, with the majority of the release occurring during the initial 2 days (75%). On the contrary, Hyt release rate from Hyt@tgel$_{10}$ significantly slowed down ($p < 0.05$) with only 10% of Hyt released after 1 h, followed by a prolonged Hyt release up to 1 week. The slow Hyt release from the hydrogel could be attributed to the densely packed inter-micellar structure due to the presence of HA. Moreover, the highly packed super-molecular structure could reduce the diffusion coefficients inside of the hydrogel leading to a prolonged drug release.

Physical stability of nanoparticles in Hyt@tgel was investigated at 4 and 25 °C over 14 days by measuring size and PDI. As reported in Table 3, Hyt NPs were stable when stored at both low and room temperatures without a significant increase in particle size and PDI. Moreover, the drug encapsulation efficiency, assessed in parallel, demonstrated no decrease in the Hyt retention rate over the 14-day period confirming the protective effect of chitosan nanoparticles on biomolecules.

Table 3. Particle size and entrapment efficiency studies of Hyt@tgel$_{10}$ before and after 14-day storage.

	Hyt@tgel$_{10}$ Before Storage	Hyt@tgel$_{10}$ after 14-Day Storage at 4 ± 1 °C	Hyt@tgel$_{10}$ after 14-Day Storage at 25 ± 1 °C
Average particle size (nm)	137.56 ± 3.13	135.32 ± 2.56 nm	139.00 ± 6.53 nm
PDI	0.16 ± 0.03	0.15 ± 0.02	0.18 ± 0.05
Entrapment efficiency (% EE)	74.18 ± 3.16	75.88 ± 4.13	77.22 ± 5.34

Note: Data were expressed as mean $\pm$ standard deviation, n = 3.

Taken together, the physicochemical behavior of Hyt@tgel$_{10}$ is consistent with a potential use as a device to be injected through a syringe, because the sol-to-gel transition temperature is between room temperature and physiological temperature.

3.3. Oxidative Damage Protection of Hyt@tgel

Several studies have concluded that OA progression is significantly related to an imbalance between the production of reactive oxygen species (ROS) and their clearance by an antioxidant defense system [52,53]. During OA pathogenesis, chondrocytes become both source and target of elevated amounts of reactive chemical species, particularly oxygen and nitrogen species triggering a vicious circle that leads to further damage of cartilage cells and matrix [54]. A wide body of evidence suggested that Hyt has antioxidant activity by inhibition and/or scavenging of reactive oxygen species (ROS) [54,55]. Moreover, a gene expression profiling study has suggested that Hyt affect the expression of genes involved in oxidative stress, inflammation, cell proliferation, or differentiation, suggesting that the beneficial effects of this molecule may be multifactorial and context-dependent [56]. The efficiency of Hyt@tgel$_{10}$ to reduce the intracellular ROS generation was assessed in C20A4 cells in the presence of hydrogen peroxide (H_2O_2) (Figure 3A,B). The stimulation of chondrocytes with H_2O_2 mimics the in vivo condition observed in OA cartilage inducing the production of cellular and mitochondrial ROS and producing proinflammatory and procatabolic responses [57,58]. A significant increase ($p < 0.001$) in chondrocyte intracellular oxidants by approximately 2.8 times was obtained with respect to untreated cells (Control) after H_2O_2 24 h treatment (Figure 3A). A short-time pre-incubation (24 h) with Hyt@tgel$_{10}$ considerably reduced ($p < 0.01$) H_2O_2-induced ROS production of about 1.4-fold with respect to the H_2O_2 group. Moreover, the protective effect of released Hyt was greatly enhanced (($p < 0.001$) by a longer pre-treatment (96 h) resulting in slight fluorescence increase with respect to control cells. MDA, a lipid peroxidation end product, is abundant in synoviocytes from patients with OA. Under oxidative stress, polyunsaturated fatty acids of cellular membrane lipids represent the prime targets of ROS attack. The lipid peroxidation leads to the formation of chemically reactive lipid aldehydes, such as MDA, capable of causing severe damage to nucleic acids and proteins, altering their functions and leading to the loss of both structural and metabolic function of cells [59]. As reported in Figure 3B, treatment of cells with H_2O_2 increased intracellular lipid peroxidation to 2-fold relative to control ($p < 0.001$). Conversely, the presence of Hyt@tgel$_{10}$ for 24 h markedly diminished ($p < 0.01$) the MDA level (1.1-fold) compared with H_2O_2 treated cells, with a marked decrease ($p < 0.001$) after 96 h leading the MDA formation to levels almost similar to control.

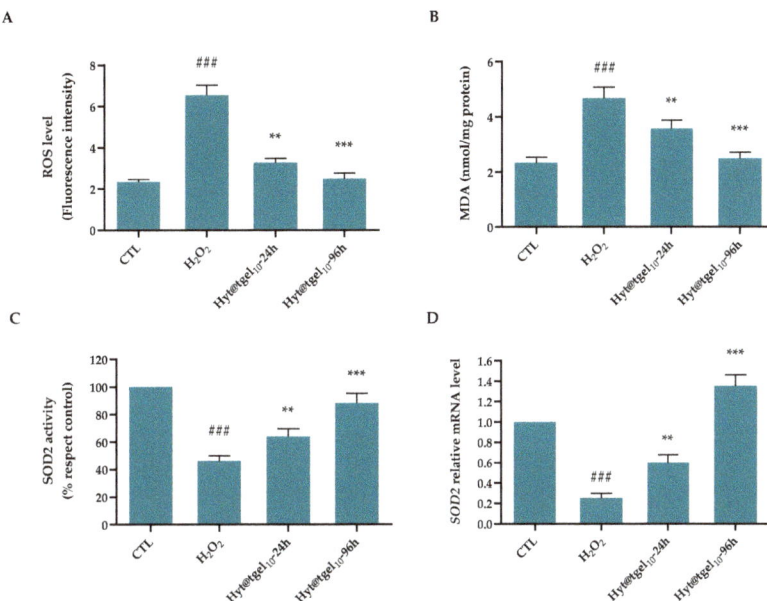

Figure 3. Antioxidant capacity of Hyt@tgel in H_2O_2-treated chondrocytes. C20A4 cells were incubated in the presence of hydrogel for 24 and 96 h and then treated with hydrogen peroxide for 24 h. (**A**) ROS release was determined by oxidized H2DCFDA (DCF). (**B**) Malondialdehyde quantity was used as a marker of lipid peroxidation. (**C**) Superoxide dismutase (SOD2) activity measured by assay kit. (**D**) SOD2 mRNA transcription level. Results are expressed as the mean of three independent experiments ± S.D (n = 3). ** $p < 0.01$, *** $p < 0.001$ versus untreated cells (control). ### $p < 0.001$ versus H_2O_2 group.

To prevent an accumulation of ROS-mediated damage, chondrocytes produce a number of antioxidant enzymes including the superoxide dismutases (SOD), catalase, and glutathione peroxidase [60]. The three SOD family members SOD1, SOD2, and SOD3 transform O_2^- into hydrogen peroxide (H_2O_2), limiting the formation of highly aggressive compounds such as $ONOO^-$ and OH^-. All SOD are expressed at lower levels in OA cartilage compared to normal control cartilage, at both the messenger RNA (mRNA) and protein level. In particular, Ruiz-Romero et al. demonstrated through a proteomics approach, a significant decrease in the major mitochondrial antioxidant protein manganese-superoxide dismutase (SOD2) in the superficial layer of OA cartilage. This SOD2 reduction makes cartilage more susceptible to ROS damage suggesting a central role of mitochondrial redox imbalance in OA pathogenesis [61]. To verify if the antioxidant actions of Hyt have been related not only to its free radical scavenging activity, but also to the ability to enhance the endogenous defense system by inducing antioxidant/detoxifying enzymes activity, SOD2 activity was assayed. As shown in Figure 3C, treatment with H_2O_2 leads to decrease in antioxidant enzyme activity of about 54% with respect to untreated cells. When chondrocytes were pre-treated with Hyt@tgel$_{10}$ for 24 and 96 h, SOD2 activity was 18% and 42%, respectively, higher than that in H_2O_2-treated cells, demonstrating a good ability to protect mitochondria from oxidative damage. Moreover, Hyt@tgel$_{10}$ pretreatment restored the SOD2 transcript to above their control levels, by significantly increasing its expression by 2.4-fold for 24 h and 5.5-fold after 96 h over the H_2O_2-depressed level (Figure 3D).

Taken together, the results reported herein confirm a key role of Hyt@tgel$_{10}$ pretreatment to effectively suppress the production of intracellular ROS and lipid peroxidation and also elevated the activity of antioxidant enzymes such as SOD, limiting oxidative stress-induced damage in the OA in vitro model.

3.4. Hyt@tgel Suppresses Inflammatory Response in Chondrocytes

Increases in the levels of the cytokines in joints plays a central role in the pathogenesis of OA by modulating oxidative stress, cartilage ECM turnover, and chondrocytes apoptosis [62,63]. The current drugs for treating OA are developed primarily to relieve pain and control symptoms, failing to cure the disease [63]. Epidemiologic studies demonstrated the lower incidence of inflammatory chronic disease, such as OA in people of the Mediterranean basin. One of the possible reasons is that Mediterranean people have a high intake of olive and olive oil rich in polyphenolic compounds with antioxidant and anti-inflammatory properties [14,64,65]. During the pathophysiological processes of OA, cytokines, hormone-like proteins, are responsible for the loss of metabolic homeostasis of tissues forming joints by promoting catabolic and destructive processes. Olive-oil-rich extracts inhibit the production of proinflammatory cytokines, including IL-1β, TNF-α, IL-6, and prostaglandin E2 in arthritic joints [66,67]. Richard and colleagues demonstrated a pivotal role of Hyt extracted from olive vegetation water in diminished secretion of cytokines (IL-1 α, IL-1 β, IL-6, IL-12, TNF-α), and chemokines (CXCL10/IP-10, CCL2/MCP-1) in murine macrophages (RAW264.7 cells) stimulated with lipopolysaccharide (LPS) [68]. Another study showed a decrease in the severity of the disease and an overall anti-IL-1β effect after treatment with olive and grape seed extract in animal models of post-traumatic OA [69]. In the present study, secreted IL-6, IL-8, and TNF-α were detected in the supernatant of chondrocytes cell line C20A4 by enzyme-linked immunosorbent assays (ELISA). As expected, incubation of cells for 24 h with Hyt@tgel$_{10}$ significantly reduces the amount of released cytokines with respect to the control in a time-dependent manner (Figure 4A–C). Consistently, the protective effects of Hyt were confirmed also by RT-qPCR analysis. As reported in Figure 4D–F, the mRNA levels of all tested cytokines (relative to the housekeeping gene) were significantly upregulated ($p < 0.01$) in H_2O_2-treated cells, compared with the control group. As expected, the H_2O_2-driven release of IL-6, IL-8, and TNF-α was decreased by Hyt@tgel$_{10}$ pre-treatment with a 50% reduction in interleukin expression levels with respect to H_2O_2-treated cells.

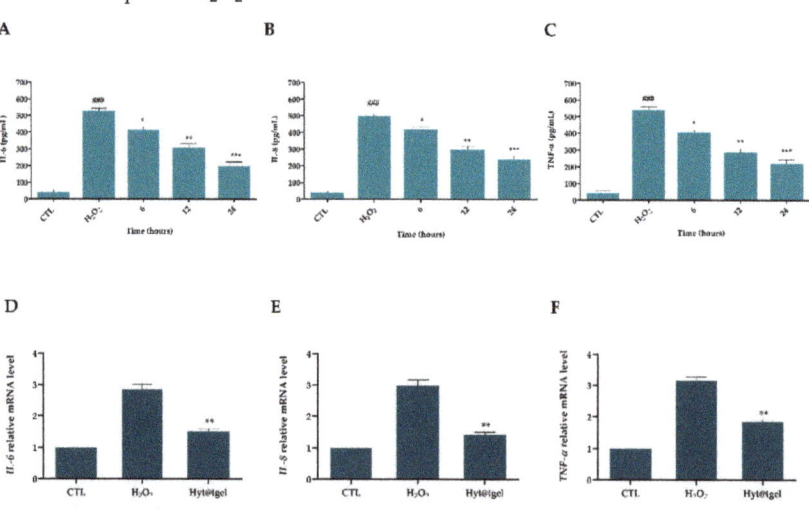

Figure 4. Hyt@tgel$_{10}$ inhibits H_2O_2-induced inflammatory response in chondrocytes. The effect of Hyt on the production of IL-6 (**A,D**), IL-8 (**B,E**), and TNF-α (**C,F**) was measured by ELISA assay (**A–C**) and qRT-PCR (**D–F**). C20A4 cells were pre-treated with Hyt@tgel$_{10}$ for 24 h, then stimulated with H_2O_2 for 24 h (ELISA assay) or 4 h (qRT-PCR). Results are expressed as the mean of three independent experiments ± S.D ($n = 3$). ### $p < 0.001$ H_2O_2 vs. CTL, * $p < 0.05$, ** $p < 0.01$, and *** $p < 0.005$ Hyt@tgel$_{10}$ vs. H_2O_2.

3.5. Hyt@tgel Protects against H_2O_2-Mediated Chondrocyte ECM Degradation

Once damaged, the cartilage is enabled to repair itself due to its special physiological structure. In the early stages of OA, the production of inflammatory mediators including cytokines and prostaglandins by the cartilage and synovial cells lead to activation of matrix metalloproteinases (MMPs) [70]. Among them, matrix metalloproteinases MMP-3 and MMP-13 can further promote cartilage inflammation, chondrocyte apoptosis, and ROS production, via a positive feedback loop [71,72]. Emerging evidence has shown that MMP13 is considered a significant biomarker to assess OA therapeutic effects and OA progression [73,74]. In this context, bioactive molecules able to suppress these inflammatory mediators or block the involved signaling pathway may help to reduce the OA pathological process [75]. To further corroborate the Hyt@tgel anti-inflammatory action, the expression of catabolic genes such as those coding for MMP-3 and MMP-13 were evaluated (Figure 5).

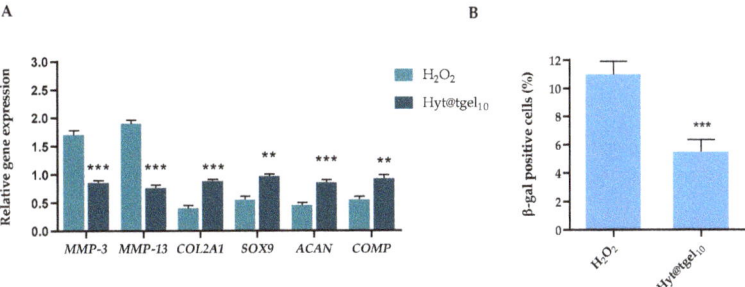

Figure 5. Hyt released by Hyt@tgel$_{10}$ prevents the expression of OA-related genes in chondrocytes treated with H_2O_2. (**A**) C20A4 chondrocytes were pre-treated with Hyt@tgel$_{10}$ for 24 h, then stimulated with H_2O_2 for 4 h ($\pm$SD, n = 3, ** p < 0.01 vs. H_2O_2 group). (**B**) Beta-galactosidase senescence assay. The graph shows the mean percentage value of senescent cells in every experimental condition ($\pm$SD, n = 3, *** p < 0.005).

Compared with the untreated group, the mRNA expression of MMP-3 and MMP-13 increased significantly (p < 0.01) in H_2O_2-treated cells, while Hyt@tgel$_{10}$ pre-treatment was able to reduce about 40% of the increases provoked by H_2O_2. These results, in line with Facchini et al. [25] corroborated the capacity of Hyt to antagonize the activation of pro-inflammatory pathways like NF-κB even in chondrocytes.

Activation of catabolic enzymes degrades proteoglycan and collagen in the articular cartilage. Moreover, inflammatory states lead to de-differentiation of chondrocytes accompanied by decreased expression of chondrocyte-specific proteins [76]. As reported in Figure 5A, the expression of SOX9 [77] (an early marker of the formation of a cartilage-like tissue matrix), COL2A1, ACAN [78], and COMP (markers of the final stage of chondrogenic differentiation) was significantly rescued by incubation with Hyt@tgel$_{10}$. These data indicate that Hyt released by Hyt@tgel influenced the ECM balance and gene expression in the chondrocytes under pathological state maintaining their metabolic activity and proliferation in their differentiated phenotype.

Although various cell types are involved in OA pathology, chondrocytes play a major role in OA induction by cellular senescence [79]. It has been shown that chondrocytes have telomere shortening with age. For this reason, chondrocyte senescence, caused by chronic stress in the cells or caused by post-traumatic inflammation, is believed to be closely related to OA [80]. Therefore, the regulation of hypertrophic or senescent chondrocytes using natural phytochemicals known to have a powerful anti-inflammatory and antioxidant activity could be a potential therapeutic target to slow or stop the progression of OA [76]. Data demonstrated that senescence was noticeably reduced in cells treated with Hyt@tgel as detected in an in-situ beta-galactosidase assay (Figure 5B). In particular, Hyt@tgel$_{10}$ treatment reduced more than two times the percentage of senescent cells compared to untreated chondrocytes.

Two different mechanisms of senescence are suggested in chondrocytes: replicative senescence and stress-induced premature senescence [81,82]. Upregulation of inflammatory cytokines expression induces senescence directly, while downregulation of chondrocyte phenotypic maintenance genes such as SOX9, BMP-2, IGF-1, and ACAN induces senescence indirectly. Thus, the association between aging and/or trauma, reduces the number of healthy and functioning chondrocytes, promoting cartilage degeneration and eventually leads to osteoarthritic pathophysiology [10]. Therefore, the reduction of this cell population lends further credit to hydroxytyrosol ability to preserve chondrocytes from senescence after Hyt@tgel$_{10}$ treatment.

4. Conclusions

OA is mainly caused by trauma induced by an external force or cartilage damage ac-cumulated during aging. This study provided new insights into the therapeutic effects of intra-articular injection of Hyt-loaded chitosan nanoparticles embedded into thermosensitive hydrogels (Hyt@tgel$_{10}$). The hydrogel exhibited a sol-gel transition behavior and a gelation time consistent with its therapeutic application. Moreover, Hyt released from hydrogel was able to protect chondrocytes from ROS damage and to revert the activation of inflammatory factors, limiting, in an in vitro model, the vicious cycle typical of OA progression. Hence, it can be concluded that the formulated hydrogel injection could be proposed for the efficient and sustained Hyt delivery for OA treatment. The next step would be the extraction of "added-value" bioactive polyphenols from by-products of the olive industry, in order to develop a green delivery system able not only to enhance human wellbeing but also to promote a sustainable environment.

Author Contributions: Conceptualization, A.V., R.C. and A.C.; investigation, A.V., R.C., F.D.C. and I.D.L.; writing—original draft preparation, A.V. and R.C.; writing—review and editing, M.B. and A.C.; supervision, M.B. and A.C.; funding acquisition, M.B., G.P. and A.C. All authors have read and agreed to the published version of the manuscript.

Funding: This work was financially supported by the PON 03 PE_00110_1/ptd1_000410 Titolo: Sviluppo di nanotecnologie Orientate alla Rigenerazione e Ricostruzione tissutale, Implantologia e Sensoristica in Odontoiatria/oculistica (SORRISO); POR Campania FESR 2014_2020 "Tecnologie abilitanti per la sintesi eco-sostenibile di nuovi materiali per la restaurativa dentale"—ABILTECH; EU funding within the Horizon 2020 Program, under the MSCA-RISE 2016 Project "VAHVISTUS" (Grant 734759); POR 2014–2020 FESR MISE Prog. n.F/200004/01-02/X45: Micro-Poli, Titolo: Micro-nanodispositivi veicolanti polifenoli isolati da scarti della filiera olivicola come nuovi integratori alimentari; Fondazione Cariplo, "grant no. 2018-1001", Economia Circolare-Ricerca per un Futuro Sostenibile" program, "High added-value bioactive polyphenols recovered from waste of olive oil production" research project.

Institutional Review Board Statement: Not applicable.

Informed Consent Statement: Not applicable.

Data Availability Statement: The data presented in this study are available in the article.

Acknowledgments: The authors gratefully acknowledge Orsolina Petillo for her assistance with cell culture (IRET-CNR).

Conflicts of Interest: The authors declare no conflict of interest.

References

1. Roos, E.; Arden, N.K. Strategies for the prevention of knee osteoarthritis. *Nat. Rev. Rheumatol.* **2016**, *12*, 92–101. [CrossRef]
2. Zheng, L.; Zhang, Z.; Sheng, P.; Mobasheri, A. The role of metabolism in chondrocyte dysfunction and the progression of osteoarthritis. *Ageing Res. Rev.* **2021**, *66*, 101249. [CrossRef]
3. Li, G.Y.; Yin, J.M.; Gao, J.J.; Cheng, T.S.; Pavlos, N.J.; Zhang, C.Q.; Zheng, M.H. Subchondral bone in osteoarthritis: Insight into risk factors and microstructural changes. *Arthritis Res. Ther.* **2013**, *15*, 223. [CrossRef]
4. Martel-Pelletier, J.; Barr, A.; Cicuttini, F.; Conaghan, P.; Cooper, C.; Goldring, M.B.; Goldring, S.R.; Jones, G.; Teichtahl, A.J.; Pelletier, J.-P. Osteoarthritis. *Nat. Rev. Dis. Prim.* **2016**, *2*, 16072. [CrossRef]

5. Ameye, L.G.; Chee, W.S. Osteoarthritis and nutrition. From nutraceuticals to functional foods: A systematic review of the scientific evidence. *Arthritis Res. Ther.* **2006**, *8*, R127. [CrossRef]
6. Bijlsma, J.W.; Berenbaum, F.; Lafeber, F.P. Osteoarthritis: An update with relevance for clinical practice. *Lancet* **2011**, *377*, 2115–2126. [CrossRef]
7. Lepetsos, P.; Papavassiliou, A.G. ROS/oxidative stress signaling in osteoarthritis. *Biochim. Biophys. Acta (BBA)-Mol. Basis Dis.* **2016**, *1862*, 576–591. [CrossRef]
8. Zahan, O.-M.; Serban, O.; Gherman, C.; Fodor, D. The evaluation of oxidative stress in osteoarthritis. *Med. Pharm. Rep.* **2020**, *93*, 12–22. [CrossRef]
9. Bolduc, J.A.; Collins, J.A.; Loeser, R.F. Reactive oxygen species, aging and articular cartilage homeostasis. *Free Radic. Biol. Med.* **2019**, *132*, 73–82. [CrossRef]
10. Ramasamy, T.S.; Yee, Y.M.; Khan, I.M. Chondrocyte Aging: The Molecular Determinants and Therapeutic Opportunities. *Front. Cell Dev. Biol.* **2021**, *9*, 625497. [CrossRef]
11. Fernandes, L.; Hagen, K.B.; Bijlsma, J.W.J.; Andreassen, O.; Christensen, P.; Conaghan, P.; Doherty, M.; Geenen, R.; Hammond, A.; Kjeken, I.; et al. EULAR recommendations for the non-pharmacological core management of hip and knee osteoarthritis. *Ann. Rheum. Dis.* **2013**, *72*, 1125–1135. [CrossRef] [PubMed]
12. Miles, E.A.; Zoubouli, P.; Calder, P. Differential anti-inflammatory effects of phenolic compounds from extra virgin olive oil identified in human whole blood cultures. *Nutrition* **2005**, *21*, 389–394. [CrossRef] [PubMed]
13. Finicelli, M.; Squillaro, T.; Di Cristo, F.; Di Salle, A.; Melone, M.A.B.; Galderisi, U.; Peluso, G. Metabolic syndrome, Mediterranean diet, and polyphenols: Evidence and perspectives. *J. Cell. Physiol.* **2019**, *234*, 5807–5826. [CrossRef] [PubMed]
14. Finicelli, M.; Squillaro, T.; Galderisi, U.; Peluso, G. Polyphenols, the Healthy Brand of Olive Oil: Insights and Perspectives. *Nutrients* **2021**, *13*, 3831. [CrossRef]
15. Valentino, A.; Di Cristo, F.; Bosetti, M.; Amaghnouje, A.; Bousta, D.; Conte, R.; Calarco, A. Bioactivity and Delivery Strategies of Phytochemical Compounds in Bone Tissue Regeneration. *Appl. Sci.* **2021**, *11*, 5122. [CrossRef]
16. Del Rio, D.; Rodriguez-Mateos, A.; Spencer, J.P.E.; Tognolini, M.; Borges, G.; Crozier, A. Dietary (Poly)phenolics in Human Health: Structures, Bioavailability, and Evidence of Protective Effects Against Chronic Diseases. *Antioxid. Redox Signal.* **2013**, *18*, 1818–1892. [CrossRef]
17. De Luca, I.; Di Cristo, F.; Valentino, A.; Peluso, G.; Di Salle, A.; Calarco, A. Food-Derived Bioactive Molecules from Mediterranean Diet: Nanotechnological Approaches and Waste Valorization as Strategies to Improve Human Wellness. *Polymers* **2022**, *14*, 1726. [CrossRef]
18. Visioli, F.; De La Lastra, C.A.; Andres-Lacueva, C.; Aviram, M.; Calhau, C.; Cassano, A.; D'Archivio, M.; Faria, A.; Favé, G.; Fogliano, V.; et al. Polyphenols and Human Health: A Prospectus. *Crit. Rev. Food Sci. Nutr.* **2011**, *51*, 524–546. [CrossRef]
19. Silva, A.F.R.; Resende, D.; Monteiro, M.; Coimbra, M.A.; Silva, A.M.S.; Cardoso, S.M. Application of Hydroxytyrosol in the Functional Foods Field: From Ingredient to Dietary Supplements. *Antioxidants* **2020**, *9*, 1246. [CrossRef]
20. Serra, A.; Rubió, L.; Borràs, X.; Macià, A.; Romero, M.-P.; Motilva, M.-J. Distribution of olive oil phenolic compounds in rat tissues after administration of a phenolic extract from olive cake. *Mol. Nutr. Food Res.* **2012**, *56*, 486–496. [CrossRef]
21. Yonezawa, Y.; Miyashita, T.; Nejishima, H.; Takeda, Y.; Imai, K.; Ogawa, H. Anti-inflammatory effects of olive-derived hydroxytyrosol on lipopolysaccharide-induced inflammation in RAW264.7 cells. *J. Veter. Med. Sci.* **2018**, *80*, 1801–1807. [CrossRef]
22. Di Meo, F.; Valentino, A.; Petillo, O.; Peluso, G.; Filosa, S.; Crispi, S. Bioactive Polyphenols and Neuromodulation: Molecular Mechanisms in Neurodegeneration. *Int. J. Mol. Sci.* **2020**, *21*, 2564. [CrossRef] [PubMed]
23. Fuccelli, R.; Fabiani, R.; Rosignoli, P. Hydroxytyrosol Exerts Anti-Inflammatory and Anti-Oxidant Activities in a Mouse Model of Systemic Inflammation. *Molecules* **2018**, *23*, 3212. [CrossRef]
24. Cetrullo, S.; D'Adamo, S.; Guidotti, S.; Borzì, R.M.; Flamigni, F. Hydroxytyrosol prevents chondrocyte death under oxidative stress by inducing autophagy through sirtuin 1-dependent and -independent mechanisms. *Biochi. Biophys. Acta (BBA)* **2016**, *1860*, 1181–1191. [CrossRef] [PubMed]
25. Facchini, A.; Cetrullo, S.; D'Adamo, S.; Guidotti, S.; Minguzzi, M.; Facchini, A.; Borzì, R.M.; Flamigni, F. Hydroxytyrosol Prevents Increase of Osteoarthritis Markers in Human Chondrocytes Treated with Hydrogen Peroxide or Growth-Related Oncogene α. *PLoS ONE* **2014**, *9*, e109724. [CrossRef] [PubMed]
26. Scalbert, A.; Williamson, G. Dietary Intake and Bioavailability of Polyphenols. *J. Nutr.* **2000**, *130*, 2073S–2085S. [CrossRef]
27. Detsi, A.; Kavetsou, E.; Kostopoulou, I.; Pitterou, I.; Pontillo, A.R.N.; Tzani, A.; Christodoulou, P.; Siliachli, A.; Zoumpoulakis, P. Nanosystems for the Encapsulation of Natural Products: The Case of Chitosan Biopolymer as a Matrix. *Pharmaceutics* **2020**, *12*, 669. [CrossRef]
28. Le, H.; Xu, W.; Zhuang, X.; Chang, F.; Wang, Y.; Ding, J. Mesenchymal stem cells for cartilage regeneration. *J. Tissue Eng.* **2020**, *11*, 2041731420943839. [CrossRef]
29. Zha, K.; Sun, Z.; Yang, Y.; Chen, M.; Gao, C.; Fu, L.; Li, H.; Sui, X.; Guo, Q.; Liu, S. Recent Developed Strategies for Enhancing Chondrogenic Differentiation of MSC: Impact on MSC-Based Therapy for Cartilage Regeneration. *Stem Cells Int.* **2021**, *2021*, 8830834. [CrossRef]
30. Meng, F.G.; Zhang, Z.Q.; Huang, G.X.; Chen, W.S.; Zhang, Z.J.; He, A.S.; Liao, W.M. Chondrogenesis of mesenchymal stem cells in a novel hyaluronate-collagen-tricalcium phosphate scaffolds for knee repair. *Eur. Cells Mater.* **2016**, *31*, 79–94. [CrossRef]

31. Arrich, J.; Piribauer, F.; Mad, P.; Schmid, D.; Klaushofer, K.; Müllner, M. Intra-articular hyaluronic acid for the treatment of osteoarthritis of the knee: Systematic review and meta-analysis. *Can. Med. Assoc. J.* **2005**, *172*, 1039–1043. [CrossRef] [PubMed]
32. Reichenbach, S.; Blank, S.; Rutjes, A.W.S.; Shang, A.; King, E.A.; Dieppe, P.A.; Jüni, P.; Trelle, S. Hylan versus hyaluronic acid for osteoarthritis of the knee: A systematic review and meta-analysis. *Arthritis Rheum.* **2007**, *57*, 1410–1418. [CrossRef] [PubMed]
33. Jung, Y.-S.; Park, W.; Park, H.; Lee, D.-K.; Na, K. Thermo-sensitive injectable hydrogel based on the physical mixing of hyaluronic acid and Pluronic F-127 for sustained NSAID delivery. *Carbohydr. Polym.* **2017**, *156*, 403–408. [CrossRef] [PubMed]
34. Calvo, P.; Remuñán-López, C.; Vila-Jato, J.L.; Alonso, M.J. Novel hydrophilic chitosan-polyethylene oxide nanoparticles as protein carriers. *J. Appl. Polym. Sci.* **1997**, *63*, 125–132. [CrossRef]
35. Conte, R.; Valentino, A.; Di Cristo, F.; Peluso, G.; Cerruti, P.; Di Salle, A.; Calarco, A. Cationic Polymer Nanoparticles-Mediated Delivery of miR-124 Impairs Tumorigenicity of Prostate Cancer Cells. *Int. J. Mol. Sci.* **2020**, *21*, 869. [CrossRef]
36. Khattab, A.; Marzok, S.; Ibrahim, M. Development of optimized mucoadhesive thermosensitive pluronic based in situ gel for controlled delivery of Latanoprost: Antiglaucoma efficacy and stability approaches. *J. Drug Deliv. Sci. Technol.* **2019**, *53*, 101134. [CrossRef]
37. Amaghnouje, A.; Mechchate, H.; Es-Safi, I.; Boukhira, S.; Aliqahtani, A.S.; Noman, O.M.; Nasr, F.A.; Conte, R.; Calarco, A.; Bousta, D. Subacute Assessment of the Toxicity and Antidepressant-Like Effects of *Origanum majorana* L. Polyphenols in Swiss Albino Mice. *Molecules* **2020**, *25*, 5653. [CrossRef]
38. Scuruchi, M.; D'Ascola, A.; Avenoso, A.; Mandraffino, G.; Campo, S.; Campo, G.M. Endocan, a novel inflammatory marker, is upregulated in human chondrocytes stimulated with IL-1 beta. *Mol. Cell. Biochem.* **2021**, *476*, 1589–1597. [CrossRef]
39. Di Cristo, F.; Valentino, A.; De Luca, I.; Peluso, G.; Bonadies, I.; Calarco, A.; Di Salle, A. PLA Nanofibers for Microenvironmental-Responsive Quercetin Release in Local Periodontal Treatment. *Molecules* **2022**, *27*, 2205. [CrossRef]
40. Musto, P.; Calarco, A.; Pannico, M.; La Manna, P.; Margarucci, S.; Tafuri, A.; Peluso, G. Hyperspectral Raman imaging of human prostatic cells: An attempt to differentiate normal and malignant cell lines by univariate and multivariate data analysis. *Spectrochim. Acta Part A Mol. Biomol. Spectrosc.* **2017**, *173*, 476–488. [CrossRef]
41. Calarco, A.; Di Salle, A.; Tammaro, L.; De Luca, I.; Mucerino, S.; Petillo, O.; Riccitiello, F.; Vittoria, V.; Peluso, G. Long-Term Fluoride Release from Dental Resins Affects STRO-1+ Cell Behavior. *J. Dent. Res.* **2015**, *94*, 1099–1105. [CrossRef] [PubMed]
42. Xia, B.; Chen, D.; Zhang, J.; Hu, S.; Jin, H.; Tong, P. Osteoarthritis Pathogenesis: A Review of Molecular Mechanisms. *Calcif. Tissue Res.* **2014**, *95*, 495–505. [CrossRef] [PubMed]
43. Anandacoomarasamy, A.; March, L. Current evidence for osteoarthritis treatments. *Ther. Adv. Musculoskelet. Dis.* **2010**, *2*, 17–28. [CrossRef] [PubMed]
44. Nowaczyk, A.; Szwedowski, D.; Dallo, I.; Nowaczyk, J. Overview of First-Line and Second-Line Pharmacotherapies for Osteoarthritis with Special Focus on Intra-Articular Treatment. *Int. J. Mol. Sci.* **2022**, *23*, 1566. [CrossRef]
45. Menazea, A.; Ahmed, M. Wound healing activity of Chitosan/Polyvinyl Alcohol embedded by gold nanoparticles prepared by nanosecond laser ablation. *J. Mol. Struct.* **2020**, *1217*, 128401. [CrossRef]
46. Ali, A.; Ahmed, S. A review on chitosan and its nanocomposites in drug delivery. *Int. J. Biol. Macromol.* **2018**, *109*, 273–286. [CrossRef]
47. Rathore, P.; Mahor, A.; Jain, S.; Haque, A.; Kesharwani, P. Formulation development, in vitro and in vivo evaluation of chitosan engineered nanoparticles for ocular delivery of insulin. *RSC Adv.* **2020**, *10*, 43629–43639. [CrossRef]
48. Campos, E.V.R.; Proença, P.L.F.; da Costa, T.G.; de Lima, R.; Fraceto, L.F.; de Araujo, D.R. Using Chitosan-Coated Polymeric Nanoparticles-Thermosensitive Hydrogels in association with Limonene as Skin Drug Delivery Strategy. *BioMed Res. Int.* **2022**, *2022*, 9165443. [CrossRef]
49. Conte, R.; De Luise, A.; Valentino, A.; Di Cristo, F.; Petillo, O.; Riccitiello, F.; Di Salle, A.; Calarco, A.; Peluso, G. Chapter 10—Hydrogel Nanocomposite Systems: Characterization and Application in Drug-Delivery Systems. In *Nanocarriers for Drug Delivery*; Elsevier: Amsterdam, The Netherlands, 2018.
50. Chen, Z.-P.; Liu, W.; Liu, D.; Xiao, Y.-Y.; Chen, H.-X.; Chen, J.; Li, W.; Cai, H.; Li, W.; Cai, B.-C.; et al. Development of brucine-loaded microsphere/thermally responsive hydrogel combination system for intra-articular administration. *J. Control. Release* **2012**, *162*, 628–635. [CrossRef]
51. Wang, S.; Wei, X.; Sun, X.; Chen, C.; Zhou, J.; Zhang, G.; Wu, H.; Guo, B.; Wei, L. A novel therapeutic strategy for cartilage diseases based on lipid nanoparticle-RNAi delivery system. *Int. J. Nanomed.* **2018**, *13*, 617–631. [CrossRef]
52. Ansari, M.Y.; Ahmad, N.; Haqqi, T.M. Oxidative stress and inflammation in osteoarthritis pathogenesis: Role of polyphenols. *Biomed. Pharmacother.* **2020**, *129*, 110452. [CrossRef] [PubMed]
53. Amrati, F.E.-Z.; Bourhia, M.; Slighoua, M.; Ibnemoussa, S.; Bari, A.; Ullah, R.; Amaghnouje, A.; Di Cristo, F.; El Mzibri, M.; Calarco, A.; et al. Phytochemical Study on Antioxidant and Antiproliferative Activities of Moroccan Caralluma europaea Extract and Its Bioactive Compound Classes. *Evid. Based Complement. Altern. Med.* **2020**, *2020*, 8409718. [CrossRef] [PubMed]
54. Zhuo, Q.; Yang, W.; Chen, J.; Wang, Y. Metabolic syndrome meets osteoarthritis. *Nat. Rev. Rheumatol.* **2012**, *8*, 729–737. [CrossRef] [PubMed]
55. Sarsour, E.H.; Kumar, M.G.; Kalen, A.L.; Goswami, M.; Buettner, G.R.; Goswami, P.C. MnSOD activity regulates hydroxytyrosol-induced extension of chronological lifespan. *AGE* **2011**, *34*, 95–109. [CrossRef]
56. Kim, Y.; Choi, Y.; Park, T. Hepatoprotective effect of oleuropein in mice: Mechanisms uncovered by gene expression profiling. *Biotechnol. J.* **2010**, *5*, 950–960. [CrossRef]

57. Ansari, M.Y.; Khan, N.M.; Ahmad, I.; Haqqi, T.M. Parkin clearance of dysfunctional mitochondria regulates ROS levels and increases survival of human chondrocytes. *Osteoarthr. Cartil.* **2018**, *26*, 1087–1097. [CrossRef]
58. Salgado, C.; Jordan, O.; Allémann, E. Osteoarthritis In Vitro Models: Applications and Implications in Development of Intra-Articular Drug Delivery Systems. *Pharmaceutics* **2021**, *13*, 60. [CrossRef]
59. Shi, Q.; Vaillancourt, F.; Côté, V.; Fahmi, H.; Lavigne, P.; Afif, H.; Di Battista, J.A.; Fernandes, J.C.; Benderdour, M. Alterations of metabolic activity in human osteoarthritic osteoblasts by lipid peroxidation end product 4-hydroxynonenal. *Arthritis Res. Ther.* **2006**, *8*, R159. [CrossRef]
60. Henrotin, Y.; Kurz, B.; Aigner, T. Oxygen and reactive oxygen species in cartilage degradation: Friends or foes? *Osteoarthr. Cartil.* **2005**, *13*, 643–654. [CrossRef]
61. Ruiz-Romero, C.; Calamia, V.; Mateos, J.; Carreira, V.; Martínez-Gomariz, M.; Fernández, M.; Blanco, F.J. Mitochondrial dysregulation of osteoarthritic human articular chondrocytes analyzed by proteomics: A decrease in mitochondrial superoxide dismutase points to a redox imbalance. *Mol. Cell. Proteom.* **2009**, *8*, 172–189. [CrossRef]
62. Kapoor, M.; Martel-Pelletier, J.; Lajeunesse, D.; Pelletier, J.-P.; Fahmi, H. Role of proinflammatory cytokines in the pathophysiology of osteoarthritis. *Nat. Rev. Rheumatol.* **2011**, *7*, 33–42. [CrossRef] [PubMed]
63. Veronesi, F.; Della Bella, E.; Cepollaro, S.; Brogini, S.; Martini, L.; Fini, M. Novel therapeutic targets in osteoarthritis: Narrative review on knock-out genes involved in disease development in mouse animal models. *Cytotherapy* **2016**, *18*, 593–612. [CrossRef] [PubMed]
64. Ravalli, S.M.; Szychlinska, M.A.; Leonardi, R.M.; Musumeci, G. Recently highlighted nutraceuticals for preventive management of osteoarthritis. *World J. Orthop.* **2018**, *9*, 255–261. [CrossRef]
65. Castrogiovanni, P.; Trovato, F.M.; Loreto, C.; Nsir, H.; Szychlinska, M.A.; Musumeci, G. Nutraceutical Supplements in the Management and Prevention of Osteoarthritis. *Int. J. Mol. Sci.* **2016**, *17*, 2042. [CrossRef]
66. Rosillo, M.; De La Lastra, C.A.; Castejón, M.L.; Montoya, T.; Cejudo-Guillén, M.; Sánchez-Hidalgo, M. Polyphenolic extract from extra virgin olive oil inhibits the inflammatory response in IL-1β-activated synovial fibroblasts. *Br. J. Nutr.* **2019**, *121*, 55–62. [CrossRef] [PubMed]
67. Chin, K.-Y.; Pang, K.-L. Therapeutic Effects of Olive and Its Derivatives on Osteoarthritis: From Bench to Bedside. *Nutrients* **2017**, *9*, 1060. [CrossRef] [PubMed]
68. Richard, N.; Arnold, S.; Hoeller, U.; Kilpert, C.; Wertz, K.; Schwager, J. Hydroxytyrosol Is the Major Anti-Inflammatory Compound in Aqueous Olive Extracts and Impairs Cytokine and Chemokine Production in Macrophages. *Planta Med.* **2011**, *77*, 1890–1897. [CrossRef]
69. Mével, E.; Merceron, C.; Vinatier, C.; Krisa, S.; Richard, T.; Masson, M.; Lesoeur, J.; Hivernaud, V.; Gauthier, O.; Abadie, J.; et al. Olive and grape seed extract prevents post-traumatic osteoarthritis damages and exhibits in vitro anti IL-1β activities before and after oral consumption. *Sci. Rep.* **2016**, *6*, 33567. [CrossRef]
70. Charlier, E.; Deroyer, C.; Ciregia, F.; Malaise, O.; Neuville, S.; Plener, Z.; Malaise, M.; de Seny, D. Chondrocyte dedifferentiation and osteoarthritis (OA). *Biochem. Pharmacol.* **2019**, *165*, 49–65. [CrossRef]
71. Musumeci, G.; Castrogiovanni, P.; Trovato, F.M.; Weinberg, A.M.; Al-Wasiyah, M.K.; Alqahtani, M.H.; Mobasheri, A. Biomarkers of Chondrocyte Apoptosis and Autophagy in Osteoarthritis. *Int. J. Mol. Sci.* **2015**, *16*, 20560–20575. [CrossRef]
72. Musumeci, G.; Aiello, F.C.; Szychlinska, M.A.; Di Rosa, M.; Castrogiovanni, P.; Mobasheri, A. Osteoarthritis in the XXIst Century: Risk Factors and Behaviours that Influence Disease Onset and Progression. *Int. J. Mol. Sci.* **2015**, *16*, 6093–6112. [CrossRef] [PubMed]
73. Wang, M.; Sampson, E.R.; Jin, H.; Li, J.; Ke, Q.H.; Im, H.-J.; Chen, D. MMP13 is a critical target gene during the progression of osteoarthritis. *Arthritis Res. Ther.* **2013**, *15*, R5. [CrossRef] [PubMed]
74. Wan, J.; Zhang, G.; Li, X.; Qiu, X.; Ouyang, J.; Dai, J.; Min, S. Matrix Metalloproteinase 3: A Promoting and Destabilizing Factor in the Pathogenesis of Disease and Cell Differentiation. *Front. Physiol.* **2021**, *12*, 663978. [CrossRef]
75. Ryu, J.-H.; Yang, S.; Shin, Y.; Rhee, J.; Chun, C.-H.; Chun, J.-S. Interleukin-6 plays an essential role in hypoxia-inducible factor 2α-induced experimental osteoarthritic cartilage destruction in mice. *Arthritis Rheum.* **2011**, *63*, 2732–2743. [CrossRef]
76. Benya, P.; Padilla, S.R.; Nimni, M.E. Independent regulation of collagen types by chondrocytes during the loss of differentiated function in culture. *Cell* **1978**, *15*, 1313–1321. [CrossRef]
77. Kawaguchi, J.; Mee, P.; Smith, A. Osteogenic and chondrogenic differentiation of embryonic stem cells in response to specific growth factors. *Bone* **2005**, *36*, 758–769. [CrossRef]
78. Setzu, A.; Lathia, J.D.; Zhao, C.; Wells, K.; Rao, M.S.; Ffrench-Constant, C.; Franklin, R.J.M. Inflammation stimulates myelination by transplanted oligodendrocyte precursor cells. *Glia* **2006**, *54*, 297–303. [CrossRef] [PubMed]
79. McCulloch, K.; Litherland, G.J.; Rai, T.S. Cellular senescence in osteoarthritis pathology. *Aging Cell* **2017**, *16*, 210–218. [CrossRef] [PubMed]
80. Brandl, A.; Hartmann, A.; Bechmann, V.; Graf, B.; Nerlich, M.; Angele, P. Oxidative stress induces senescence in chondrocytes. *J. Orthop. Res.* **2011**, *29*, 1114–1120. [CrossRef]
81. Rim, Y.A.; Nam, Y.; Ju, J.H. The Role of Chondrocyte Hypertrophy and Senescence in Osteoarthritis Initiation and Progression. *Int. J. Mol. Sci.* **2020**, *21*, 2358. [CrossRef]
82. Vinatier, C.; Dominguez, E.; Guicheux, J.; Caramés, B. Role of the Inflammation-Autophagy-Senescence Integrative Network in Osteoarthritis. *Front. Physiol.* **2018**, *9*, 706. [CrossRef] [PubMed]

Article

Preparation, Characterization, and Antioxidant Activity of Nanoemulsions Incorporating Lemon Essential Oil

Ting Liu [1,2], Zhipeng Gao [3], Weiming Zhong [3], Fuhua Fu [1,2], Gaoyang Li [1,2], Jiajing Guo [1,2,*] and Yang Shan [1,2,*]

[1] Longping Branch, Graduate School of Hunan University, Changsha 410125, China; ltchangsha98@163.com (T.L.); fhfu686@163.com (F.F.); lgy7102@163.com (G.L.)
[2] International Joint Lab on Fruits & Vegetables Processing, Quality and Safety, Hunan Key Lab of Fruits & Vegetables Storage, Processing, Quality and Safety, Hunan Agriculture Product Processing Institute, Hunan Academy of Agricultural Sciences, Changsha 410125, China
[3] College of Animal Science and Technology, Hunan Agricultural University, Changsha 410128, China; gaozhipeng627@163.com (Z.G.); zhongweiming2021@163.com (W.Z.)
* Correspondence: guojiajing1986@163.com (J.G.); sy6302@sohu.com (Y.S.); Tel.: +86-(0)731-8469-8915 (J.G.); +86-(0)731-8469-1289 (Y.S.)

Citation: Liu, T.; Gao, Z.; Zhong, W.; Fu, F.; Li, G.; Guo, J.; Shan, Y. Preparation, Characterization, and Antioxidant Activity of Nanoemulsions Incorporating Lemon Essential Oil. *Antioxidants* 2022, 11, 650. https://doi.org/10.3390/antiox11040650

Academic Editors: Li Liang and Hao Cheng

Received: 11 February 2022
Accepted: 25 March 2022
Published: 28 March 2022

Publisher's Note: MDPI stays neutral with regard to jurisdictional claims in published maps and institutional affiliations.

Copyright: © 2022 by the authors. Licensee MDPI, Basel, Switzerland. This article is an open access article distributed under the terms and conditions of the Creative Commons Attribution (CC BY) license (https://creativecommons.org/licenses/by/4.0/).

Abstract: Lemon essential oil (LEO) is a kind of citrus essential oil with antioxidant, anti-inflammatory, and antimicrobial activities, but low water solubility and biological instability hinder its industrial application. In this study, LEO was nanoemulsified to solve these problems. The preparation procedure of lemon essential oil nanoemulsions (LEO-NEs) was optimized, and the physicochemical characterization and antioxidant activities were explored. Single-factor experiments (SFEs) and response surface methodology (RSM) were conducted for the effects on the mean droplet size of LEO-NEs. Five factors of SFE which may influence the droplet size were identified: HLB value, concentration of essential oil, concentration of surfactant, ultrasonic power, and ultrasonic time. On the basis of the SFE, the RSM approach was used to optimize the preparation procedure to obtain LEO-NEs with the smallest droplet size. LEO-NEs exhibited good antioxidant activity when the HLB value was 13, content of surfactant was 0.157 g/mL, ultrasonic time was 23.50 min, and ultrasonic power was 761.65 W. In conclusion, these results can provide a good theoretical basis for the industrial application of lemon essential oil.

Keywords: lemon essential oil; nanoemulsions; ultrasonication; response surface methodology; antioxidant activities

1. Introduction

Essential oils (EOs), as volatile products of secondary plant metabolism, are well known for their antioxidant [1,2], anti-inflammatory [3], and antimicrobial [4] activities. Citrus EOs have high yield and demand in EO, which are major by-products of citrus processing. Lemon essential oil (LEO) is a kind of citrus EOs, which is commonly used for flavoring and fragrance. The FDA has also deemed LEO safe for use as a preservative or flavoring agent [5]. Furthermore, some researchers reported that LEO had antioxidant activity using DPPH, ABTs, and β-Carotene bleaching assays [6,7]. The antioxidant activity of the LEO is related to the preservation of food and the prevention of diseases. Thus, it has the prospect to replace synthetic preservatives [8].

However, the greatest impediment to the widespread use of LEO is its insolubility in water, and other disadvantages include volatility, low stability, and sensitivity to the environment. LEO could be encapsulated in emulsions to reduce its hydrophobicity, but conventional emulsions are thermodynamically unstable and the components tend to separate from each other [9]. These problems can be solved by nanoemulsions (NEs) prepared using emerging nanotechnology [10]. A NE is a type of drug delivery system with a simple preparation process and stable formulation quality. It has a certain kinetic and thermodynamic stability [10], which can effectively improve the stability of the drug after emulsification on the one hand, and reduce the irritation of drug delivery on the other [11]. The droplet size of the NE is relatively small (20–200 nm) [12,13]. Meanwhile, the particle size of NEs determines its surface and interface properties. NEs with small particle size have a low particle weight and high surface-to-volume ratio, and the Brownian motion of small particle NEs can overcome gravity, which can reduce the occurrence of coalescence, aggregation, and flocculation [14,15]. However, the small droplets in oil-in-water nanoemulsions are mainly composed of oil and dispersed in water with surfactant, whereby their minimum particle size is limited by the oil [16]. Currently, the methods for preparing NEs are classified into high-energy emulsification and low-energy emulsification methods according to the physicochemical mechanism of droplet rupture. Ultrasound is a widely used high-energy process to prepare NEs. It consumes less surfactant with smaller particles compared to the low-energy method [17]. Meanwhile, it provides better control of the system and has a lower production cost than other high-energy methods (microfluidization, high-pressure homogenization) [18].

Thus, the main purpose of this study was to employ the ultrasonic method for preparing LEO-NEs with small particle size, good stability, and high antioxidant activity. SFEs and RSM were employed to prepare optimized LEO-NEs and investigate the individual effects of the independent variables on the droplet size. The findings can provide a basis for formulating and rationalizing the application of LEO-NEs and lay the foundation for their scale-up production in the cosmetics and the food industries.

2. Materials and Methods

2.1. Materials and Chemicals

Lemon was obtained from Sichuan Province. Tween-80 and Span-80 were purchased from Sinopharm Chemical Reagent Co., Ltd. (Shanghai, China). the total antioxidant capacity assay kits with DPPH and ABTS were purchased from Suzhou Comin Biotechnology Co., Ltd. (Suzhou, China). Ultrapure water (MILLI Q) was used in the experiments.

2.2. Methods

2.2.1. Extraction and GC-MS Analysis of Lemon Essential Oil (LEO)

The method of extracting essential oil referred to Guo et al. [4]. LEO was extracted from a mixture of lemon peel and water by steam distillation. Sodium chloride was added in the extraction process, and anhydrous sodium sulfate was added to dry the essential oil after extraction. The determination of LEO components was determined according to procedures reported earlier [19]. LEO was analyzed by GC–MS using an Agilent 7890A GC with a Gerstel MPS autosampler and an Agilent 5975C MSD detector. The carrier gas was helium with a flow rate of 1 mL/min. The temperature was programmed as follows: the initial temperature of 40 °C was maintained for 1 min; the temperature was increased to 220 °C at a rate of 3 °C/min for 25 min; the final temperature of 250 °C was reached at a rate of 5 °C/min for 10 min. MS conditions were 70 eV EI and an ion source temperature of 230 °C. The mass-to-charge (m/z) range was set to 35–350 atomic units. The National Institute of Standards and Technology (NIST 08) was used to compare the data of the LEO components.

2.2.2. Preparation of Lemon Oil-Based Nanoemulsions (LEO-NEs)

The LEO-NEs were formed from LEO, a mixture of two surfactants (Tween-80 and Span-80), and deionized water. A procedure for the oil–water mixtures was followed to obtain 20 mL; the pre-emulsion was centrifuged for 5 min at 10,000 rpm using a high-speed homogenizer (F6/10, Jingxin, Shanghai, China). The homogenate was processed further by an ultrasonicator (JY92-11D, Jingxin, Shanghai, China). During the ultrasonication process, samples were put in ice water for a low-temperature environment.

2.3. Optimization and Statistical Design of LEO-NEs
2.3.1. Single-Factor Experiments (SFE)

Single-factor experiments were designed to investigate the effects of hydrophilic–lipophilic balance (HLB) value, content of Span-80 and Tween-80 (STmix), concentration of essential oil, ultrasonic time, and ultrasonic power on the mean droplet size, which can also provide a reasonable data range for the design of the response surface methodology. Specific parameters are presented in Table 1. The HLB value represented the combined affinity of hydrophilic and oleophilic groups in emulsifier molecules for oil or water [20]. Different HLB values of surfactants can contribute to the formation of two types of emulsions: water-in-oil (W/O) emulsion and oil-in-water (O/W) emulsion. To prepare the O/W LEO-NEs with hydrophilicity, an oil-in-water emulsifier with a high HLB value (8–15) was chosen. According to Nirmal et al. [21], different combinations of STmix were used to create surfactant HLB values ranging from 8–15, as shown in Table 2.

Table 1. Variables of single-factor experiments (SFE) and Response surface methodology (RSM).

Factors of SFE	Variables						
HLB value of STmix	8	9	10	11	12	13	14	15
concentration of LEO (g/mL)	0.05	0.06	0.07	0.08	0.09	0.1		
STmix content (g/mL)	0.0125	0.025	0.05	0.1	0.2			
ultrasonic time (min)	0	10	20	30	40			
ultrasonic power (W)	100	300	500	700	900			
Independent Variables of RSM	Levels		
	−1	0	1
A: HLB value of STmix	11	12	13
B: content of STmix (g/mL)	0.05	0.125	0.2
C: ultrasonic time (min)	10	20	30
D: ultrasonic power (W)	500	700	900

Table 2. Different combinations of Span 80 and Tween 80 used to create surfactant HLB value.

HLB	Span 80 (%)	Tween 80 (%)
8	65.4	34.6
9	56.9	43.1
10	46.7	53.3
11	37.4	62.6
12	28	72
13	18.7	81.3
14	9.3	90.7
15	0	100

2.3.2. Response Surface Methodology (RSM) Design

The levels of the independent variables to be used in the Box–Behnken designs were determined by the results of the SFE. The RSM explored the effects of the selection factor over 29 runs. In this work, the BBD with four variables (factor A was the HLB value, factor B was the STmix content, factor C was the ultrasonic time, and factor D was the ultrasonic power) at three levels (−1, 0, 1) was carried out to evaluate the effect on the dependent variable. The mean droplet size (Y) was the response value. The optimum formulation was chosen by the analysis of the RSM. The specific parameters are shown in Table 1.

2.4. Characterization of LEO-NEs
2.4.1. Mean Droplet Size and Polydispersity Index (PDI) of LEO-NEs

The mean droplet size, particle size distribution, and PDI were measured using an NS-90 nano-granularity analyzer (Malvern Instruments Ltd., Malvern, UK). The average diameter of the particles indicated the average particle size. The intensity of particles of different diameters indicated the particle size distribution. To avoid bubbles and multiple light scattering, the LEO-NE was diluted 50-fold with ultrapure water. Three sets of measurements were performed in each sample to determine the mean droplet size and PDI of LEO-NEs in 1 mL of the diluted samples.

2.4.2. Transmission Electron Microscopy (TEM) Images of LEO-NEs

The particle morphology of the LEO-NEs with a 20-fold dilution was observed by TEM (Hitachi HT-7700, Tokyo, Japan). Dilution was undertaken to prevent inter-particle aggregation.

2.4.3. DPPH Radical-Scavenging Activity

The DPPH scavenging assay using 0.5 g/mL of LEO and LEO-NEs (stored for 7 days) was measured following the kit instructions. The EOs were diluted with extraction buffer in the kit. Firstly, 380 µL of Reagent 1 was added to 20 µL of sample and then shaken vigorously for 20 min. The change in absorbance was measured at 515 nm by the microplate reader (Thermo Scientific, Waltham, MA, USA). The percentage inhibition free radical scavenging rate of DPPH was calculated as follows:

$$\text{DPPH scavenging activity (Inhibition\%)} = [(A_{control} - A_{sample})/A_{control}] \times 100 \quad (1)$$

2.4.4. ABTs Radical-Scavenging Activity

The ABTs scavenging assay of 0.5 g/mL of LEO and LEO-NEs (stored for 7 days) was performed following the kit instructions. The change in absorbance was measured at 734 nm by the microplate reader. The percentage inhibition of ABTs was calculated with the following formula:

$$\text{ABTs scavenging activity (Inhibition\%)} = [(A_{control} - A_{sample} + A_{blank})/A_{control}] \times 100 \quad (2)$$

2.5. Data Analysis

The results of the single-factor experiments were analyzed by Graphpad Prism version 8 software. The statistical analysis of the results of the response surface test was performed by Design-Expert version 13 software. All of the components, as well as their probable interactions, were examined using statistical parameters for analyses of variance (ANOVAs), such as degrees of freedom, F-ratios, and p-values. The model with a good fit to the data was selected ($p < 0.05$).

3. Results

3.1. Chemical Composition of the Lemon Essential Oil

A lemon-like odor liquid oil isolated by steam distillation from lemon peels was transparent and colorless. The components of the LEO identified are given in Table 3. Analysis of the volatile constituents of the LEO compounds by GC–MS identified 15 compounds that accounted for more than 0.5%, with a total of 96.36%. The major components detected in LEO were limonene (48.54%), α-pinene (30.90%), β-citral (3.65%), and β-myrcene (3.01%). As seen, the main constituents of the EO in this study were composed of monoterpene hydrocarbons (83.53%), including limonene, α-pinene, β-myrcene, and terpinolene.

Table 3. Chemical composition (%) of the essential oil isolated from lemon peels.

No	Main Component	Content (%)	Classification
1	Limonene	48.54	Monoterpene Hydrocarbons
2	α-Pinene	30.9	Monoterpene Hydrocarbons
3	β-Citral	3.65	Monoterpene aldehydes
4	β-Myrcene	3.01	Monoterpene Hydrocarbons
5	Neryl Acetate	1.74	Oxygenated Terpenes
6	β-Bisabolene	1.31	Sesquiterpene Hydrocarbons
7	α-Terpineol	1.11	Oxygenated Terpenes
8	Terpinolene	1.08	Monoterpene Hydrocarbons
9	α-bergamotene	0.97	Sesquiterpene Hydrocarbons
10	Thujane	0.85	Monoterpene alkanes
11	Caryophyllene	0.72	Sesquiterpene Hydrocarbons
12	4-Terpineol	0.68	Oxygenated Terpenes
13	Geraniol	0.68	Oxygenated Terpenes
14	Nerol	0.61	Oxygenated Terpenes
15	Valencene	0.51	Sesquiterpene Hydrocarbons

3.2. Single-Factor Experiments

The effects of parameters on the mean droplet size of LEO-NEs were investigated using single-factor experiments, including the HLB value (8, 9, 10, 11, 12, 13, 14 and 15), concentration of essential oil (0.05, 0.06, 0.07, 0.08, 0.09 and 1 g/mL), concentration of surfactant (0.0125, 0.025, 0.05, 0.1 and 0.2 g/mL), ultrasonic power (100, 300, 500, 700 and 900 W) and ultrasonic time (0, 10, 20, 30 and 40 min). The ranges for parameter values of RSM were set to the right and left of the optimum values.

3.2.1. Effect of HLB Value on the Mean Droplet Size of LEO-NEs

The HLB value of the surfactant can assist in identifying the best-suited stabilizer. When the HLB value of the STmix couple matches the HLB value required for the EO to form nanoemulsions, NEs with a small droplet size can be produced [22]. It was a crucial step to select an appropriate HLB value to obtain LEO-NEs with the smallest particle size. In the present work, the impact of the HLB value on the mean droplet size of LEO-NEs was studied first. As indicated in Figure 1a, when the HLB value changed from 8–12, the mean droplet size progressively declined, while the mean droplet size exhibited an upward trend when the HLB value was above 12. Furthermore, the particle size of LEO-NEs grew considerably when the HLB value increased from 14 to 15. Therefore, the optimum HLB value for the smallest droplet of LEO-NEs was 12.

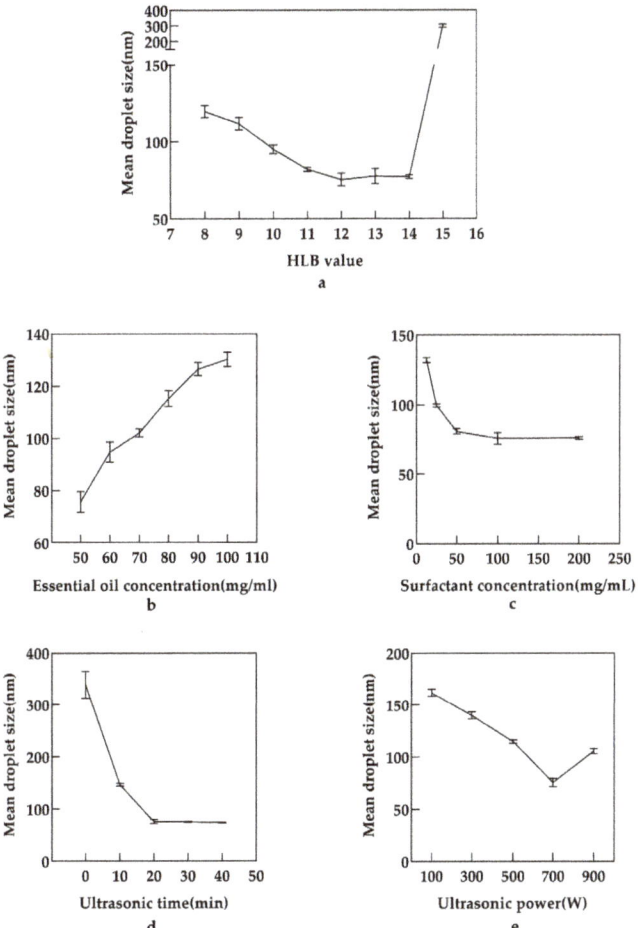

Figure 1. Effects of HLB value (**a**), essential oil concentration (**b**), Surfactant concentration (**c**), ultrasound time (**d**) and ultrasonic power (**e**) on the mean droplet size of NEO-NEs.

3.2.2. Effect of Essential Oil Concentration on the Mean Droplet Size of LEO-NEs

To explore the effect of essential oil concentration on the mean droplet size of LEO-NEs, the formulation was performed with different LEO concentrations (ranging from 0.05 to 0.1 g/mL). As shown in Figure 1b, a significant increase in the mean droplet size was observed when the LEO content was changed from 0.05 g/mL to 0.1 g/mL. The nanoemulsion with a low concentration of LEO was more suitable for production applications. According to our results, the essential oil concentration of 0.05 g/mL in LEO-NEs was finally chosen for the subsequent experiments.

3.2.3. Effect of Surfactant Concentration on the Mean Droplet Size of LEO-NEs

STmix with different concentrations was used in the NEs system. As shown in Figure 1c, a sharp decrease in the mean droplet size from 133.71 to 75.66 nm was observed when STmix concentration increased from 0.0125 to 0.1 g/mL. On the other hand, it remained almost constant when increasing the surfactant concentration from 0.1 to 0.2 g/mL. Therefore, the 0.1 g/mL surfactant concentration was selected for subsequent experiments.

3.2.4. Effect of Ultrasonic Time on the Mean Droplet Size of LEO-NEs

Various ultrasonic time was used to prepare the NEs, with the aim of investigating the effects on the mean droplet size of LEO-NEs. As shown in Figure 1d, when the ultrasonic time was 0, which means that the emulsion was not treated by ultrasound, the mean droplet size fluctuated over a wide range, and the repeatability of the experiment was poor. Meanwhile, a layering phenomenon was observed after staying still at room temperature overnight. The smallest particle size was observed at the ultrasonic time of 20 min. The increase in ultrasonic time can promote the integration of water and oil. Longer ultrasonic times, on the other hand, may result in higher degradation or disintegration of bioactive chemicals in LEO, as well as energy waste [23]. Therefore, ultrasonic time of 20 min was selected for the subsequent studies considering both saving energy and achieving the best results.

3.2.5. Effect of Ultrasonic Power on the Mean Droplet Size of LEO-NEs

To study the effects of ultrasonic power on the mean droplet size of LEO-NEs, the preparation process was carried out with different ultrasonic powers ranging from 100 to 900 W. As shown in Figure 1e, the value of droplet size decreased with the increase in ultrasonic power from 100 to 700 W. However, the particle size increased instead when the ultrasonic power was increased from 700 to 900 W. Excessive ultrasonic power may induce a rise in the number of bubbles in solvents during cavitation, lowering the efficiency of the ultrasound energy delivered into the medium [24]. As a result, the ultrasonic power of 700 W was chosen for further experiments.

3.3. Response Surface Optimization of LEO-NEs

The preparation process was further optimized using BBD experiments to obtain the best experimental parameters. The BBD of the response surface was used to optimize the formulation and preparation of LEO-NEs. The following regression equation model was obtained by regression analysis:

$$Y = 88.50 - 14.85\,A + 6.87\,B - 31.66\,C - 10.15\,D - 5.40\,AB + 4.88\,AC + 2.60\,AD \\ - 17.62\,BC + 6.21\,BD + 3.01\,CD + 3.93\,A^2 + 19.65\,B^2 + 14.46\,C^2 + 3.34\,D^2 - 26.38\,A^2B \\ + 10.80\,A^2C - 0.1079\,A^2D + 13.10\,AB^2 + 13.07\,AC^2 + 2.67\,B^2C - 11.58\,B^2D + 2.48\,BC^2 \quad (3)$$

As shown in Table 4, the generation of a model with no significant lack of fit implied its suitability. The $R^2 > 90\%$ indicated that the model could accurately reflect the change in the response value when the fitness was high. The coefficient of variance (CV), which is the ratio of the estimated standard error to the mean of the observed responses, is related to the model reproducibility. Our model had a CV of 9.97% (<10%), which is usually considered to be sufficiently reproducible. The ratio of adequate precision reflects the ratio of response to the deviation, and its value was 11.840 (>4), indicating an adequate signal.

An analysis of variance of the regression coefficient revealed that C was extremely significantly different ($p < 0.01$), while A significantly differed in its linear effect ($p < 0.05$), remaining factors indicated a non-significant difference. The main effect relationship of each factor could be ranked as ultrasonic time > HLB value > ultrasonic power > surfactant content. Among interaction effects, B^2 had extremely significant differences ($p < 0.01$), while BC and C^2 significantly differed ($p < 0.05$). The remaining effects were not significant ($p > 0.1$). The experimental values of mean droplet size of nanoemulsions are presented in Table 5.

Table 4. ANOVA of RSM outcome [α].

Source	Sum of Squares	df	Mean Square	F-Value	p-Value	
Model	19,794.11	22	899.73	8.11	0.0077	significant
A-HLB	881.89	1	881.89	7.95	0.0304	
B-Surfactant content	188.81	1	188.81	1.7	0.2398	
C-Ultrasonic time	4009.95	1	4009.95	36.14	0.001	
D-Ultrasonic power	411.79	1	411.79	3.71	0.1023	
AB	116.53	1	116.53	1.05	0.3449	
AC	95.39	1	95.39	0.8598	0.3896	
AD	26.95	1	26.95	0.243	0.6396	
BC	1241.27	1	1241.27	11.19	0.0155	
BD	154.36	1	154.36	1.39	0.2828	
CD	36.29	1	36.29	0.3271	0.5881	
A^2	100.41	1	100.41	0.905	0.3782	
B^2	2505.85	1	2505.85	22.59	0.0032	
C^2	1357.05	1	1357.05	12.23	0.0129	
D^2	72.53	1	72.53	0.6538	0.4496	
Residual	665.65	6	110.94			
Lack of Fit	514.75	2	257.38	6.82	0.0514	not significant

[α] $R^2 = 0.97$; adj. $R^2 = 0.85$; C.V. (%) = 9.97; adequate precision = 11.84.

Table 5. Experimental values of mean droplet size of nanoemulsions obtained from BBD experimental design.

Run	HLB	Surfactant Content (g/mL)	Ultrasound Time (min)	Ultrasound Power (W)	Mean Droplet Size (nm)
1	12	0.2	20	500	132.53
2	12	0.2	20	900	101.51
3	12	0.125	10	900	119.93
4	11	0.2	20	700	94.85
5	11	0.125	20	900	104.0
6	13	0.125	30	700	87.79
7	11	0.125	20	500	129.70
8	12	0.05	20	900	75.35
9	13	0.125	20	900	79.50
10	13	0.125	20	500	94.81
11	12	0.125	30	500	76.90
12	13	0.125	10	700	119.75
13	12	0.125	10	500	146.25
14	12	0.05	20	500	131.22
15	12	0.125	20	700	94.79
16	13	0.2	20	700	80.56
17	11	0.125	30	700	81.59
18	12	0.125	20	700	92.41
19	12	0.2	10	700	184.80
20	12	0.125	20	700	91.35
21	12	0.05	10	700	130.87
22	12	0.125	20	700	83.11
23	12	0.125	30	900	62.63
24	11	0.05	20	700	123.07
25	12	0.05	30	700	108.12
26	13	0.05	20	700	130.37
27	12	0.125	20	700	80.830
28	12	0.2	30	700	91.587
29	11	0.125	10	700	133.083

Finally, the response surface was plotted using Design-expert 13. The effect of the two-factor interaction on the size of the mean droplet is intuitively shown in Figure 2. The response surface slope is steeper in Figure 2d, indicating that the interplay of surfactant content and ultrasound time had a bigger impact on LEO-NE particle size. The gradient of the response surface was moderate, as shown in Figure 2c, showing that the interaction of HLB value and ultrasonic power had less of an effect on the droplet size.

3.4. Physicochemical Properties and Stability of LEO-NEs

The above regression model was used to generate the optimum process parameters and validation results. When the HLB value (A) was 13, surfactant content (B) was 0.157 g/mL, ultrasonic time (C) was 23.50 min, and ultrasonic power (D) was 761.65 W, the predicted minimum mean droplet size was 66.82 nm. The actual mean particle diameter was 64.60 nm, and the PDI was 0.255. The particle size distribution of NEO-NEs is shown in Figure 3a. Then, the morphological changes and the changes in particle size of NEs during the storage period were observed, and the difference in antioxidant activity between emulsion and essential oil was compared.

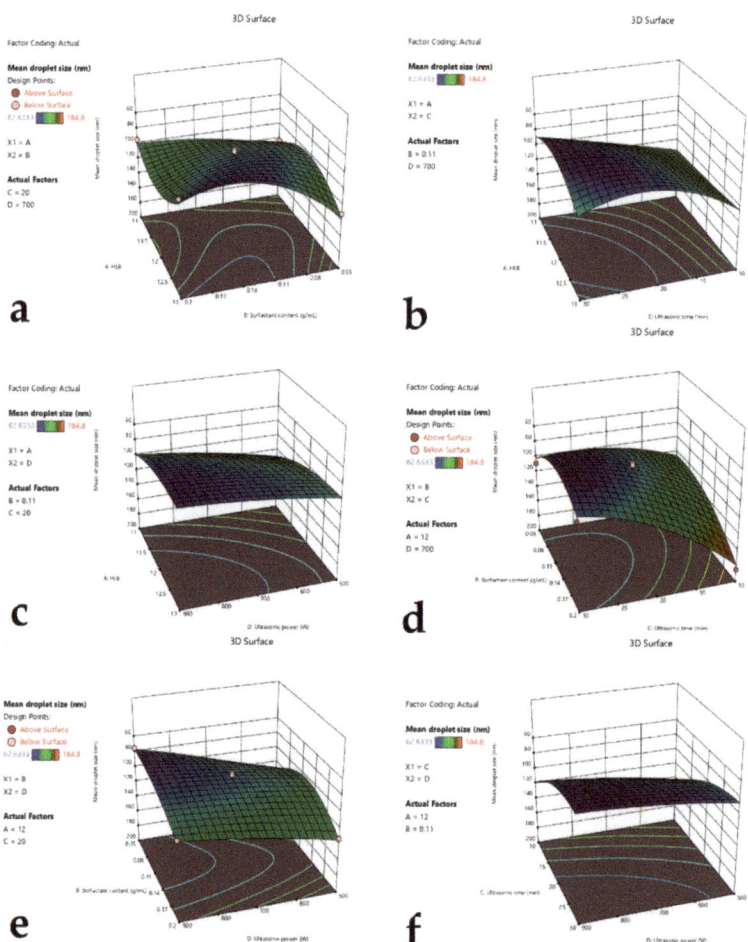

Figure 2. Response surface plot showing the significant ($p < 0.05$) interaction effect for mean droplet size as a function of (**a**) HLB value of STmix and STmix content, (**b**) HLB value of STmix and ultrasonic time, (**c**) HLB value of STmix and ultrasonic power, (**d**) STmix content and ultrasonic time, (**e**) STmix content and ultrasonic power, and (**f**) ultrasonic time and ultrasonic power.

3.4.1. Morphological Observation of LEO-NEs

TEM was used to describe the morphology of the LEO-NEs, as shown in Figure 3a, the droplets were well distributed and spherical. However, the diameters of the particles in the NEs were not the same, ranging from 50–100 nm, in line with those reported by the particle size meter.

3.4.2. Changes in Particle Size of LEO-NEs during the Storage Period

Figure 3b demonstrates the changes in the mean droplet of LEO-NEs for 1 week at a storage temperature of 25 °C. The particle size of the LEO-NEs changed little during a week, ranging from 62.96 nm to 64.60 nm.

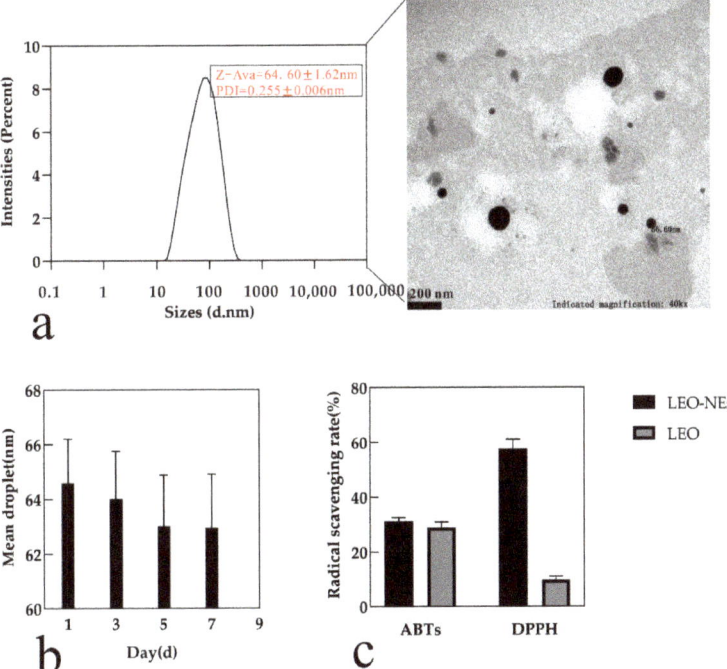

Figure 3. Physicochemical Properties and Stability of LEO-NEs: (**a**) particle size distribution and transmission electron microscopy (TEM) of droplets of LEO-NEs; (**b**) effect of storage time on particle size of LEO-NEs; (**c**) antioxidant activity comparison between LEO-NEs and LEO.

3.4.3. Antioxidant Activity of LEO-NEs

The DPPH and ABTs assays were used to measure the free-radical-scavenging potential to compare the difference between LEO-NEs and EO at the same concentration (0.05 g/mL). It has been shown that EOs exhibit antioxidant activities due to large amounts of polyphenol compounds. Figure 3c shows that the antioxidant activity of the LEO-NEs was higher than that of LEO with the same concentration, whereby the inhibition of DPPH free radicals by LEO-NEs (57.61%) was much better than that by LEO (8.74%), but the inhibition of ABTs free radicals by LEO-NEs (31.74%) was similar to that by LEO (30.61%).

4. Discussion

Coinciding with the reports by other Hirai et al. [25], Aguilar et al. [26], Perdones et al. [27] and Campolo et al. [28], limonene was the most abundant component in LEO while its content may vary. LEO was rich in constituents with monoterpene structure (limonene, α-pinene, etc.) which have been proven to possess antioxidant activity [29]. For example, limonene was shown to prevent neuronal suffering [30], oxidative stress on lymphocytes, and mitochondrial dysfunction [29] through its antioxidant activity. In addition, LEO components present in other studies were not detected in this experiment such as β-phellandrene [25], camphene, and sabinene [26]. Differences in LEO composition may be due to differences in geographic location, environmental factors, plant age, developmental stage, harvest time, extraction site, and extraction method [31].

In this study, LEO-NEs were prepared by the ultrasonic method using STmix as an emulsifier. The influence of each factor was studied by SFEs. Firstly, when the HLB values of STmix ranged from 8–14, the mean droplet size of NEs was less than 200 nm, while NEs could not be formed when HLB was 15. The large range of suitable HLB values means that many kinds of surfactants can be used to prepare LEO-NEs. Tween-80 was not suitable for this experiment; however, it was also used to make lemon LEO-NEs in other experiments. Mossa et al. [13] reported that the droplet size of LEO-NEs was 131.9 nm, while the particle size of the LEO-NEs was 181.5 nm in the study of Yazgan [32]. Although these studies were able to form NEs with Tween-80, the particle sizes were larger than 100 nm. Furthermore, the droplet diameter of LEO-NEs was 91 nm [33] and 135 nm [34] when produced with Tween-80 using the high-pressure homogenizer method. These results indicate that different essential oil components and different emulsification methods may lead to different particle sizes when constructing NEs.

When the concentration of essential oil was 0.05–0.1 g/mL, the particle size increases with the increase in concentration of essential oil, indicating its greater impact on particle size. However, previous studies showed that the particle size would not increase when the essential oil exceeds a certain amount if the concentration of surfactant micelles remains not changed, because of a saturation with lemon oil, whereby any further lemon oil droplets added to the nanoemulsions would not dissolve [35]. This phenomenon did not occur in our experiments because the concentration range of LEO was not large enough. An increase in surfactant concentration can also lead to a decrease in particle size. The surfactants can affect inter-particle interactions in emulsions, whereby a the higher surfactant concentration results in weaker inter-particle interactions and smaller droplets formed [36]. The effect of surfactant concentration on mean particle size may be related to the surfactant dose required to cover the surface of the formed droplets, whereby self-emulsification would be more dependent on surfactant concentration [37]. In the process of ultrasonic preparation of NEs, the particle size did not decrease when the surfactant concentration increased to a certain amount. In addition, the dependence of the mean particle size on surfactant concentration did not depend strongly on storage time and temperature [12].

Ultrasonic cavitation is a feasible and energy-efficient method for preparing NEs, which offers improvements in terms of stability and decreases the Ostwald ripening rate. During the ultrasonication processes, soundwave energy causes cavities and sinusoidal pressure variations in the liquid–liquid interphase, resulting in a shockwave action on the particle surface and a reduction in particle size [38]. The particle size of nanoemulsions prepared with the ultrasonic method is generally determined by the sonication time and sonication power, but is insensitive to ultrasonication amplitude [39]. Understanding the dynamic routes is critical for reducing processing time and avoiding energy oversupply. When the ultrasonic time reached a certain value, the particle size reached the minimum. Increasing the ultrasonic time would not lead to a significant change in particle size. The increase in ultrasonic power led to a decrease and then increase in particle size, coinciding with the report of Kentish et al. [40]. In addition, Floris et al. [41] reported that high ultrasonic power may destroy bioactive substances.

The small size of the particles in NEs would result in less agglomeration or precipitation and higher stability of the system [42,43]. RSM was used to optimize NEs to obtain the smallest droplet size. The interaction between surfactant concentration and ultrasonic time had the greatest effect on particle size. However, the particle size did not decrease indefinitely, as it was limited by the ingredients of the essential oil. The optimal preparation conditions obtained by RSM were similar to those obtained by SFE, and the conditions predicted by RSM were relatively more precise.

Due to the mass transfer of oil molecules, droplets in NEs change from smaller droplets to larger droplets through an intermediate water phase, which is called Ostwald ripening. Ostwald ripening leads to droplet growth and phase separation [44]. From the TEM image and the particle size change during storage, particle diameter does not exceed 200 nm; hence, the ripening phenomenon was not serious in LEO-NEs. However, the TEM images revealed that the diameters of the particles in the nanoemulsion were not the same. The TEM image was similar to that presented by Kaur et al. [45] and Zhong et al. [46]. In previous studies, the structure of NE was presented a spherical substance consisting of several small spherical packets [46]. The particle size of LEO-NEs had the tendency to decrease in 1 week, possibly due to the EOs in the NEs undergoing a small amount of evaporation, thereby reducing the content of essential oil. In the study of Zhong et al. [46], there was a tendency for the particle size to increase with storage time, which may have been due to Ostwald ripening.

The prepared NE was not only stable but also had sustained-release activities. The study of antioxidant activities is essential as reflected in the reduction in reactive oxygen species (ROS) in the food and cosmetics industries. We could find that the encapsulation of essential oils in NEs helped to enhance their antioxidant activities when comparing the antioxidant activity of essential oils and NEs. DPPH scavenging activity refers to the ability to reduce the stable DPPH free radical to its reduced form DPPH-H [47]. ABTs scavenging activity refers to the ability to decolorize the radical cation (ABTS$^{\bullet+}$) [48]. Due to the different principles of determination, the two results are not necessarily related. The different methods employed to indicate antioxidant activity can comprehensively profile the antioxidant activities of LEO-NEs. According to a previous study [6], the DPPH radical scavenging activity and ABTs radical-scavenging activity of pure LEO were 32.85% and 41.57% respectively. These results are close to the antioxidant capacity of LEO-NEs in our study, but the composition and determination method of the essential oil had an impact on the results. In addition, the antioxidant activity of LEO and LEO-NEs may be due to components in LEO with antioxidant activity. However, The antioxidant activity of LEO should not only consider the primary constituents [49]. The main antioxidant components in lemon essential oil need to be further studied.

5. Conclusions

This study explored the optimum preparation procedure of LEO-NEs using SFEs and RSM. The optimal parameters were as follows: HLB value of 13, surfactant content of 0.157 g/mL, ultrasonic time of 23.50 min, and ultrasonic power of 761.65 W. The optimized mean droplet size was 64.60 nm. In addition, the TEM images and storage results demonstrated the good dispersion and stability of LEO-NEs. The antioxidant activity experiments showed that LEO-NEs had better antioxidant capacity than essential oils. Some of the characteristics of LEO-NEs investigated in this study and future endeavors may lay the foundation for the practical application of antioxidant activity and other biological activities of LEO-NEs.

Author Contributions: Conceptualization, J.G.; methodology, T.L.; software, T.L. and W.Z.; writing—review and editing, T.L., Y.S. and J.G.; visualization, T.L., W.Z. and Z.G.; supervision, G.L., F.F. and Y.S.; project administration and funding acquisition, Y.S. All authors contributed to the article and approved the submitted version. All authors have read and agreed to the published version of the manuscript.

Funding: This work was financially supported by the Agricultural Science and Technology Innovation Project of Hunan Province, China, (2021CX05, the Changsha Municipal Natural Science Foundation (kq2202332, kq2014070), the Agricultural Science and Technology Innovation Fund of Hunan (2020CX47), the Key Laboratory of Agro-Products Processing, Ministry of Agriculture and Rural Affairs of China (S2021KFKT-22), the National Natural Science Foundation of China (32073020), and the Hunan innovative province construction project (2019NK2041).

Institutional Review Board Statement: Not applicable.

Informed Consent Statement: Not applicable.

Data Availability Statement: Data is contained within the article.

Conflicts of Interest: The authors declare no conflict of interest.

References

1. Ilic, Z.S.; Milenkovic, L.; Tmusic, N.; Stanojevic, L.; Stanojevic, J.; Cvetkovic, D. Essential oils content, composition and antioxidant activity of lemon balm, mint and sweet basil from Serbia. *LWT* **2022**, *153*, 112210. [CrossRef]
2. Herrera-Calderon, O.; Chacaltana-Ramos, L.J.; Huayanca-Gutierrez, I.C.; Algarni, M.A.; Alqarni, M.; Batiha, G.E.S. Chemical Constituents, In Vitro Antioxidant Activity and In Silico Study on NADPH Oxidase of *Allium sativum* L. (Garlic) Essential Oil. *Antioxidants* **2021**, *10*, 1844. [CrossRef] [PubMed]
3. Pucci, M.; Raimondo, S.; Zichittella, C.; Tinnirello, V.; Corleone, V.; Aiello, G.; Moschetti, M.; Conigliaro, A.; Fontana, S.; Alessandro, R. Biological Properties of a Citral-Enriched Fraction of Citrus limon Essential Oil. *Foods* **2020**, *9*, 1290. [CrossRef] [PubMed]
4. Guo, J.J.; Gao, Z.P.; Li, G.Y.; Fu, F.H.; Liang, Z.E.N.; Zhu, H.; Shan, Y. Antimicrobial and antibiofilm efficacy and mechanism of essential oil from Citrus Changshan-huyou Y. B. chang against Listeria monocytogenes. *Food Control* **2019**, *105*, 256–264. [CrossRef]
5. Sharma, N.; Tripathi, A. Effects of *Citrus sinensis* (L.) Osbeck epicarp essential oil on growth and morphogenesis of *Aspergillus niger* (L.) Van Tieghem. *Microbiol. Res.* **2008**, *163*, 337–344. [CrossRef] [PubMed]
6. Guo, J.J.; Gao, Z.P.; Xia, J.L.; Ritenour, M.A.; Li, G.Y.; Shan, Y. Comparative analysis of chemical composition, antimicrobial and antioxidant activity of citrus essential oils from the main cultivated varieties in China. *LWT* **2018**, *97*, 825–839. [CrossRef]
7. Ben Hsouna, A.; Ben Halima, N.; Smaoui, S.; Hamdi, N. Citrus lemon essential oil: Chemical composition, antioxidant and antimicrobial activities with its preservative effect against Listeria monocytogenes inoculated in minced beef meat. *Lipids Health Dis.* **2017**, *16*, 146. [CrossRef]
8. Fernandez-Lopez, J.; Viuda-Martos, M. Introduction to the Special Issue: Application of Essential Oils in Food Systems. *Foods* **2018**, *7*, 56. [CrossRef]
9. Aghababaei, F.; Cano-Sarabia, M.; Trujillo, A.J.; Quevedo, J.M.; Ferragut, V. Buttermilk as Encapsulating Agent: Effect of Ultra-High-Pressure Homogenization on Chia Oil-in-Water Liquid Emulsion Formulations for Spray Drying. *Foods* **2021**, *10*, 1059. [CrossRef]
10. Lawrence, M.J.; Rees, G.D. Microemulsion-based media as novel drug delivery systems. *Adv. Drug Deliv. Rev.* **2012**, *64*, 175–193. [CrossRef]

11. Zainol, S.; Basri, M.; Bin Basri, H.; Shamsuddin, A.F.; Abdul-Gani, S.S.; Karjiban, R.A.; Abdul-Malek, E. Formulation Optimization of a Palm-Based Nanoemulsion System Containing Levodopa. *Int. J. Mol. Sci.* **2012**, *13*, 13049–13064. [CrossRef] [PubMed]
12. Walker, R.M.; Decker, E.A.; McClements, D.J. Physical and oxidative stability of fish oil nanoemulsions produced by spontaneous emulsification: Effect of surfactant concentration and particle size. *J. Food Eng.* **2015**, *164*, 10–20. [CrossRef]
13. Mossa, A.T.H.; Mohafrash, S.M.M.; Ziedan, E.S.H.E.; Abdelsalam, I.S.; Sahab, A.F. Development of eco-friendly nanoemulsions of some natural oils and evaluating of its efficiency against postharvest fruit rot fungi of cucumber. *Ind. Crop. Prod.* **2021**, *159*, 113049. [CrossRef]
14. Falleh, H.; Ben Jemaa, M.; Neves, M.A.; Isoda, H.; Nakajima, M.; Ksouri, R. Peppermint and Myrtle nanoemulsions: Formulation, stability, and antimicrobial activity. *LWT* **2021**, *152*, 112377. [CrossRef]
15. Tadros, T.; Izquierdo, P.; Esquena, J.; Solans, C. Formation and stability of nano-emulsions. *Adv. Colloid Interface Sci.* **2004**, *108–109*, 303–318. [CrossRef]
16. Ngan, C.L.; Basri, M.; Lye, F.F.; Masoumi, H.R.F.; Tripathy, M.; Karjiban, R.A.; Abdul-Malek, E. Comparison of process parameter optimization using different designs in nanoemulsion-based formulation for transdermal delivery of fullerene. *Int. J. Nanomed.* **2014**, *9*, 4375–4386. [CrossRef]
17. Yukuyama, M.N.; Ghisleni, D.D.M.; Pinto, T.J.A.; Bou-Chacra, N.A. Nanoemulsion: Process selection and application in cosmetics—A review. *Int. J. Cosmet. Sci.* **2016**, *38*, 13–24. [CrossRef]
18. Sepahvand, S.; Amiri, S.; Radi, M.; Akhavan, H.R. Antimicrobial Activity of Thymol and Thymol-Nanoemulsion Against Three Food-Borne Pathogens Inoculated in a Sausage Model. *Food Bioprocess Technol.* **2021**, *14*, 1936–1945. [CrossRef]
19. Gao, Z.P.; Zhong, W.M.; Chen, K.Y.; Tang, P.Y.; Guo, J.J. Chemical composition and anti-biofilm activity of essential oil from *Citrus medica* L. var. sarcodactylis Swingle against Listeria monocytogenes. *Ind. Crop. Prod.* **2020**, *144*, 112036. [CrossRef]
20. Sun, Y.; Liu, Z.L.; Wang, X.M.; Zhang, F.J.; Huang, X.; Li, J.R.; Sun, X.; Guo, Y.M.; Han, X.B. Effect of HLB value on the properties of chitosan/zein/lemon essential oil film-forming emulsion and composite film. *Int. J. Food Sci. Technol.* **2021**, *56*, 4925–4933. [CrossRef]
21. Nirmal, N.P.; Mereddy, R.; Li, L.; Sultanbawa, Y. Formulation, characterisation and antibacterial activity of lemon myrtle and anise myrtle essential oil in water nanoemulsion. *Food Chem.* **2018**, *254*, 1–7. [CrossRef] [PubMed]
22. Abdelhameed, M.F.; Asaad, G.F.; Ragab, T.I.M.; Ahmed, R.F.; El Gendy, A.G.; Abd El-Rahman, S.S.; Elgamal, A.M.; Elshamy, A.I. Oral and Topical Anti-Inflammatory and Antipyretic Potentialities of Araucaria bidiwillii Shoot Essential Oil and Its Nanoemulsion in Relation to Chemical Composition. *Molecules* **2021**, *26*, 5833. [CrossRef] [PubMed]
23. Luo, Q.; Zhang, J.R.; Li, H.B.; Wu, D.T.; Geng, F.; Corke, H.; Wei, X.L.; Gan, R.Y. Green Extraction of Antioxidant Polyphenols from Green Tea (*Camellia sinensis*). *Antioxidants* **2020**, *9*, 785. [CrossRef] [PubMed]
24. Maran, J.P.; Priya, B. Ultrasound-assisted extraction of polysaccharide from *Nephelium lappaceum* L. fruit peel. *Int. J. Biol. Macromol.* **2014**, *70*, 530–536. [CrossRef]
25. Hirai, M.; Ota, Y.; Ito, M. Diversity in principal constituents of plants with a lemony scent and the predominance of citral. *J. Nat. Med.* **2022**, *76*, 254–258. [CrossRef]
26. Aguilar-Hernandez, M.G.; Sanchez-Bravo, P.; Hernandez, F.; Carbonell-Barrachina, A.A.; Pastor-Perez, J.J.; Legua, P. Determination of the Volatile Profile of Lemon Peel Oils as Affected by Rootstock. *Foods* **2020**, *9*, 241. [CrossRef]
27. Perdones, A.; Escriche, I.; Chiralt, A.; Vargas, M. Effect of chitosan-lemon essential oil coatings on volatile profile of strawberries during storage. *Food Chem.* **2016**, *197*, 979–986. [CrossRef]
28. Campolo, O.; Romeo, F.V.; Algeri, G.M.; Laudani, F.; Malacrino, A.; Timpanaro, N.; Palmeri, V. Larvicidal Effects of Four Citrus Peel Essential Oils Against the Arbovirus Vector Aedes albopictus (Diptera: Culicidae). *J. Econ. Entomol.* **2016**, *109*, 360–365. [CrossRef]
29. Roberto, D.; Micucci, P.; Sebastian, T.; Graciela, F.; Anesini, C. Antioxidant Activity of Limonene on Normal Murine Lymphocytes: Relation to H_2O_2 Modulation and Cell Proliferation. *Basic Clin. Pharmacol.* **2010**, *106*, 38–44. [CrossRef]
30. Piccialli, I.; Tedeschi, V.; Caputo, L.; Amato, G.; De Martino, L.; De Feo, V.; Secondo, A.; Pannaccione, A. The Antioxidant Activity of Limonene Counteracts Neurotoxicity Triggered byA $\beta_{(1-42)}$ Oligomers in Primary Cortical Neurons. *Antioxidants* **2021**, *10*, 937. [CrossRef]
31. Ozogul, Y.; El Abed, N.; Ozogul, F. Antimicrobial effect of laurel essential oil nanoemulsion on food-borne pathogens and fish spoilage bacteria. *Food Chem.* **2022**, *368*, 130831. [CrossRef] [PubMed]
32. Yazgan, H.; Ozogul, Y.; Kuley, E. Antimicrobial influence of nanoemulsified lemon essential oil and pure lemon essential oil on food-borne pathogens and fish spoilage bacteria. *Int. J. Food Microbiol.* **2019**, *306*, 108266. [CrossRef] [PubMed]
33. Walker, R.M.; Gumus, C.E.; Decker, E.A.; McClements, D.J. Improvements in the formation and stability of fish oil-in-water nanoemulsions using carrier oils: MCT, thyme oil, & lemon oil. *J. Food Eng.* **2017**, *211*, 60–68. [CrossRef]
34. Ziani, K.; Fang, Y.; McClements, D.J. Fabrication and stability of colloidal delivery systems for flavor oils: Effect of composition and storage conditions. *Food Res. Int.* **2012**, *46*, 209–216. [CrossRef]
35. Ziani, K.; Fang, Y.; McClements, D.J. Encapsulation of functional lipophilic components in surfactant-based colloidal delivery systems: Vitamin E, vitamin D, and lemon oil. *Food Chem.* **2012**, *134*, 1106–1112. [CrossRef]
36. Smejkal, G.B.; Ting, E.Y.; Nambi, K.N.A.; Schumacher, R.T.; Lazarev, A.V. Characterization of Astaxanthin Nanoemulsions Produced by Intense Fluid Shear through a Self-Throttling Nanometer Range Annular Orifice Valve-Based High-Pressure Homogenizer. *Molecules* **2021**, *26*, 2856. [CrossRef]

37. Gulotta, A.; Saberi, A.H.; Nicoli, M.C.; McClements, D.J. Nanoemulsion-Based Delivery Systems for Polyunsaturated (ω-3) Oils: Formation Using a Spontaneous Emulsification Method. *J. Agric. Food Chem.* **2014**, *62*, 1720–1725. [CrossRef]
38. Fathordoobady, F.; Sannikova, N.; Guo, Y.G.; Singh, A.; Kitts, D.D.; Pratap-Singh, A. Comparing microfluidics and ultrasonication as formulation methods for developing hempseed oil nanoemulsions for oral delivery applications. *Sci. Rep.* **2021**, *11*, 72. [CrossRef]
39. Gupta, A.; Eral, H.B.; Hatton, T.A.; Doyle, P.S. Controlling and predicting droplet size of nanoemulsions: Scaling relations with experimental validation. *Soft Matter* **2016**, *12*, 1452–1458. [CrossRef]
40. Kentish, S.; Wooster, T.J.; Ashokkumar, M.; Balachandran, S.; Mawson, R.; Simons, L. The use of ultrasonics for nanoemulsion preparation. *Innov. Food Sci. Emerg. Technol.* **2008**, *9*, 170–175. [CrossRef]
41. Floris, A.; Meloni, M.C.; Lai, F.; Marongiu, F.; Maccioni, A.M.; Sinico, C. Cavitation effect on chitosan nanoparticle size: A possible approach to protect drugs from ultrasonic stress. *Carbohyd. Polym.* **2013**, *94*, 619–625. [CrossRef] [PubMed]
42. Moazeni, M.; Davari, A.; Shabanzadeh, S.; Akhtari, J.; Saeedi, M.; Mortyeza-Semnani, K.; Abastabar, M.; Nabili, M.; Moghadam, F.H.; Roohi, B.; et al. In vitro antifungal activity of Thymus vulgaris essential oil nanoemulsion. *J. Herb. Med.* **2021**, *28*, 100452. [CrossRef]
43. Aziz, Z.A.A.; Nasir, H.M.; Ahmad, A.; Setapar, S.H.M.; Ahmad, H.; Noor, M.H.M.; Rafatullah, M.; Khatoon, A.; Kausar, M.A.; Ahmad, I.; et al. Enrichment of Eucalyptus oil nanoemulsion by micellar nanotechnology: Transdermal analgesic activity using hot plate test in rats' assay. *Sci. Rep.* **2019**, *9*, 13678. [CrossRef] [PubMed]
44. Romes, N.B.; Abdul Wahab, R.; Abdul Hamid, M.; Oyewusi, H.A.; Huda, N.; Kobun, R. Thermodynamic stability, in-vitro permeability, and in-silico molecular modeling of the optimal Elaeis guineensis leaves extract water-in-oil nanoemulsion. *Sci. Rep.* **2021**, *11*, 20851. [CrossRef]
45. Kaur, K.; Kumar, R.; Arpita; Goel, S.; Uppal, S.; Bhatia, A.; Mehta, S.K. Physiochemical and cytotoxicity study of TPGS stabilized nanoemulsion designed by ultrasonication method. *Ultrason. Sonochem.* **2017**, *34*, 173–182. [CrossRef]
46. Zhong, W.M.; Tang, P.Y.; Liu, T.; Zhao, T.Y.; Guo, J.J.; Gao, Z.P. Linalool Nanoemulsion Preparation, Characterization and Antimicrobial Activity against Aeromonas hydrophila. *Int. J. Mol. Sci.* **2021**, *22*, 11003. [CrossRef]
47. Azizkhani, M.; Kiasari, F.J.; Tooryan, F.; Shahavi, M.H.; Partovi, R. Preparation and evaluation of food-grade nanoemulsion of tarragon (*Artemisia dracunculus* L.) essential oil: Antioxidant and antibacterial properties. *J. Food Sci. Technol.* **2021**, *58*, 1341–1348. [CrossRef]
48. Aouf, A.; Ali, H.; Al-Khalifa, A.R.; Mahmoud, K.F.; Farouk, A. Influence of Nanoencapsulation Using High-Pressure Homogenization on the Volatile Constituents and Anticancer and Antioxidant Activities of Algerian Saccocalyx satureioides Coss. et Durieu. *Molecules* **2020**, *25*, 4756. [CrossRef]
49. da Costa, J.S.; Barroso, A.S.; Mourao, R.H.V.; da Silva, J.K.R.; Maia, J.G.S.; Figueiredo, P.L.B. Seasonal and Antioxidant Evaluation of Essential Oil from *Eugenia uniflora* L., Curzerene-Rich, Thermally Produced in Situ. *Biomolecules* **2020**, *10*, 328. [CrossRef]

Article

Engineering of Liposome Structure to Enhance Physicochemical Properties of *Spirulina plantensis* Protein Hydrolysate: Stability during Spray-Drying

Maryam Mohammadi [1,2,*], Hamed Hamishehkar [3], Marjan Ghorbani [4], Rahim Shahvalizadeh [1,2], Mirian Pateiro [5,*] and José M. Lorenzo [5,6]

[1] Drug Applied Research Center, Student Research Committee, Tabriz University of Medical Sciences, Tabriz 51656-65811, Iran; rahimshahvalizadeh@gmail.com
[2] Department of Food Science, Faculty of Agriculture, University of Tabriz, Tabriz 51666-16471, Iran
[3] Drug Applied Research Center, Tabriz University of Medical Sciences, Tabriz 51656-65811, Iran; hamishehkarh@tbzmed.ac.ir
[4] Stem Cell Research Center, Tabriz University of Medical Sciences, Tabriz 51666-14766, Iran; Ghorbani.marjan65@yahoo.com
[5] Centro Tecnológico de la Carne de Galicia, Avd. Galicia No. 4, Parque Tecnológico de Galicia, 32900 Ourense, Spain; jmlorenzo@ceteca.net
[6] Facultade de Ciencias, Universidade de Vigo, Área de Tecnoloxía dos Alimentos, 32004 Ourense, Spain
* Correspondence: ma.mohammadi@tabrizu.ac.ir (M.M.); mirianpateiro@ceteca.net (M.P.)

Abstract: Encapsulating hydrolysates in liposomes can be an effective way to improve their stability and bioactivity. In this study, *Spirulina* hydrolysate was successfully encapsulated into nanoliposomes composed of different stabilizers (cholesterol or γ-oryzanol), and the synthesized liposomes were finally coated with chitosan biopolymer. The synthesized formulations were fully characterized and their antioxidant activity evaluated using different methods. Then, stabilization of coated nanoliposomes (chitosomes) by spray-drying within the maltodextrin matrix was investigated. A small mean diameter and homogeneous size distribution with high encapsulation efficiency were found in all the formulations, while liposomes stabilized with γ-oryzanol and coated with chitosan showed the highest physical stability over time and preserved approximately 90% of their initial antioxidant capacity. Spray-dried powder could preserve all characteristics of peptide-loaded chitosomes. Thus, spray-dried hydrolysate-containing chitosomes could be considered as a functional food ingredient for the human diet.

Keywords: *Spirulina platensis*; bioactive peptides; encapsulation; liposomes; chitosome

1. Introduction

The microalga *Spirulina* has gained more attention in areas such as the pharmaceutical, food, poultry, and aquaculture industries for its nutritional and health benefits [1]. Certain therapeutic effects of *Spirulina* (reduced hyperlipidemia, obesity, and blood cholesterol; antioxidant and anticancer activity; immune system improvement; and increased beneficial intestinal bacteria) have been proven by pre-clinical and clinical studies, which are related to their bioactive constitution, e.g., phycocyanins, carotenoids, phenolic compounds, and polyunsaturated fatty acids. The green-blue microalgae are a rich source of proteins (60–70% of dry matter) and, due to the absence of cellulose in the cell wall, are very digestible. Thus, they have gained more attention in recent years as a food supplement, especially for athletes and vegetarians [2].

However, the undesirable taste, low digestibility, and high allergenicity of algae-based protein isolates for monogastric animals and humans have limited their application in the food industry. Protein hydrolysates and peptides can be excellent alternatives to overcome the problems associated with the direct consumption of protein isolates, as they have nutritional and health-promoting features and act as natural antioxidant agents in food preservation [3,4]. Although algae-derived peptides have various advantages, their instability during storage and under harsh conditions (e.g., in the gastrointestinal tract), low absorption efficiency, bitterness, reaction with the food matrix, and possible inactivation inhibit the application of hydrolysates in foods and beverages. Incorporating these bioactive compounds in lipid-based nanocarriers such as liposomes can be an appropriate solution to cover all of these problems and increase their efficacy under different conditions [5].

Liposomes have an enclosed vesicular structure and are able to accommodate both water-soluble and hydrophobic compounds in their internal aqueous core and bilayer space, respectively [6]. Moreover, the typical constituents of liposomes are completely natural, and their nontoxicity has led to the broad application of these vesicular systems in the encapsulation of various bioactive compounds [7,8]. However, the major drawbacks of this versatile carrier are the fluidity of the intravesicular space and the flexible bilayer structure, which can lead to the physical instability of vesicles (aggregation/flocculation and fusion/coalescence), resulting in changes in size and loss of liposome-incorporated bioactive materials over time. A possible solution to this problem is to engineer the liposomal structure [9]. Mostly, cholesterol has been applied as a stabilizing factor in vesicular systems because it can increase the packing of phospholipid molecules, reduce the fluidity of intravesicular space, and consequently create a more rigid and stable structure over time and under severe shear stress [10]. Moreover, the above-mentioned problems can be improved by depositing an oppositely charged biopolymer such as chitosan around the liposome surface through electrostatic interaction [11,12]. Several studies on improving hydrolysate stability during storage and processing using chitosan coating have been conducted [13,14].

To make a formulation that is more stable over time and more appropriate for industrial application, it can be transformed into powder form by spray-drying or freeze-drying. Freeze-drying is a more expensive technology and requires more time and energy compared with spray-drying. Thus, spray-drying technology is an economical strategy to make powdered liposomal dispersions, and it has a wide range of use in the food industry compared to other drying techniques. However, there have been no studies regarding the simultaneous use of different stabilizers and coating materials to increase the stability and bioactivity of liposome-containing *Spirulina* hydrolysates. Therefore, the aims of this research were as follows: (1) to explore the effect of cholesterol and γ-oryzanol as stabilizing agents and chitosan as a coating material on the mean diameter and encapsulation efficiency of *Spirulina* hydrolysate-loaded liposome, (2) to examine the antioxidant capacity of synthesized liposomes using the different methods, and (3) to estimate the stability of synthesized liposomes during storage. Finally, the optimum formulation for easy usage was converted into powder form and its physicochemical and structural properties and antioxidant activity were evaluated.

2. Materials and Methods

2.1. Materials and Reagents

Pepsin from powdered porcine gastric mucosa (activity ≥ 250 units/mg solid), 1,1-diphenyl-2-picrylhydrazyl (DPPH), 2,2′-azino-bis(3-ethylbenzothiazoline-6-sulfonic acid) diammonium salt (ABTS), 2,4,6(tripyridyl)-1,3,5-triazine (TPTZ), cholesterol, and Coomassie brilliant blue (G250) were purchased from Sigma-Aldrich (St. Louis, MO, USA); potassium persulfate, iron (III) chloride hexahydrate, trichloroacetic acid (TCA), ferrous chloride, γ-oryzanol, iron sulfate, and maltodextrin were obtained from Merck (Darmstadt, Germany). *Spirulina platensis* powder was purchased from Noor Daro Gonbad (Gonbad Kavous, Iran).

2.2. Protein Hydrolysis

For protein hydrolysis, lyophilized *Spirulina platensis* protein was dispersed in distilled water (DW) to achieve a protein concentration of 3% (w/v). Protein hydrolysates were produced by pepsin protease for 240 min. The hydrolysis conditions were set as follows:

pH 2, temperature 37 °C, enzyme-to-substrate (E/S) ratio of 6% (w/w). The hydrolysis reaction was performed in a shaker incubator (Unimax 1010; Heidolph, Schwabach, Germany), then the enzymes were thermally inactivated (90 °C, 10 min), and the solution was cooled down to room temperature. The hydrolysate solution was then centrifugated at 4550× g for 10 min and the supernatant was collected for further analysis [4].

2.3. Degree of Hydrolysis (DH)

The extent of enzymatic hydrolysis can be defined by the degree of hydrolysis (DH), which is a key factor determining the chain length of peptides, and thereby their functional properties. A higher DH corresponds to mean shorter peptide length and vice versa. This index is significantly influenced by hydrolysis time and the type of enzyme used. To determine DH, 1 mL of hydrolysate was added to 1 mL of TCA (0.44 M), followed by centrifugation at 7800× g for 10 min at 4 °C. The collected supernatants were analyzed for soluble protein by the Bradford assay. DH was estimated using the following equation [15]:

$$DH\ (\%) = \frac{TCA - \text{Soluble Protein of Hydrolysate}}{\text{Total Protein of Sample (non - hydrolyzed)}} \times 100 \quad (1)$$

2.4. Amino Acid Profile

An RP-HPLC apparatus (Young Lin Acme 9000, YL Instruments, Anyang, Korea) equipped with a reverse-phase column (150 mm × 4.6 mm × 5 μm; RP-C18 ODS-A, Barcelona, Spain), a fluorescence detector (LC305; Lab Alliance, State College, PA, USA), and a mobile phase of acetate buffer (50 mM at pH 3.4, with a flow rate of 1.3 mL/min) were used to determine the amino acids in *Spirulina* protein hydrolysates. For this purpose, the hydrolysate was intensively treated with HCl (6 M) at 110 °C for 24 h. The digested sample was derivatized with orthophthaldehyde and injected to the HPLC column. The amount of amino acids in hydrolysates was expressed as mg/100 g protein. The biological value (*BV*) and amino acid score (*AAS*) of hydrolysates, as nutritional parameters, were determined using the following equations [16]:

$$AAS = \frac{\%\ \text{Essential amino acids in sample}}{\%\ \text{Essential amino acids recommended by FAO}} \quad (2)$$

$$BV\ (\%) = 10^{2.15} \times Lys^{0.41} \times (Phe + Tyr)^{0.6} \times (Met + Cys)^{0.77} \times Thr^{0.24} \times Trp^{0.21} \quad (3)$$

where each amino acid symbol is expressed as % amino acid in sample/% amino acid FAO pattern.

2.5. Preparation of Spirulina Hydrolysate (HS)-Loaded Liposomes

HS-encapsulated nanoliposomes were prepared using the thin layer hydration method as described by Mohammadi et al. [7] with slight modifications. For this procedure, 1.2 % (w/v) Phospholipon 90 G (soybean lecithin of ~90% phosphatidylcholine; Lipoid GmbH, Ludwigshafen, Germany) and 2 stabilizing agents (cholesterol or 0.15% (w/v) γ-oryzanol) were dissolved in 15 mL of 96% ethanol and stirred on a hotplate at 50 °C for complete solubilization. Subsequently, the solvent was evaporated using a rotary evaporator (Heidolph, Germany) at 50 °C until a thin film was formed in the round-bottomed flasks. The resulting lipid films were hydrated with 15 mL of DW containing HS at 0.3% (w/v) with continuous agitation on a rotary evaporator at 55 °C, followed by sonication using a sonication probe (130 W, 20 kHz; Vibra-Cell Sonics & Materials, Newtown, CT, USA) at 80% sonication strength for 10 min. During sonication, the sample was placed into an ice bath to avoid overheating of dispersion. To prepare the empty liposomes, the same method was applied, except HS was excluded in the hydration step and the thin layer was hydrated only with DW.

2.6. Preparation of Chitosan-Coated Nanoliposomes

For coating with chitosan, prepared nanoliposomes were added to chitosan solution (0.4%, w/v) dissolved in acetic acid (1% v/v) in a drop-wise manner with a volume ratio of 1:1 and stirred for 2 h.

2.7. Characterization of HS-Loaded Liposomes

2.7.1. Measurement of Particle Size and Zeta Potential (ζ)

Liposome dispersions were diluted 1:10 with DW before analysis by a zetasizer (Zetasizer Nano ZS, Malvern Instruments Ltd., Malvern, UK).

2.7.2. Encapsulation Efficiency

Encapsulation efficiency (EE) was determined by separating encapsulated hydrolysates from free ones using an Amicon filter (Amicon Ultra-15, with molecular weight cutoff of 30 kDa; Millipore Corp., Cork, Ireland), followed by centrifugation at 3000 rpm for 10 min. Free and total hydrolysates were determined by calculating the protein amount using the Bradford method as described previously.

EE was determined according to following equation:

$$EE = \frac{\text{Total protein content} - \text{Amount of free hydrolysate}}{\text{Total protein content}} \times 100 \qquad (4)$$

2.7.3. Scanning Electron Microscopy (SEM)

To investigate the morphological features of the vesicles, γ-oryzanol liposome (with and without chitosan coating) was dispersed onto the laboratory lamel and dried at 37 °C, then transferred to adhesive-coated aluminum pin stubs. The stubs were coated with a thin layer of gold and examined using a scanning electron microscope (MIRA3, TESCAN, Brno, Czech Republic) [17].

2.7.4. Transmission Electron Microscopy (TEM) Measurements

For TEM measurement, 5 µL of each sample was placed onto a copper grid coated with carbon film for 3 min before being blotted off using filter paper. After that, 10 µL of contrast dye containing 2% uranyl acetate was placed onto the grid, left for 2 min, and blotted off with filter paper. Finally, the grids were loaded onto a specimen holder and then into a transmission electron microscope (100 Kv; LEO 906, Zeiss, Oberkochen, Germany).

2.7.5. Fourier-Transform Infrared Spectroscopy (FT-IR)

To determine the functional groups of liposomes, lyophilized samples (ALPHA 1–4 LD freeze dryer, Martin Christ, Osterode am Harz, Germany) were formed into KBr pellets with a mass ratio of 1:100 [17]. The samples were analyzed using FTIR (4300, Shimadzu, Kyoto, Japan) from 4000 to 400 cm^{-1} with a minimum of 256 scans/spectrum and a constant scan speed of 4°/s.

2.7.6. Determination of Total Phenolic Content (TPC)

TPC was performed according to the Folin–Ciocalteu method described by de Araujo et al. [18]. Briefly, 300 µL of sample was mixed with 125 µL of Folin–Ciocalteu reagent and 1825 µL of DW. After the mixture was vortexed for 5 min at ambient temperature, 250 µL of sodium carbonate solution (20%, w/v) was added and it was vortexed for another 5 min. Then, the mixture was placed in a water bath at 40 °C for 30 min. The samples were then centrifuged at 10,000 rpm for 10 min, and the absorbance of the upper phase was measured at 765 nm by a spectrophotometer (Ultrospec 2000; Scinteck, Cambridge, UK). The results were expressed as mg gallic acid per g sample using the following formula [18]:

$$C = c\frac{V}{m} \qquad (5)$$

where C is the total phenolic content (mg GAE/g dry extract), c is the concentration of gallic acid obtained from the calibration curve (mg/mL), V is the volume of extract (mL), and m is the mass of the extract (g).

2.7.7. Antioxidant Activity of Protein Hydrolysates

2.7.7.1. DPPH Radical Scavenging Activity

To measure DPPH radical scavenging activity, 1 mL of each concentration of hydrolysates was added to 1 mL of DPPH solution (0.1 mM), followed by incubation for 30 min in the dark. The absorbance was read at 517 nm and the DPPH radical scavenging activity was calculated by the following equation [19]:

$$\text{DPPH } radical\ scavenging\ activity\ (\%) == \frac{A_{control} - A_{sample}}{A_{control}} \times 100 \qquad (6)$$

where $A_{control}$ and A_{sample} are the absorbance of the control and sample, respectively.

2.7.7.2. ABTS Radical Scavenging Activity

The mixture of ABTS (7 mM) and potassium persulfate (2.45 mM) with a volume ratio of 1:1 generates a green-blue reagent (ABTS$^+$) after 12–16 h incubation in the dark, and has maximum absorption at 734 nm. When this cationic radical is exposed to the hydrogen donating compound, the green-blue is decolorized and the color intensity is measured at 734 nm. To measure ABTS radical scavenging activity, 40 µL of the prepared hydrolysate concentration was mixed with 4 mL of diluted ABTS solution, vortexed vigorously for 30 s, and incubated in the dark for 6 min. The absorbance was measured at 734 nm. The ABTS radical scavenging activity was calculated by the following equation [20]:

$$\text{ABTS } radical\ scavenging\ activity\ (\%) = \frac{A_{control} - A_{sample}}{A_{control}} \times 100 \qquad (7)$$

2.7.7.3. Ferric Reducing/Antioxidant Power (FRAP) Assay

The capability of hydrolysate to reduce Fe^{+3} ions present in the complex to a Fe^{+2} form with 2,4,6-tri (2-pyridyl)-s-triazine (TPTZ) was determined by the ferric ion reducing capacity (FRAP) assay as described previously. The FRAP reagent was freshly prepared by mixing TPTZ (10 mM) dissolved in 40 mM HCL, iron (III) chloride hexahydrate (20 mM) dissolved in water, and acetate buffer (0.3 mM) at pH 3.6 at a ratio of 1:1:10 (v/v/v) and warming it to 37 °C. Then, 900 µL of the working solution was mixed with 100 µL of different concentrations of hydrolysate and the corresponding nano-formulated system, followed by incubation of the mixture at 37 °C for 30 min, and the resulting blue color absorbance at 595 nm was recorded. Different concentrations of $FeSO_4$, in the range 0–1 mM, were used as the calibration curve [21].

2.8. Storage Stability of HS-Loaded Liposome and Chitosome

In order to perform this test, HS-loaded liposomes (stabilized with γ-oryzanol) and the corresponding chitosomes were stored at 4 °C for 1 month. The mean diameter, PDI, ζ-potential, and precipitation were monitored during storage by DLS as described in Section 2.7. The residual antioxidant activity of selected formulations was controlled by ABTS assay.

2.9. Spray-Drying of Chitosomes

Before spray drying, the γ-oryzanol-stabilized chitosome dispersion was mixed with maltodextrin solution (40% w/v) at a mass ratio of 40:60, and stirred overnight at room temperature. Then, the resulting dispersions were spray-dried using a Mini Spray Dryer (Büchi Labortechnik, Flawil, Switzerland). The inlet and outlet air temperature were 130 and 75 °C, respectively. Dried powders were stored in airtight containers and placed in a desiccator at room temperature.

2.10. Characterization of Spray-Dried Powder

Production yield was measured by calculating the mass ratio of the produced powder to the total solid content in the feed. Other physical properties of the spray-dried powder, such as moisture content, bulk density, and solubility, were computed using the methods described by Sarabandi et al. [20].

The particle morphology of spray-dried powder was evaluated by scanning electron microscopy (SEM; MIRA3, TESCAN, Brno, Czech Republic).

To determine the size, polydispersity, and ζ of reconstituted nanoliposomes, the powder was dissolved in an appropriate concentration and its particle size and ζ potential were determined by a zetasizer.

2.11. Statistical Analysis

Statistical analysis was performed using SPSS software (version 24.0, IBM, Chicago, IL, USA). Normal distribution and variance homogeneity had been previously tested (Shapiro–Wilk). Data of 3 repetitions were subjected to analysis of variance (ANOVA), followed by Tukey's test at a 5% significance level.

3. Results and Discussion

3.1. Characterization of Hydrolyzed Spirulina Protein (HS)

Extracted *Spirulina* isolate was hydrolyzed by pepsin enzyme. The hydrolysis degree was found to be 16.5% over 4 h hydrolysis time. The solubility of hydrolysate under harsh acidic conditions was improved after enzymatic hydrolysis, but the highest solubility was obtained under alkaline pH conditions.

Figure 1 shows the amino acid composition of pepsin-hydrolyzed peptides. According to the obtained profile, the hydrolysate was rich in acidic amino acids (aspartic and glutamic acid), arginine, valine, lysine, alanine, glycine, threonine, and leucine. All the essential amino acids (except sulfur-containing amino acids) were present in the hydrolysate at concentrations higher than the FAO recommended levels for adults. Moreover, the hydrolysate showed good nutritional value as determined by amino acid score (72%) and biological value (78%), and good antioxidant activity (IC_{50} 1 mg/mL).

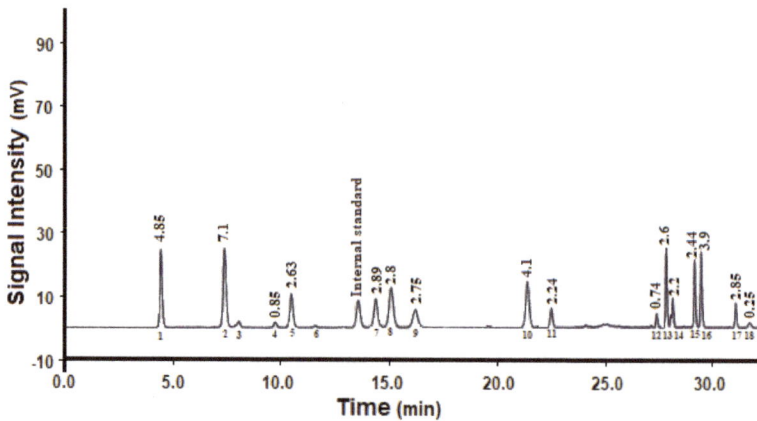

Figure 1. Amino acid composition of pepsin-hydrolyzed peptides: (1) aspartic acid; (2) glutamic acid; (3) asparagine; (4) histidine; (5) serine; (6) glutamine; (7) arginine; (8) glycine; (9) threonine; (10) alanine; (11) tyrosine; (12) methionine; (13) valine; (14) phenylalanine; (15) isoleucine; (16) leucine; (17) lysine; (18) tryptophan.

Compared to native protein (with a DPPH IC_{50} value of 3 mg/mL), the hydrolysates had a significantly lower IC_{50} value of 1.0 mg/mL, indicating their effectiveness in scavenging DPPH radicals. The ABTS IC_{50} of hydrolysate was estimated to be 2 mg/mL. Compared with native protein (with an IC_{50} value of 4.5 mg/mL), the hydrolysates had lower IC_{50} values, indicating their effectiveness against ABTS radicals.

Encapsulating these bioactive compounds in lipid-based nanocarriers such as liposomes can be an appropriate solution to cover all of the mentioned problems and increase their efficacy under different conditions.

3.2. Characterization of Uncoated and Chitosan-Coated Liposomal Dispersions

In this study, HS was encapsulated into the liposomal carrier and its experimental characteristics were investigated. Two stabilizers (cholesterol and γ-oryzanol) were applied in the preparation of primary liposome dispersion, then the resulting nanoliposomes were coated with cationic chitosan polymer at a final concentration of 0.2% w/v (this concentration resulted in the smallest particle size and highest surface charge on the coated liposome dispersions), and their effects on the physicochemical properties of the resulting formulations were examined. The mean particle diameter and ζ of primary and chitosan-coated nanoliposomes (chitosomes) are shown in Table 1. Liposomes stabilized by cholesterol and γ-oryzanol had a small particle size and homogeneous size distribution, and their surface charge was between −11 and −14 mV. Following the addition of the chitosan polymer, the particle size and PDI of nanoliposomes increased and ζ changed from negative to positive values (approximately 29 mV), confirming that cationic chitosan successfully covered the primary liposomes. These findings are in accordance with those of Altin et al. [22], who reported that surface coating of primary liposomes containing phenolic extract from cocoa hull waste with cationic chitosan by electrostatic deposition increased the particle size of liposomes and the secondary liposomes had a positive charge.

SEM and TEM images of primary liposomes (γ-oryzanol-liposomes) and the corresponding chitosomes are shown in Figure 2. The cholesterol-liposome and γ-oryzanol-liposome had similar shape and morphology. The results obtained from the zetasizer apparatus were somewhat confirmed by SEM and TEM. The SEM images show spherical particles with a small particle size < 100 nm and narrow distribution. In TEM images, the spherical structure and monodispersed distribution of primary liposomes are very clear [23].

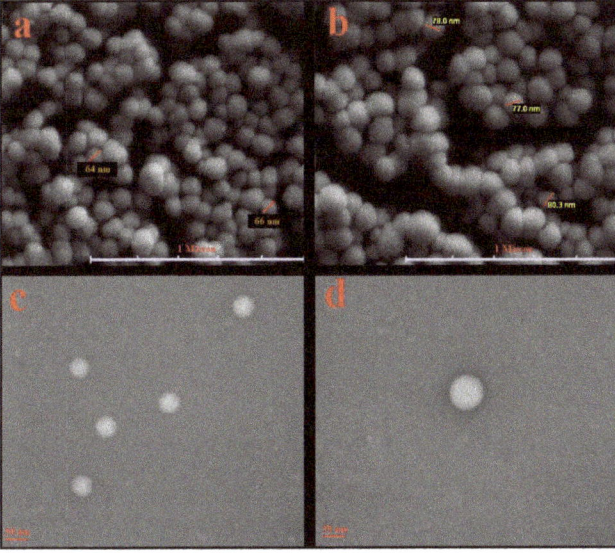

Figure 2. (**a,b**) SEM and (**c,d**) TEM images of primary liposomes stabilized with γ-oryzanol (**a,c**) and corresponding γ-oryzanol-chitosomes (**b,d**).

Table 1. Characteristics of hydrolysate-loaded γ-oryzanol-liposome and chitosome.

	Z-Average (nm)	PDI	ζ (mV)	Encapsulation Efficiency (%)
Cholesterol-liposome	56.6 ± 2.1 [c]	0.17 ± 0.02 [d]	−11.1 ± 1.5 [b]	85.0 ± 1.2 [a]
γ-Oryzanol-liposome	63.9 ± 1.8 [b]	0.18 ± 0.02 [c]	−14.8 ± 1.1 [b]	87.0 ± 1.3 [a]
Cholesterol-chitosome	66.8 ± 2.4 [b]	0.27 ± 0.03 [b]	24.8 ± 1.2 [a]	88.0 ± 1.0 [a]
γ-Oryzanol-chitosome	78.4 ± 1.9 [a]	0.28 ± 0.01 [a]	26.9 ± 1.5 [a]	89.0 ± 2.0 [a]

Means in same column with different superscripts (a, b, c, d) are statistically different ($p < 0.05$).

3.3. Encapsulation Efficiency (EE)

The EE of the primary HS-loaded liposomes stabilized by cholesterol and γ-oryzanol is given in Table 1. Overall, both liposomes showed high EE > 85%, indicating that both stabilizing agents had good potential for encapsulation of HS. Incorporating sterol compounds in the liposome structure significantly increased the rigidity of the liposome membrane; thus, the system could encapsulate a larger amount of hydrophilic bioactive material. In another study, high EE was reported for fish hydrolyzed collagen-loaded liposomes stabilized by cholesterol and glycerol [24]. In another study, orange seed protein hydrolysates were produced using alcalase and pepsin enzymes, which were incorporated into uncoated liposomes and chitosome systems. The hydrolysates produced with alcalase showed a higher EE than those produced with pepsin. The authors suggested that this difference may be related to the higher DH of alcalase hydrolysate (approximately 24%), the lower molecular weight of resulting peptides compared to pepsin hydrolysate, and the easy incorporation of alcalase hydrolysate into the aqueous core of liposomes [14]. The authors also claimed that incorporating peptides into the chitosomes led to increased EE of vesicles compared to plain liposomes. This may be related to occupying pores in the surface of the liposome surface preventing the leakage of incorporated bioactive materials. These findings were consistent with those reported by [14,20].

3.4. Determination of Total Phenolic Content and Antioxidant Activity of Liposomes and Chitosomes

Liposomal nanocarriers can be applied for encapsulation of both liposoluble and hydrophilic antioxidant and phenolic compounds to improve their bioavailability. The total phenolic and antioxidant capacity of uncoated liposomes and chitosome dispersions are shown in Table 2. There was no significant difference ($p < 0.05$) between HS and HS-loaded cholesterol-liposomes by TPC, DPPH, ABTS, or FRAP assay, indicating that the phenolic compounds and, subsequently, the antioxidant properties of HS were properly preserved in the nanoliposomal carrier. Preservation of the antioxidant activity of anthocyanin-rich black carrot extract after 21 days of storage by encapsulating in liposomes has been reported [25].

Table 2. Antioxidant activity and total phenol content of hydrolysate-loaded γ-oryzanol-liposomes and chitosomes.

	Total Phenol (mg Gallic Acid/g Extract)	FRAP Assay (μM FeSO$_4$/mL)	Inhibition of ABTS Radical (%)	Inhibition of DPPH Radical (%)
Pepsin hydrolysate	51.0 ± 1.4 [b]	400.0 ± 3.1 [b]	53.0 ± 1.2 [b]	50.0 ± 1.5 [b]
Cholesterol-liposome	53.6 ± 1.6 [b]	395.0 ± 3.5 [b]	52.0 ± 2.1 [b]	47.0 ± 2.0 [b]
γ-Oryzanol-liposome	152.9 ± 2.1 [a]	650.0 ± 2.8 [a]	85.0 ± 2.5 [a]	90.0 ± 1.8 [a]
Cholesterol-chitosome	55.6 ± 2.4 [b]	410.0 ± 3.4 [b]	54.0 ± 2.1 [b]	48.0 + 2.1 [b]
γ-Oryzanol-chitosome	155.0 ± 1.9 [a]	655.0 ± 3.1 [a]	86.0 ± 1.9 [a]	88.0 ± 2.3 [a]

Means in same column with different superscripts (a, b) are statistically different ($p < 0.05$).

After the cholesterol-liposome surface was coated with chitosan, the antioxidant activity remained unchanged as compared to the uncoated liposome. The antioxidant activity of chitosan has been reported by others [22]. The authors suggested that the phenolic bioactive material could be partially located on the surface of the liposome, and consequently chitosan–phenolic compound conjugates might be formed, and these couples synergistically improve the antioxidant activity [22].

Conversely, in another study, after coating the surface of sour cherry extract-loaded liposomes with cationic chitosan, the TPC content decreased from 38.19 to 31.23 mg/L. The authors suggested that the available chitosan on the liposome surface might block the availability of phenolic compounds on the surface of uncoated liposomes [26].

γ-Oryzanol, a plant sterol with a structure similar to cholesterol, has a wide capacity for scavenging free radicals, consequently preventing lipid oxidation. The HS-loaded γ-oryzanol-liposomes showed higher TPC and antioxidant capacity. This was attributed to the cooperative scavenging capacity of γ-oryzanol with HS in a liposome system [27]. This cooperative antioxidative effect was reported by Li et al. [28]. Sage extract (SE) and zein hydrolysate in combination showed higher antioxidant activity than the simple sum of their individual effects [28].

3.5. Fourier-Transform Infrared Spectroscopy (FTIR)

The structural changes in synthesized liposomes with and without HS and the successful chitosan coating were confirmed by FTIR spectroscopy. From the IR spectrum of the HS (Figure 3a), a broadband at 3300 cm^{-1} was attributed to O–H and N–H stretching and two bands at 2926 and 2853 cm^{-1} were related to CH_2 stretching vibrations of aliphatic chains. The amide region bands (1658 cm^{-1} corresponding to protein amide I, 1550 cm^{-1} corresponding to protein amide II, and 1247 cm^{-1} related to protein amide III) were clearly visible in the IR spectrum of the HS [29].

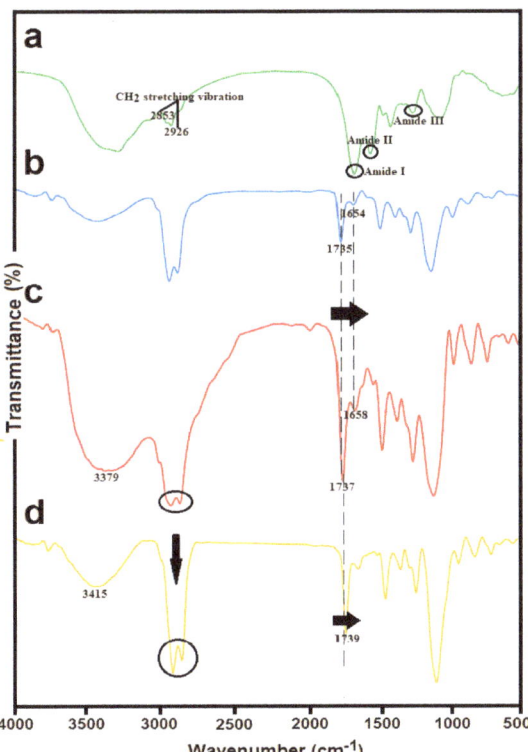

Figure 3. FTIR spectra of (**a**) *Spirulina* hydrolysate (HS); (**b**) blank nanoliposomes stabilized with γ-oryzanol; (**c**) HS-loaded γ-oryzanol nanoliposomes; and (**d**) HS-loaded γ-oryzanol-chitosomes.

Blank nanoliposomes (Figure 3b) were observed at the following wavenumbers: 3438 cm^{-1} related to hydroxyl stretch vibration, 2925 cm^{-1} attributed to stretch vibrations of a methylene group, 1735 cm^{-1} mainly related to the stretching vibration of the polar head ester groups of phospholipids, 1654 cm^{-1} related to C=C stretching vibrations, 1246 and 1109 cm^{-1} corresponding to symmetric and antisymmetric stretch vibrations of a phosphate group, and 958 cm^{-1} related to asymmetrical stretch vibrations of N$^+$/CH$_3$.

Comparing the IR spectra of HS-loaded γ-oryzanol-liposomes (Figure 3c) and corresponding blank liposomes (Figure 3b), a great similarity between their spectra can be observed. The incorporation of hydrolysate into the liposomal carrier resulted in a shift in some frequencies. The most important of these changes were slight shifts at 1735 and 1654 cm^{-1} to 1737 and 1658 cm^{-1} for HS-loaded liposomes, which may correspond to the possible interaction of HS with carbonyl ester groups at the interfacial part of the liposomal bilayers [30].

For the HS-loaded γ-oryzanol-chitosomes (Figure 3d), a shift from 3379 to 3415 cm^{-1} was detected, which may correspond to hydrogen bonding between hydroxyl or amino groups of chitosan and carboxylic acid or amino groups of hydrolysates. Moreover, after chitosan coating, significant changes in the absorption bands of acyl chains (3000–2800 cm^{-1}) were detected. These peaks were converted into two narrow and intense peaks at 2862 and 2924 cm^{-1} in the case of chitosan-coated liposomes. Further evidence for electrostatic conjugation of chitosan on liposome surface was a considerable shift to higher frequencies in the carbonyl group (from 1737 to 1739 cm^{-1}). This indicates that the carbonyl groups are involved with cationic groups of chitosan, resulting in the destruction of some hydrogen bonds [31].

3.6. Storage Stability of HS-Loaded Liposomes and Chitosomes

Regarding our previous research on the effect of temperature on the physical stability of vitamin D$_3$-loaded liposomes and the instability and aggregation of the formulation at ambient temperature due to higher fluidity of the lipid bilayer and higher loss of encapsulated bioactive material, refrigerator temperature (4 °C) was selected to examine the physical stability of selected formulations [7]. Liposomes as vesicular droplets have permeable and flexible bilayers and a high tendency to fuse and aggregate, resulting in the release of encapsulated bioactive materials during storage. In this study, stabilizing agents (cholesterol and γ-oryzanol) and coating material (chitosan) were tested to enhance the physical stability of liposomes. HS-loaded γ-oryzanol-liposomes and the corresponding chitosomes had no marked difference ($p < 0.05$) in magnitude, PDI, and ζ after 30 days of storage at 4 °C (Table 3). Moreover, both samples preserved more than 90% of their initial antioxidant activity in ABTS radical scavenging activity ($p < 0.05$). When the storage time was expanded, slight lipid oxidation may have occurred in unsaturated fatty acids of the phospholipid bilayers, leading to decreased antioxidative activity of the tested formulations. On the other hand, no significant precipitate was observed, especially in the HS-loaded chitosome system, during storage. When storage time was expanded to 2 months, the chitosome system showed higher physical stability compared to the uncoated liposomes. In contrast, the stability of HS-loaded cholesterol-liposomes in terms of mean diameter, PDI, and ζ was much lower than that of HS-loaded γ-oryzanol-liposomes at 4 °C. When storage time was expanded, the mean diameter and PDI of these formulations significantly increased ($p < 0.05$), and observable precipitates were separated into round-bottomed Falcon tubes.

Table 3. Physical stability of hydrolysate-loaded γ-oryzanol-liposomes and chitosomes during one month at 4 °C.

	Z-Average (nm)	PDI	ζ (mV)	Antioxidant Activity (ABTS Assay)
Pepsin hydrolysate	-	-	-	25.0 ± 1.8 [b]
Cholesterol-liposome	350.6 ± 2.1 [a]	0.31 ± 0.02 [b]	−9.1 ± 1.5 [b]	28.0 ± 2.1 [b]
γ-Oryzanol-liposome	90.9 ± 1.8 [d]	0.19 ± 0.02 [d]	−13.8 ± 1.1 [b]	75.0 ± 2.5 [a]
Cholesterol-chitosome	220.8 ± 2.4 [b]	0.38 ± 0.03 [a]	20.8 ± 1.2 [a]	38.0 ± 2.1 [b]
γ-Oryzanol-chitosome	120.4 ± 1.9 [c]	0.29 ± 0.01 [c]	24.9 ± 1.5 [a]	76.0 ± 1.9 [a]

Means in same column with different superscripts (a, b, c, d) are statistically different ($p < 0.05$).

On the other hand, the residual antioxidant activity of HS-loaded cholesterol-liposomes was 50% of its initial antioxidant activity. This observed instability of HS-loaded cholesterol-liposomes could have resulted from the poor coating of liposomal space cores with cholesterol stabilizing agents, and possible replacement of empty spaces with hydrophilic domain residues (glycine, arginine, and lysine) of hydrolysates. Nevertheless, some hydrophobic domains of *Spirulina* hydrolysates could interact with acyl chains of the lipid bilayer of liposome through hydrophobic binding, leading to more flexibility and fluidity of liposomes [32]. HS-loaded γ-oryzanol-chitosomes showed better results in terms of particle size, PDI, aggregation, and antioxidant activity compared to HS-loaded cholesterol-liposomes during storage. In summary, HS-loaded γ-oryzanol-liposomes and HS-loaded γ-oryzanol-chitosomes showed promising storage stability with no precipitate or change in mean diameter and a slight decrease in antioxidant activity at 4 °C. Thus, γ-oryzanol as a stabilizing agent and chitosan polymer as a coating material were able to reduce membrane fluidity and flexibility, contributing to stability.

In another study, different stabilizers (cholesterol and glycerol) were used to encapsulate peptides obtained from defatted Asian sea bass skin. Regarding the results, both formulations showed small particle size and high encapsulation efficiency. However, after the lyophilization process, hydrolysate-loaded cholesterol-liposomes showed higher stability and higher antioxidant activity in the gastrointestinal tract than lyophilized glycine-liposomes during storage at 25 °C for 28 days [24]. It was reported that curcumin-loaded chitosomes showed higher physical stability than curcumin-loaded uncoated liposomes due to the lower flexibility of chitosan-coated liposomes and, as a result, lower membrane fusion between droplets [33].

3.7. Production Yield and Physicochemical Properties of Spray-Dried Chitosomes

The powder moisture content was found to be 4.65 ± 0.61%, which is lower than the specified minimum moisture content of many powders used in food applications to inhibit microbiological spoilage and lipid oxidation and extend the product's shelf life.

The bulk density of the powder was 0.31 ± 0.01 g cm^{-3}. A higher bulk density has several advantages, including more convenient storage conditions due to lower space requirements for storage and greater protection against oxidation during storage due to the presence of less air in powder. In addition, the solubility and production yield of powder were found to be 96% and 57%, respectively.

3.8. Microstructure and Particle Size Distribution

The morphology of dried particles is mostly affected by the evaporation rate and viscoelastic properties of shell material. SEM images (Figure 4) show mostly spherical particles with little evidence of roughness or fracturing in the microcapsules. Moreover, most of the spray-dried powders showed a well-defined spherical shape with a particle size of around 1–3 µm, which is smaller than the threshold diameter used in food fortification (10–50 µm) [34]. The coarse powder results in a sandy and unfavorable mouthfeel in the fortified food products. These results are in accordance with those reported by Sarabandi et al. [13].

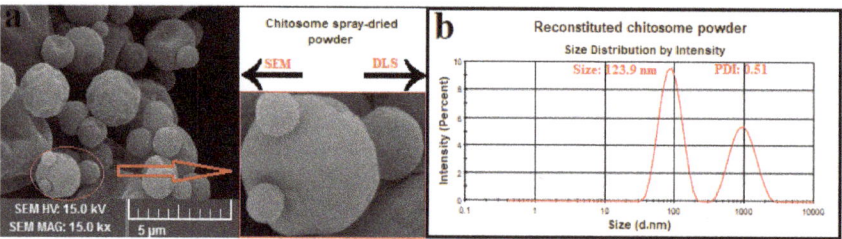

Figure 4. (a) SEM image of spray-dried hydrolysate-loaded γ-oryzanol-chitosomes; (b) particle size result of reconstituted γ-oryzanol-chitosome powder.

3.9. Properties of Reconstituted Chitosomes

3.9.1. Physical Properties

It is essential to conserve the physicochemical properties of nanoliposomes after the spray-drying process. The effect of spray-drying on particle size, PDI, and ζ of reconstituted chitosomes was investigated (Table 4). As expected, the chitosan coating increased the physical stability of nanocarriers after spray-drying, and slight changes were observed in the size and ζ of reconstituted chitosomes, but the PDI for these systems changed from 0.33 to 0.51. In a previously reported study, the effect of spray-drying on the physical properties of reconstituted liposomal powders was investigated. Regarding those results, the uncoated liposome systems were immediately unstable, as the maltodextrin was added to the dispersion as a wall matrix, i.e., these systems were not spray-dried. However, the coated systems did not significantly affect the spray-drying process and the diameter of the liposomal powder was smaller than that of liposomal dispersion before thermal processing. The authors suggested that when coated liposomal systems are added to a solution containing salts or sugars, which have the potential to induce an osmotic driving force and reduce water activity, migration of water molecules happens from the core to the liposomal surface, reducing the concentration gradient between the internal and external aqueous phases of the liposomes, thus reducing the size of the liposomal powder [35]. Moreover, the addition of maltodextrin as a hydrophilic nonionic polysaccharide had no effect on the ζ of coated liposomes.

Table 4. Physicochemical properties of reconstituted hydrolysate-loaded γ-oryzanol-chitosomes.

	Z-Average (nm)	PDI	ζ (mV)	Residual Antioxidant Activity (%)
HS-loaded chitosome dispersion	96.9 ± 1.8 [b]	0.33 ± 0.02 [b]	31.4 ± 1.5 [a]	86.0 ± 1.2 [a]
HS-loaded chitosome powder	123.9 ± 2.1 [a]	0.51 ± 0.02 [a]	33.2 ± 1.1 [a]	83.0 ± 1.3 [a]

Means in same column with different superscripts (a, b) are statistically different ($p < 0.05$).

3.9.2. Retention of Antioxidant Activity (AA)

The effect of spray-drying and thermal stress on the retention of ABTS radical scavenging activity in the chitosome system is shown in Table 4. There was no significant difference between these two indices in chitosome system retention before and after the spray-drying process ($p < 0.05$), which shows the positive effect of chitosan coating on the preservation of the biological activity of peptide fractions against thermal stresses during spray-drying. Our results are consistent with those reported by Sarabandi et al. [13]. In another study, about 39% of total phenols, 30% of flavonoids, and 47% of radical scavenging activity of extract were maintained after the spray-drying of nanoliposomes [22]. The spray-drying of nanoliposomes loaded with black mulberry extract resulted in the preservation of approximately 69% of total phenolic and 56% of anthocyanin compounds in chitosan-coated nanoliposomes [36].

4. Conclusions

Spirulina plantensis hydrolysate was successfully encapsulated into nanoliposomes using the thin-layer hydration method of sonication. This study shows that stabilizing the liposome structure with γ-oryzanol and covering the liposomes with a polycationic chitosan polymer provided long-term physical stability, and the system preserved its antioxidant capacity over time. Thus, this study suggests that chitosomes stabilized with γ-oryzanol could be used as a promising delivery system to protect against the loss of hydrolysate under processing or storage conditions.

Author Contributions: Methodology, M.M.; formal analysis, M.M. and R.S.; data curation and preparation, M.M.; Project administration, H.H.; writing—original draft preparation, M.M. and R.S.; Review and editing, H.H., M.G., M.P., and J.M.L. All authors have read and agreed to the published version of the manuscript.

Funding: This research received no external funding.

Institutional Review Board Statement: Not applicable.

Informed Consent Statement: Not applicable.

Data Availability Statement: Data is contained within the article.

Acknowledgments: The authors would like to acknowledge the Drug Applied Research Center, Tabriz University of Medical Sciences, for financial support.

Conflicts of Interest: The authors declare no conflict of interest.

References

1. Nakata, H.; Nakayama, S.M.M.; Kataba, A.; Yohannes, Y.B.; Ikenaka, Y.; Ishizuka, M. Evaluation of the ameliorative effect of Spirulina (*Arthrospira platensis*) supplementation on parameters relating to lead poisoning and obesity in C57BL/6J mice. *J. Funct. Foods* **2021**, *77*, 104344. [CrossRef]
2. Chronakis, I.S.; Galatanu, A.N.; Nylander, T.; Lindman, B. The behaviour of protein preparations from blue-green algae (*Spirulina platensis* strain Pacifica) at the air/water interface. *Colloids Surf. A Physicochem. Eng. Asp.* **2000**, *173*, 181–192. [CrossRef]
3. Sedighi, M.; Jalili, H.; Darvish, M.; Sadeghi, S.; Ranaei-Siadat, S.O. Enzymatic hydrolysis of microalgae proteins using serine proteases: A study to characterize kinetic parameters. *Food Chem.* **2019**, *284*, 334–339. [CrossRef]
4. Fan, X.; Cui, Y.; Zhang, R.; Zhang, X. Purification and identification of anti-obesity peptides derived from *Spirulina platensis*. *J. Funct. Foods* **2018**, *47*, 350–360. [CrossRef]
5. Costa, A.M.; Bueno, K.T.L.; da Rosa, A.P.C.; Costa, J.A.V. The antioxidant activity of nanoemulsions based on lipids and peptides from *Spirulina* sp. LEB18. *LWT* **2019**, *99*, 173–178. [CrossRef]
6. Mohammadi, M.; Jafari, S.M.; Hamishehkar, H.; Ghanbarzadeh, B. Phytosterols as the core or stabilizing agent in different nanocarriers. *Trends Food Sci. Technol.* **2020**, *101*, 73–88. [CrossRef]
7. Mohammadi, M.; Ghanbarzadeh, B.; Hamishehkar, H. Formulation of nanoliposomal vitamin d3 for potential application in beverage fortification. *Adv. Pharm. Bull.* **2014**, *4*, 569–575. [PubMed]
8. Wang, Y.; Ding, R.; Zhang, Z.; Zhong, C.; Wang, J.; Wang, M. Curcumin-loaded liposomes with the hepatic and lysosomal dual-targeted effects for therapy of hepatocellular carcinoma. *Int. J. Pharm.* **2021**, *602*, 120628. [CrossRef]
9. Briuglia, M.L.; Rotella, C.; McFarlane, A.; Lamprou, D.A. Influence of cholesterol on liposome stability and on in vitro drug release. *Drug Deliv. Transl. Res.* **2015**, *5*, 231–242. [CrossRef]
10. Kaddah, S.; Khreich, N.; Kaddah, F.; Charcosset, C.; Greige-Gerges, H. Cholesterol modulates the liposome membrane fluidity and permeability for a hydrophilic molecule. *Food Chem. Toxicol.* **2018**, *113*, 40–48. [CrossRef] [PubMed]
11. Esposto, B.S.; Jauregi, P.; Tapia-Blácido, D.R.; Martelli-Tosi, M. Liposomes vs. chitosomes: Encapsulating food bioactives. *Trends Food Sci. Technol.* **2021**, *108*, 40–48. [CrossRef]
12. Wang, X.X.; Cheng, F.; Wang, X.X.; Feng, T.; Xia, S.; Zhang, X. Chitosan decoration improves the rapid and long-term antibacterial activities of cinnamaldehyde-loaded liposomes. *Int. J. Biol. Macromol.* **2021**, *168*, 59–66. [CrossRef]
13. Sarabandi, K.; Jafari, S.M. Effect of chitosan coating on the properties of nanoliposomes loaded with flaxseed-peptide fractions: Stability during spray-drying. *Food Chem.* **2020**, *310*, 125951. [CrossRef]
14. Mazloomi, S.N.; Mahoonak, A.S.; Ghorbani, M.; Houshmand, G. Physicochemical properties of chitosan-coated nanoliposome loaded with orange seed protein hydrolysate. *J. Food Eng.* **2020**, *280*, 109976. [CrossRef]
15. Daliri, H.; Ahmadi, R.; Pezeshki, A.; Hamishehkar, H.; Mohammadi, M.; Beyrami, H.; Khakbaz Heshmati, M.; Ghorbani, M. Quinoa bioactive protein hydrolysate produced by pancreatin enzyme- functional and antioxidant properties. *LWT* **2021**, *150*, 111853. [CrossRef]

16. Venuste, M.; Zhang, X.; Shoemaker, C.F.; Karangwa, E.; Abbas, S.; Kamdem, P.E. Influence of enzymatic hydrolysis and enzyme type on the nutritional and antioxidant properties of pumpkin meal hydrolysates. *Food Funct.* **2013**, *4*, 811–820. [CrossRef]
17. Khosh manzar, M.; Mohammadi, M.; Hamishehkar, H.; Piruzifard, M.K. Nanophytosome as a promising carrier for improving cumin essential oil properties. *Food Biosci.* **2021**, *42*, 101079. [CrossRef]
18. De Araújo, J.S.F.; de Souza, E.L.; Oliveira, J.R.; Gomes, A.C.A.; Kotzebue, L.R.V.; da Silva Agostini, D.L.; de Oliveira, D.L.V.; Mazzetto, S.E.; da Silva, A.L.; Cavalcanti, M.T. Microencapsulation of sweet orange essential oil (*Citrus aurantium* var. *dulcis*) by liophylization using maltodextrin and maltodextrin/gelatin mixtures: Preparation, characterization, antimicrobial and antioxidant activities. *Int. J. Biol. Macromol.* **2020**, *143*, 991–999. [CrossRef]
19. Yavari Maroufi, L.; Ghorbani, M.; Mohammadi, M.; Pezeshki, A. Improvement of the physico-mechanical properties of antibacterial electrospun poly lactic acid nanofibers by incorporation of guar gum and thyme essential oil. *Colloids Surf. A Physicochem. Eng. Asp.* **2021**, *622*, 126659. [CrossRef]
20. Sarabandi, K.; Jafari, S.M.; Mohammadi, M.; Akbarbaglu, Z.; Pezeshki, A.; Khakbaz Heshmati, M. Production of reconstitutable nanoliposomes loaded with flaxseed protein hydrolysates: Stability and characterization. *Food Hydrocoll.* **2019**, *96*, 442–450. [CrossRef]
21. Sundararajan, B.; Moola, A.K.; Vivek, K.; Kumari, B.D.R. Formulation of nanoemulsion from leaves essential oil of Ocimum basilicum L. and its antibacterial, antioxidant and larvicidal activities (*Culex quinquefasciatus*). *Microb. Pathog.* **2018**, *125*, 475–485. [CrossRef]
22. Altin, G.; Gültekin-Özgüven, M.; Ozcelik, B. Chitosan coated liposome dispersions loaded with cacao hull waste extract: Effect of spray drying on physico-chemical stability and in vitro bioaccessibility. *J. Food Eng.* **2018**, *223*, 91–98. [CrossRef]
23. Tan, C.; Feng, B.; Zhang, X.; Xia, W.; Xia, S. Biopolymer-coated liposomes by electrostatic adsorption of chitosan (chitosomes) as novel delivery systems for carotenoids. *Food Hydrocoll.* **2016**, *52*, 774–784. [CrossRef]
24. Chotphruethipong, L.; Battino, M.; Benjakul, S. Effect of stabilizing agents on characteristics, antioxidant activities and stability of liposome loaded with hydrolyzed collagen from defatted Asian sea bass skin. *Food Chem.* **2020**, *328*, 127127. [CrossRef]
25. Guldiken, B.; Gibis, M.; Boyacioglu, D.; Capanoglu, E.; Weiss, J. Physical and chemical stability of anthocyanin-rich black carrot extract-loaded liposomes during storage. *Food Res. Int.* **2018**, *108*, 491–497. [CrossRef] [PubMed]
26. Akgün, D.; Gültekin-Özgüven, M.; Yücetepe, A.; Altin, G.; Gibis, M.; Weiss, J.; Özçelik, B. Stirred-type yoghurt incorporated with sour cherry extract in chitosan-coated liposomes. *Food Hydrocoll.* **2020**, *101*, 105532. [CrossRef]
27. Liu, R.R.; Xu, Y.; Chang, M.; Tang, L.; Lu, M.; Liu, R.R.; Jin, Q.; Wang, X. Antioxidant interaction of α-tocopherol, γ-oryzanol and phytosterol in rice bran oil. *Food Chem.* **2021**, *343*, 128431. [CrossRef] [PubMed]
28. Li, Y.; Liu, H.; Han, Q.; Kong, B.; Liu, Q. Cooperative antioxidative effects of zein hydrolysates with sage (*Salvia officinalis*) extract in a liposome system. *Food Chem.* **2017**, *222*, 74–83. [CrossRef] [PubMed]
29. Sepúlveda, C.T.; Alemán, A.; Zapata, J.E.; Montero, M.P.; Gómez-Guillén, M.C. Characterization and storage stability of spray dried soy-rapeseed lecithin/trehalose liposomes loaded with a tilapia viscera hydrolysate. *Innov. Food Sci. Emerg. Technol.* **2021**, *71*, 102708. [CrossRef]
30. Hasan, M.; Ben Messaoud, G.; Michaux, F.; Tamayol, A.; Kahn, C.J.F.; Belhaj, N.; Linder, M.; Arab-Tehrany, E. Chitosan-coated liposomes encapsulating curcumin: Study of lipid–polysaccharide interactions and nanovesicle behavior. *RSC Adv.* **2016**, *6*, 45290–45304. [CrossRef]
31. Ramezanzade, L.; Hosseini, S.F.; Nikkhah, M. Biopolymer-coated nanoliposomes as carriers of rainbow trout skin-derived antioxidant peptides. *Food Chem.* **2017**, *234*, 220–229. [CrossRef] [PubMed]
32. Chotphruethipong, L.; Aluko, R.E.; Benjakul, S. Hydrolyzed collagen from porcine lipase-defatted seabass skin: Antioxidant, fibroblast cell proliferation, and collagen production activities. *J. Food Biochem.* **2019**, *43*, 1–13. [CrossRef] [PubMed]
33. Peng, S.; Zou, L.; Liu, W.; Li, Z.; Liu, W.; Hu, X.; Chen, X.; Liu, C. Hybrid liposomes composed of amphiphilic chitosan and phospholipid: Preparation, stability and bioavailability as a carrier for curcumin. *Carbohydr. Polym.* **2017**, *156*, 322–332. [CrossRef]
34. Sarabandi, K.; Gharehbeglou, P.; Jafari, S.M. Spray-drying encapsulation of protein hydrolysates and bioactive peptides: Opportunities and challenges. *Dry. Technol.* **2020**, *38*, 577–595. [CrossRef]
35. Liu, W.; Wei, F.; Ye, A.; Tian, M.; Han, J. Kinetic stability and membrane structure of liposomes during in vitro infant intestinal digestion: Effect of cholesterol and lactoferrin. *Food Chem.* **2017**, *230*, 6–13. [CrossRef] [PubMed]
36. Gültekin-Özgüven, M.; Karadağ, A.; Duman, Ş.; Özkal, B.; Özçelik, B. Fortification of dark chocolate with spray dried black mulberry (*Morus nigra*) waste extract encapsulated in chitosan-coated liposomes and bioaccessability studies. *Food Chem.* **2016**, *201*, 205–212. [CrossRef]

Article

Yogurt Fortification by the Addition of Microencapsulated Stripped Weakfish (*Cynoscion guatucupa*) Protein Hydrolysate

Karina Oliveira Lima [1], Meritaine da Rocha [2], Ailén Alemán [3], María Elvira López-Caballero [3,*], Clara A. Tovar [4], María Carmen Gómez-Guillén [3], Pilar Montero [3,*] and Carlos Prentice [1,†]

1. Laboratory of Food Technology, School of Chemistry and Food, Federal University of Rio Grande (FURG), Rio Grande 96203-900, RS, Brazil; karinah_ol@hotmail.com (K.O.L.); dqmprent@furg.br (C.P.)
2. Laboratory of Microbiology, School of Chemistry and Food, Federal University of Rio Grande (FURG), Santo Antônio da Patrulha 95500-000, RS, Brazil; meritaine@gmail.com
3. Institute of Food Science, Technology and Nutrition (ICTAN-CSIC), 28040 Madrid, Spain; ailen@ictan.csic.es (A.A.); cgomez@ictan.csic.es (M.C.G.-G.)
4. Department of Applied Physics, University of Vigo, As Lagoas, 32004 Ourense, Spain; tovar@uvigo.es
* Correspondence: elvira.lopez@ictan.csic.es (M.E.L.-C.); mpmontero@ictan.csic.es (P.M.)
† In memoriam.

Abstract: The aim of the present work was to fortify yogurt by adding a stripped weakfish (*Cynoscion guatucupa*) protein hydrolysate obtained with the enzyme Protamex and microencapsulated by spray drying, using maltodextrin (MD) as wall material. The effects on the physicochemical properties, syneresis, texture, viscoelasticity, antioxidant and ACE inhibitory activities of yogurt after 1 and 7 days of storage were evaluated. In addition, microbiological and sensory analyses were performed. Four yogurt formulations were prepared: control yogurt (without additives, YC), yogurt with MD (2.1%, YMD), with the free hydrolysate (1.4%, YH) and the microencapsulated hydrolysate (3.5%, YHEn). Yogurts to which free and microencapsulated hydrolysates were added presented similar characteristics, such as a slight reduction in pH and increased acidity, with a greater tendency to present a yellow color compared with the control yogurt. Moreover, they showed less syneresis, the lowest value being that of YHEn, which also showed a slight increase in cohesiveness and greater rheological stability after one week of storage. All yogurts showed high counts of the microorganisms used as starters. The hydrolysate presence in both forms resulted in yogurts with antioxidant activity and potent ACE-inhibitory activity, which were maintained after 7 days of storage. The incorporation of the hydrolysate in the microencapsulated form presented greater advantages than the direct incorporation, since encapsulation masked the fishy flavor of the hydrolysate, resulting in stable and sensorily acceptable yogurts with antioxidant and ACE inhibitory activities.

Keywords: fish protein hydrolysate; microencapsulation; yogurt; physicochemical properties; antioxidant activity; antihypertensive activity

1. Introduction

Milk and dairy products, such as yogurt, are widely appreciated and consumed worldwide because of their sensory and nutritional characteristics [1,2]. Compared with milk, yogurt is more nutritious and an excellent source of protein, calcium, phosphorus, riboflavin, thiamine, vitamin B12, niacin, magnesium, and zinc [3]. However, despite their beneficial health effects, these products are generally not considered to be an important source of bioactive compounds [2].

There are some studies related to the addition to yogurt of bioactive compounds, such as *Spirulina platensis* [4,5], rice bran [6], strawberry pulp [1], mushroom extracts (*Agaricus bisporus*) [7], fish collagen [8], monk fruit extract (*Siraitia grosvenorii*) [9] and *Ficus glomerata* Roxb fruit extract [10], in which the addition of these compounds increased the antioxidant and/or angiotensin-converting enzyme (ACE) inhibitory capacity. This may be an ideal strategy for facilitating the consumption of bioactive compounds and increasing the functionality of food, as people of all ages widely accept and consume yogurt.

Correspondingly, proteins derived from fish by-products represent a very interesting source of bioactive peptides due to their low cost and the established requirement of reducing agro-industrial waste [11]. Fish protein hydrolysates are sources of peptides with diverse bioactivities, including antioxidant and antihypertensive actions [12]. Some peptides present antihypertensive activity as they are able to inhibit the angiotensin-converting enzyme (ACE), which has a fundamental role in the regulation of blood pressure [12–14]. Moreover, antioxidants can also delay oxidative stress, involved in the occurrence of various diseases, including hypertension and aging [15].

To improve bioactivity, protein hydrolysates and bioactive peptides can be incorporated into different foods, such as dairy products, for their functional properties. However, scarce information about the incorporation of fish hydrolysates into yogurt is available. In one study, bovine and fish skin hydrolysates induced greater syneresis and less firmness and viscoelasticity in skimmed bovine milk yogurt than caseinate hydrolysate, which could be attributed to their hindering effects on yogurt acidification [16]. Moreover, the incorporation of fish protein hydrolysates can be difficult due to their high hygroscopicity, bitter taste, chemical instability, interaction with the food matrix, incompatibility and limited bioavailability [17–19]. Microencapsulation is a process in which the compounds of interest are coated or incorporated into a protective matrix, and is considered effective for overcoming the limitations mentioned above [20].

Among the numerous encapsulation techniques, spray drying is widely used in the food industry [21,22]. This technique consists of atomizing a formulation containing the protective matrix and bioactive compounds in small drops, followed by subsequent rapid drying through a stream of hot air to produce dry microparticles [14,23]. In previous studies, Lima et al. [24] used spray drying to encapsulate a stripped weakfish hydrolysate in maltodextrin. The microencapsulated hydrolysate was characterized, and the stability and biological properties were evaluated in vitro and in vivo in a model of *Caenorhabditis elegans*; the results showed improvements in growth and reproduction rate as well as a protective effect on nematodes exposed to oxidative stress upon consumption of the encapsulates.

Based on the above-mentioned information, the objective of this study was to develop a functional yogurt by incorporating a microencapsulated protein hydrolysate from stripped weakfish (*Cynoscion guatucupa*) in the formulation. For this purpose, sensory and microbiological analyses were performed to confirm that a quality product was obtained. Subsequently, the effects on the physicochemical and rheological properties, texture, and bioactivity of the yogurts after 1 and 7 days of storage were evaluated.

2. Materials and Methods

2.1. Materials

The by-products of stripped weakfish (*Cynoscion guatucupa*) were obtained from a fishing company in the city of Rio Grande (RS, Brazil). The carcasses and trimmings were processed in a meat–bone separator (High Tech, HT250C, Chapecó, Brazil), discarding the skin and bones. The resulting muscle was packed in plastic bags and stored at −18 °C until use. Protamex enzyme purchased from Sigma-Aldrich (St. Louis, MO, USA) was used in the hydrolysis process. Maltodextrin with a dextrose equivalent (DE) of 5 was purchased from Manuel Riesgo S.A. (Madrid, Spain). All chemical reagents used in this study were of analytical grade.

2.2. Protein Hydrolysate

The fish muscle protein hydrolysate, with a degree of hydrolysis of 5%, was produced using the enzyme Protamex at 50 °C and pH 7 as previously described in Lima et al. [24]. The lyophilized hydrolysate was stored at −18 °C until use. Hydrolysate characteristics (amino acid profile, Fourier transform infrared spectroscopy, morphological and biological activity) were previously described in Lima et al. [24,25].

2.3. Microencapsulation

The hydrolysate was microencapsulated with maltodextrin (MD) by spray drying as described by Sarabandi et al. [21], with some modifications. The MD and the hydrolysate (60:40 w/w) were dissolved in distilled water constituting 10% of the total solids and stirred for 3 h at room temperature. Subsequently, the solutions were atomized in a Mini Spray Dryer B-290 (Büchi, Switzerland) at 0.3 L/h with an inlet temperature of 130 °C, aspiration rate 100% and an outlet temperature of 70 ± 2 °C. Thus, a microencapsulated hydrolysate (microcapsules composed of maltodextrin and hydrolysate) was obtained. The characteristics and stability of the encapsulation have been previously described in Lima et al. [24].

2.4. Yogurt Preparation

Yogurts were made with UHT (Ultra High Temperature) whole cow milk (fat 3.6%; protein 3.1% and carbohydrates 4.6%) and natural yogurt was purchased from a local market (Madrid, Spain). Yogurts were prepared in a Thermomix (Vorwerk & Co., Wuppertal, Germany). Firstly, the milk was heated to 40–45 °C while stirring for 3 min, followed by the addition of natural yogurt and stirring for 5 min. Later, the tested ingredients were added, and the mixture was subsequently stirred for another 5 min to constitute the different samples. The temperature was maintained during the mixing steps. For each 100 mL of milk, 12.5 g of natural yogurt and different concentrations of the tested ingredients were added, in order to maintain the same concentration of wall material and hydrolysate present in the microencapsulated hydrolysate. Four lots of yogurts were then obtained: control yogurt (without additional ingredients; lot YC); yogurt with the addition of 2.1 g maltodextrin (wall material; lot YMD); yogurt with the addition of 1.4 g of free protein hydrolysate (lot YH) and yogurt with the addition of 3.5 g of microencapsulated hydrolysate (lot YHEn, composed of 2.1g MD and 1.4 g of hydrolysate). Later, the formulations were transferred to disposable plastic cups of 100 mL and incubated at 43 °C until reaching pH 4.6 [26]. The yogurts were stored under refrigeration (6 ± 1 °C) for 7 days.

2.5. Yogurt Characterization

2.5.1. Proximal Analyses, Physicochemical Analysis and Color

Proximate composition (moisture, ash, protein, and fat) was evaluated following the official methods AOAC 19,927.05, AOAC 945.46, AOAC 991.22, and AOAC 905.02, respectively [27]. Carbohydrate content was estimated by difference. Protein content was estimated by nitrogen determination using the factor of 6.25. The pH was measured by a digital pH-meter (Methrom 827, Herisau, Switzerland), previously calibrated, at room temperature. The titratable acidity (TA) of the yogurts was determined by titrating 9 g of the sample with 0.1 M NaOH using phenolphthalein as an indicator, which was expressed as % lactic acid. Both determinations were performed in triplicate. The color of the yogurts was measured in a Konica Minolta CM-3500d spectrophotometer (Konica Minolta Sensing, Inc., Osaka, Japan) from ten measurements of each sample, and the CIELAB color space was used to obtain the color coordinates. The color was expressed by the parameters L*, a* and b*. The whiteness index (WI) was determined according to Equation (1):

$$WI = 100 - \sqrt{(100 - L)^2 + a^2 + b^2} \qquad (1)$$

2.5.2. Syneresis

The syneresis of the yogurts was determined according to Santillán-Urquiza et al. [28]. Approximately 10 g of yogurt was centrifuged at 176× g for 20 min at 10 °C. The syneresis, expressed as percentage, was performed in triplicate and estimated as the weight of the supernatant released over the weight of the initial yogurt × 100.

2.5.3. Texture Analysis

Firmness and cohesiveness parameters were determined using a texture analyzer (TA.XTplus, Stable Micro Systems) as described by Santillán-Urquiza et al. [28], with modifications. Briefly, the compression force (N) in 50 mL of yogurt was measured using a 36 mm diameter × 50 mm high cylindrical body (P36R), with a speed of 0.5 mm/s and reaching a depth of 20 mm. Three different yogurt cups were measured for each lot and the firmness results were expressed in N.

2.5.4. Rheological Analyses

Oscillatory shear measurements were performed on a Bohlin CVO rheometer (Bohlin Instruments Ltd., Gloucestershire, UK) using a cone-plate geometry (4° angle, 40 mm diameter, 0.15 mm gap). The temperature in the lower plate was 5 °C. Frequency sweep tests were carried out over a range of angular frequencies between 0.63 and 63 rad/s with an oscillation strain of 5%, selected from the linear viscoelastic region (LVER). The storage modulus (G') and loss modulus (G'') were plotted as a function of angular frequency (ω).

To understand the viscoelastic properties of the different yogurts, flow and viscoelastic properties of the additive aqueous suspensions were examined. Thus, three aqueous suspensions at the same concentration (1.5% w/v) were prepared: maltodextrin (MD), free hydrolysate (H), and microencapsulated hydrolysate (HEn). For determining the viscoelastic moduli, time sweeps at 20 °C for 1800 s at 0.05 Hz and small stress (σ = 1.5 Pa) were performed to minimize structural changes.

Flows were characterized by the step test, with three intervals starting with a pre-shear interval to homogenize the suspensions (100 s^{-1} for 300 s). The three steps were: (1) reference interval (150 s^{-1} for 90 s); (2) high shear-rate interval (1000 s^{-1} for 45 s) to damage the internal structure; (3) regeneration interval (150 s^{-1}, 600 s) at 20 °C. Each rheological test was repeated five times.

2.6. Microbiological Analysis

Yogurts were evaluated for microbiological analysis according to the method described by Arancibia et al. [29]. Aliquots of approximately 10.0 ± 0.2 g were weighed and transferred to sterile polyethylene bags (Sterilin, Stone, UK) with 90 mL of 0.1% (w/v) sterile peptone water (Oxoid, Unipath Ltd., Basingstoke, UK) and homogenized for 1 min medium speed in a Stomacher (Colworth 400, Seward, UK) at room temperature. Then, appropriate dilutions were prepared for the following bacteriological determinations: (i) *Streptococcus thermophilus*, on spread plates of ESTY agar + lactose (0.5%) (Pronadisa, Spain) and (ii) *Lactobacillus delbruckii* sp. *bulgaricus*, on spread plates of MRS Agar (Pronadisa, Spain) + tween 80 (0.1%) (Sigma-Aldrich, Darmstadt, Germany) + Cysteine (0.05%) (Sigma-Aldrich), both incubated at 42 °C for 24 h in an anaerobic jar (Oxoid), (iii) *Enterobacteriaceae*, on pour double-layered plates of Violet Red Bile Glucose Agar—VRBG (Oxoid), incubated at 30 °C for 48 h, and (iv) molds and yeasts on spread plates of potato dextrose agar (Scharlab, Spain) incubated at 25 °C for 72 h. These analyses were carried out on the day the yogurts were prepared to confirm that a suitable product was obtained. All determinations were conducted in duplicate.

2.7. Yogurt Bioactivity

For the analysis of antioxidant and antihypertensive activities, the supernatant of yogurt was used, which was prepared according to Zhang et al. [30]. For that, 10 g of yogurt samples were centrifuged at 4330× g for 5 min at 4 °C, and then the supernatants were recentrifuged under the same conditions.

2.7.1. Antioxidant Activity
ABTS Radical Scavenging Activity

The 2,2'-azinobis-(3-ethylbenzothiazoline-6-sulfonic acid) (ABTS) radical scavenging activity was determined using the methods detailed by Zheng et al. [31], with modifications.

The stock solution of the ABTS radical was generated by incubating 7 mM ABTS with 140 mM potassium persulfate in the dark for 16 h at room temperature. Before use, the stock solution was diluted with phosphate-buffered saline (PBS, pH 7.4) to an absorbance of 0.70 ± 0.02 at 734 nm. A 50 µL aliquot of the yogurt supernatant (diluted 1:5 in PBS) was homogenized with 150 µL of ABTS, and after reaction at 30 °C for 6 min in the dark, the absorbance was measured at 734 nm. The ABTS radical scavenging activity was calculated as follows: $[(A_c - A_s)/A_c] \times 100$, where A_c is the absorbance of the control, reagents without sample, and A_s is absorbance with sample. The determination was carried out in triplicate.

Reducing Power

The reducing power was assessed as described by Canabady-Rochelle et al. [32], with modifications. An aliquot of 70 µL of yogurt supernatant (diluted 1:5 in phosphate buffer pH 6.6, 200 mM) was homogenized with 35 µL of potassium ferricyanide solution (1% w/v), and incubated at 50 °C for 20 min. Subsequently, 135 µL of distilled water, 33 µL of trichloroacetic acid (10% w/v) and 27 µL of ferric chloride (0.1% w/v) were added and after 10 min, the absorbance was measured at 700 nm. The reducing power of the sample was shown as the absorbance at 700 nm after subtracting the absorbance value from the blank, where a higher absorbance value indicates greater reducing power. The determination was carried out in triplicate.

2.7.2. ACE Inhibitory Activity

The ability of the sample to inhibit the angiotensin-converting enzyme (ACE) was determined according to Alemán et al. [33], with some modifications. The reaction was composed of 50 µL of 5 mM Hipuryl-histidyl-leucine (HHL), 80 µL of ACE (0.025 U/mL) and 20 µL of the yogurt supernatant (diluted 1:5 in 100 mM potassium phosphate buffer, containing 300 mM NaCl, pH 8.3). The determination was performed by reverse phase High Performance Liquid Chromatography (RP-HPLC) (model SPE-MA10AVP, Shimadzu, Kyoto, Japan). The injection volume was 50 µL and the flow rate 0.8 mL/ min, using an acetonitrile gradient from 20% to 60% in 0.1% trifluoroacetic acid (TFA) (v/v) for 26 min. The results were expressed as % of ACE-inhibitory activity.

2.8. Sensory Analysis

The sensory analysis of the formulated yogurts was conducted at the Instituto de Ciencia y Tecnología de Alimentos y Nutrición (ICTAN-CSIC, Madrid, Spain) with 10 semi-trained judges of both sexes, selected from among students and researchers of the Institute. The specific attributes that the panel was asked to classify in the yogurts were the following: color, flavor, odor, texture and acidity. The scale values ranged from 0 to 9. Number 9 was assigned to the most positive terms, while number 0 was assigned to the least desirable attributes/characteristics. Regarding texture and acidity, number 9 was assigned to firm and very acid, respectively. Sensory tests were performed in individual booths with lighting control; samples were served in disposable plastic cups, encoded with 3-digit numbers obtained from a table of random numbers. Mineral water was also provided for cleansing the palate between the evaluations of the different samples of yogurt.

2.9. Statistical Analysis

The data were submitted to analysis of variance (ANOVA) and the means compared by the *t*-Student test (to compare between day 1 and day 7) or the Tukey test with a significance level of 5% using Statistica software (StatSoft, Inc., Tulsa, OK, USA).

3. Results and Discussion

3.1. Visual Appearance, Proximal Analyses and Physicochemical Properties of Yogurts

Figure 1 shows the visual appearance of the different yogurts. To the naked eye, all yogurts presented a similar color, with similar gel-like appearance and no evidence of whey separation. The proximate composition revealed that the yogurts were similar, with significant differences ($p < 0.05$) mainly in carbohydrate and protein content due to the incorporation of maltodextrin and hydrolysate in some yogurts.

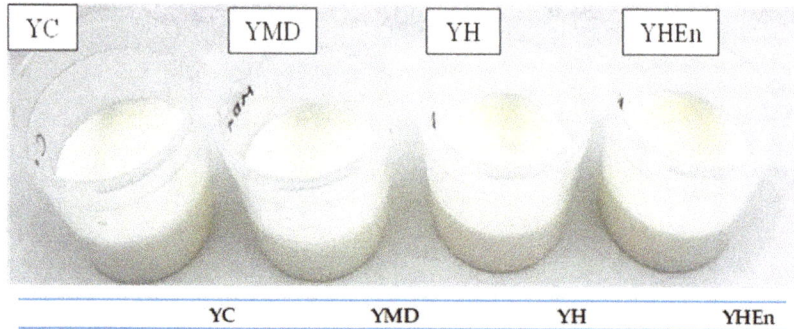

	YC	YMD	YH	YHEn
Carbohydrates	4.6[a]	6.56[b]	4.54[a]	6.47[c]
Protein	3.1[a]	3.04[a]	4.44[b]	4.35[b]
Fat	3.6[a]	3.53[bc]	3.55[bc]	3.48[c]
Ash	0.23[a]	0.23[a]	0.23[a]	0.22[a]
Moisture	88.47[a]	86.65[bc]	87.25[ab]	85.48[c]

Figure 1. Visual appearance and proximate composition of the different yogurts. YC: control yogurt; YMD: yogurt with addition of maltodextrin; YH: yogurt with addition of free protein hydrolysate; YHEn: yogurt with addition of microencapsulated protein hydrolysate. Different letters (a,b,c) indicate significant differences between samples within the same line ($p < 0.05$).

The pH, titratable acidity (TA), color and syneresis of the yogurts after 1 and 7 days of storage under refrigeration, are shown in Table 1. The lowest pH value after 1 and 7 days of storage was observed in the yogurt with the addition of the microencapsulated hydrolysate (YHEn) while the highest pH observed was in the yogurt with the addition of MD (YMD), ($p < 0.05$). The pH values of MD, the free hydrolysate and microencapsulated hydrolysate were 5.7, 7.03 and 6.91, respectively. Hence, it seems that the pH of the yogurts was probably not affected by the addition of these ingredients. Moreover, a decrease in pH was observed in all yogurts throughout storage, which can be mostly attributed to the production of microbial metabolites. Similarly, a decrease in pH was also observed by Abdel-Hamid et al. [9] when supplementing probiotic yogurt with 1% and 2% monk fruit extract (*Siraitia grosvenorii*). These authors reported that supplementation with the extract may have stimulated bacterial growth. Notably, fermented products with pH in the range of 4.2 and 4.4 are preferred by consumers [34]. In the present work, these pH ranges were achieved in yogurts after 7 days of storage (Table 1).

The titratable acidity of the yogurts varied from 0.73% to 1.09% (Table 1), the YH and YHEn yogurts showing the highest values, both after 1 and 7 days of storage. In addition, with the exception of YC, there was an increase in the titratable acidity with storage time ($p < 0.05$). Similar results were reported by Córdova-Ramos et al. [35] when studying the addition of jumbo squid powder (*Dosidicus gigas*) using maltodextrin as an encapsulating agent by spray drying (MD with different DE (11) and drying conditions) in yogurt. The authors verified an increase in acidity as squid powder concentration increased (1%, 3%, 5%, 7% and 10%), ranging from 0.76% to 1.05% in the yogurt without squid powder and with the highest concentration, respectively. An increase in titratable acidity was also observed in yogurts supplemented with pineapple peel powder or inulin after 28 days of storage [36]. According to the authors, the supplementation increased the acidifying capacity of the starter cultures during storage.

Color is the first characteristic perceived by consumers, and it can influence their preference, so it is an important attribute to be evaluated [37]. The color parameters (L*, a* and b*) of the different formulated yogurts are shown in Table 1. A decrease in lightness values (L*) was observed for yogurts with the addition of the different tested ingredients, and therefore, the highest values observed for YC also reflected a higher whiteness index. As also observed by Silva et al. [5] such a phenomenon may have occurred because this batch did not have any powder ingredients. Conversely, on the same evaluation day, the YH and YHEn yogurts tended to show higher values for yellowness (b*) ($p < 0.05$), probably due to the color of the hydrolysate. Carmona et al. [38] verified that encapsulating yellow–orange cactus pear *Opuntia ficus-indica* pulp, using maltodextrin (MD) as an encapsulating agent, protected the quality of the pigment (thus allowing its use as a yellow colorant for yogurt) and that, additionally, maltodextrin did not negatively affect b* and a* parameters in the yogurt when compared with other treatments.

Table 1. Physicochemical properties of yogurts during storage.

Parameters	Time (Days)	Samples			
		YC	YMD	YH	YHEn
pH	1	4.58 ± 0.01 bA	4.61 ± 0.01 aA	4.57 ± 0.01 bA	4.53 ± 0.02 cA
	7	4.38 ± 0.01 bB	4.43 ± 0.01 aB	4.37 ± 0.01 bB	4.35 ± 0.01 cB
TA (% lactic acid)	1	0.73 ± 0.01 bA	0.73 ± 0.01 bB	0.94 ± 0.02 aB	0.95 ± 0.01 aB
	7	0.77 ± 0.04 bA	0.75 ± 0.01 bA	1.08 ± 0.01 aA	1.09 ± 0.02 aA
L*	1	83.25 ± 0.12 aB	82.64 ± 0.10 bB	82.37 ± 0.07 bcB	82.09 ± 0.19 cA
	7	83.59 ± 0.04 aA	83.30 ± 0.15 bA	82.80 ± 0.11 cA	82.45 ± 0.03 dA
a*	1	−1.53 ± 0.06 abA	−1.57 ± 0.03 bA	−1.42 ± 0.03 aA	−1.47 ± 0.04 abA
	7	−1.49 ± 0.02 bcA	−1.54 ± 0.03 cA	−1.43 ± 0.03 abA	−1.42 ± 0.02 aB
b*	1	6.11 ± 0.08 bA	6.21 ± 0.10 bA	7.16 ± 0.04 aA	7.22 ± 0.08 aA
	7	6.20 ± 0.06 bA	6.20 ± 0.07 bA	7.26 ± 0.10 aA	7.26 ± 0.05 aA
WI	1	82.10 ± 0.12 aB	81.50 ± 0.11 bA	80.92 ± 0.06 cB	80.63 ± 0.21 cA
	7	82.39 ± 0.06 aA	82.12 ± 0.16 aA	81.28 ± 0.14 bA	80.96 ± 0.05 cA
Syneresis (%)	1	13.97 ± 0.63 aA	11.0 ± 0.45 bA	11.10 ± 0.43 bA	9.25 ± 0.19 cA
	7	13.25 ± 1.12 aA	9.89 ± 1.12 bcA	10.53 ± 0.47 bB	7.98 ± 0.79 cA

TA: titratable acidity; L*: lightness; a*: red-green axis; b*: blue-yellow axis; WI: whiteness index; YC: control yogurt; YMD: yogurt with addition of maltodextrin; YH: yogurt with addition of free protein hydrolysate; YHEn: yogurt with addition of microencapsulated protein hydrolysate. Different lowercase letters (a,b,c,d) indicate a significant difference between samples for the same day ($p < 0.05$). Different uppercase letters (A,B) indicate a significant difference for the same sample on different days ($p < 0.05$).

The YMD sample showed differences ($p < 0.05$) in a* values compared with YH after 1 day of storage and with YH and YHEn after 7 days of storage, but was similar to YC ($p > 0.05$). Regarding storage, in general the L* parameter showed a slight increase over time, while a* and b* values did not change, except in YHEn, which did not present any change in relation to the L* parameter and showed a slight decrease in a* values. The whiteness index (WI) was in the same range in all samples, decreasing less than 2% compared with the control ($p < 0.05$).

Syneresis is an important parameter in yogurt since it can affect its quality during storage through the accumulation of serum on the surface, influencing the acceptability of the product [4,28,30]. The addition of the different ingredients provided a decrease in syneresis, with the lowest values observed in YHEn, independently of storage time ($p < 0.05$) (Table 1). In general, syneresis was not influenced by storage time, with the exception of YH, which showed a small decrease after day 7 ($p < 0.05$). Córdova-Ramos et al. [35] reported a decrease in syneresis in yogurts with the addition of different concentrations (1, 3, 5, 7 and 10 g/100 mL) of jumbo squid powder (*Dosidicus gigas*) obtained by spray drying using MD as encapsulating agent. The decrease in syneresis was associated with the functional properties of the protein, since protein and MD facilitate water retention, thus preventing this phenomenon. These authors reported that the lowest syneresis values were observed in yogurts with 7 and 10 g/100 mL of the powder (4.20% and 1.0%, respectively) while the addition of 3 g/100 mL presented a syneresis percentage (9.10%) similar to that obtained in the present study in YHEn (Table 1). In addition, the lowest syneresis was found in YHEn. This result was similar to that reported by Demirci et al. [6] in yogurts with the addition of different concentrations of rice bran (1, 2 and 3%), which ranged from 9.79% to 8.80%, while the control yogurt presented 10.29%. The authors related the decrease in syneresis to the water-retention capacity of rice bran dietary fibers. In the present work, the reduction in the syneresis of formulated yogurts may be associated with the water-retention capacity of the protein hydrolysate by ion-dipole interactions. This fact may be due to the amino acid profile, which has a predominance of hydrophilic amino acids (57.4%) and negatively charged amino acids (64.4%). Moreover, the reduction in syneresis was enhanced in the yogurt to which the microencapsulated hydrolysate was added, possibly because MD also contributed to increase water retention by hydrogen bonding and dipole–dipole interactions.

3.2. Textural and Viscoelastic Properties

Texture is an important attribute regarding the quality of yogurt [37]. The firmness and cohesiveness of the different yogurts after 1 and 7 days of storage are shown in Table 2. The addition of the hydrolysate to the yogurts, both free and microencapsulated, resulted in a slight decrease in firmness compared to the YC sample ($p < 0.05$), representing approximately 11% (YH) and 10% (YHEn) after 1 day, and 30% (YH) and 19% (YHEn) after 7 days of storage. On the other hand, YH and YHEn showed increases in cohesiveness ($p < 0.05$) of approximately 12% and 15%, respectively, after 1 day and ~92% and 18%, respectively, after 7 days of storage compared with YC. After 7 days YH showed the highest cohesiveness values ($p < 0.05$). In general, the firmness and cohesiveness of the different yogurts were not significantly influenced by storage time, except for YC and YH, which showed greater firmness and greater cohesiveness, respectively, after 7 days of storage.

Table 2. Textural and viscoelastic parameters (Equations (2) and (3)) of yogurts during refrigerated storage.

Parameters	Time (Days)	Samples			
		YC	YMD	YH	YHEn
Firmness (N)	1	1.11 ± 0.02 [aB]	1.13 ± 0.02 [aA]	0.99 ± 0.03 [bA]	1.00 ± 0.02 [bA]
	7	1.34 ± 0.09 [aA]	1.21 ± 0.10 [abA]	0.94 ± 0.03 [cA]	1.09 ± 0.08 [bcA]
Cohesiveness	1	0.69 ± 0.02 [bA]	0.71 ± 0.01 [bA]	0.77 ± 0.00 [aB]	0.79 ± 0.03 [aA]
	7	0.67 ± 0.02 [bA]	0.69 ± 0.04 [bA]	1.29 ± 0.04 [aA]	0.79 ± 0.11 [bA]
G_0' (Pa)	1	40.71 ± 0.02	48.24 ± 0.06	27.62 ± 0.08	35.08 ± 0.07
	7	101.94 ± 0.09	85.00 ± 0.06	31.14 ± 0.11	36.98 ± 0.07
n'	1	0.116 ± 0.001	0.122 ± 0.001	0.144 ± 0.003	0.131 ± 0.002
	7	0.105 ± 0.001	0.117 ± 0.004	0.129 ± 0.004	0.121 ± 0.002
R^2 (Equation (2))	1	0.998	0.988	0.958	0.972
	7	0.986	0.993	0.922	0.975
G_0'' (Pa)	1	13.11 ± 0.06	14.71 ± 0.09	8.81 ± 0.08	11.24 ± 0.08
	7	30.72 ± 0.06	23.33 ± 0.08	9.39 ± 0.10	11.05 ± 0.09
n''	1	0.214 ± 0.003	0.208 ± 0.005	0.236 ± 0.006	0.226 ± 0.005
	7	0.177 ± 0.002	0.191 ± 0.003	0.239 ± 0.008	0.232 ± 0.006
R^2 (Equation (3))	1	0.977	0.947	0.937	0.946
	7	0.988	0.976	0.913	0.946
tanδ	1	0.355 ± 0.007 [aA]	0.343 ± 0.008 [aA]	0.356 ± 0.007 [aA]	0.358 ± 0.011 [aA]
	7	0.333 ± 0.012 [aB]	0.338 ± 0.005 [aA]	0.343 ± 0.004 [aB]	0.345 ± 0.011 [aA]

YC: control yogurt; YMD: yogurt with addition of maltodextrin; YH: yogurt with addition of free protein hydrolysate; YHEn: yogurt with addition of microencapsulated protein hydrolysate. Different lowercase letters (a,b,c) indicate a significant difference between samples for the same day ($p < 0.05$). Different uppercase letters (A,B) indicate a significant difference for the same sample on different days ($p < 0.05$).

Barkallah et al. [4] studied the addition of different concentrations of *Spirulina platensis* (0.25%, 0.5%, 0.75% and 1.0%) to yogurt and observed greater firmness in the control yogurt and with the addition of 0.25% of *Spirulina platensis* (0.67 and 0.62 N, respectively). The addition of higher concentrations resulted in yogurts with lower firmness values which, according to the authors, was due to the interruption of gel formation as concentrations of microalga were increased. In the same study, the authors found no differences in relation to cohesiveness for the samples. It is worth noting that both firmness and cohesiveness values were lower than those observed in the present study.

Additionally, Öztürk et al. [39], when studying the fortification of set-type yogurts with peeled or unpeeled oleaster (*Elaeagnus angustifolia* L.) flours (1% and 2%), reported a reduction in firmness and an increase in cohesiveness on the first day of storage compared to the control yogurt. According to the authors, the control yogurt showed a longer fermentation time resulting in greater firmness, while the increase in cohesiveness would occur due to the water retention capacity of the flours added into the protein matrix. During fermentation, the formation of lactic acid by the action of microorganisms occurs, and the pH drop induces the aggregation of casein and the formation of disulfide bonds between denatured whey proteins and k–casein, resulting in the characteristic gel formation, texture and properties of yogurt [37,40,41].

In the present study, the yogurts with the addition of the free and microencapsulated hydrolysate required less time to ferment (about 60–90 min) compared with the control yogurt or that with MD; this factor may have influenced the lower firmness values found. Moreover, the addition of these ingredients may have caused an interruption of gel formation, reducing firmness. Parallel to this, the greater cohesiveness may be associated with the water retention capacity, mainly due to the functional properties of protein hydrolysates [42].

The viscoelastic properties of the yogurt samples were evaluated by small-amplitude oscillatory shear (SAOS) tests. Mechanical spectra of the different yogurts after 1 and 7 days of storage showed typical gel-like behaviour, with $G' > G''$ along the whole frequency range (Figure 2). The power law model may be used to fit both G' and G'' with the angular frequency ω (Equations (2) and (3)):

$$G' = G_0' \cdot \omega^{n'} \qquad (2)$$

$$G'' = G_0'' \cdot \omega^{n''} \qquad (3)$$

where G_0' and G_0'' are the respective storage and loss moduli at 1 rad/s, and n' and n'' exponents denote the viscoelastic response in terms of the time stability of both G' and G'' at short time scales. The viscoelastic parameters resulting from Equations (2) and (3) (G_0' and G_0'') are both a measurement of the gel strength of samples since they provide the complete (elastic and viscous) deformation resistance [43].

After day 1, the yogurts with the free hydrolysate (YH) and microencapsulated hydrolysate (YHEn) exhibited the lowest values for G_0' and G_0'' (Table 2). This result indicates that the hydrolysate, in both forms, reduced the gel strength of the casein matrix, maintaining a similar degree of viscoelasticity as was evidenced by the similar values for the loss factor at 1 rad/s (*tanδ*) (Table 2). This result could be explained by the fact that the hydrolysate introduces a negative electric charge, consequently increasing the electrostatic repulsive forces within the micellar structure, which expand the network, enhancing the hydration of the gel network [44]. This result is consistent with the observed decrease in firmness in YH and YHEn as compared with YC from textural analysis.

The yogurt with maltodextrin (YMD) exhibited slightly higher values for G_0' and G_0'' than those in YC after day 1. The electrostatic neutral character of these glucose polymers would favor the mutual associations between maltodextrin and casein by hydrogen bonding and dipole–dipole interactions, showing a certain binder role for MD in the yogurt matrix. This result is consistent with the lower *tanδ* value for YMD vs. YC (Table 2), which suggests greater strength in the intermolecular interactions and consequently a longer bond lifetime in the gel matrix [45].

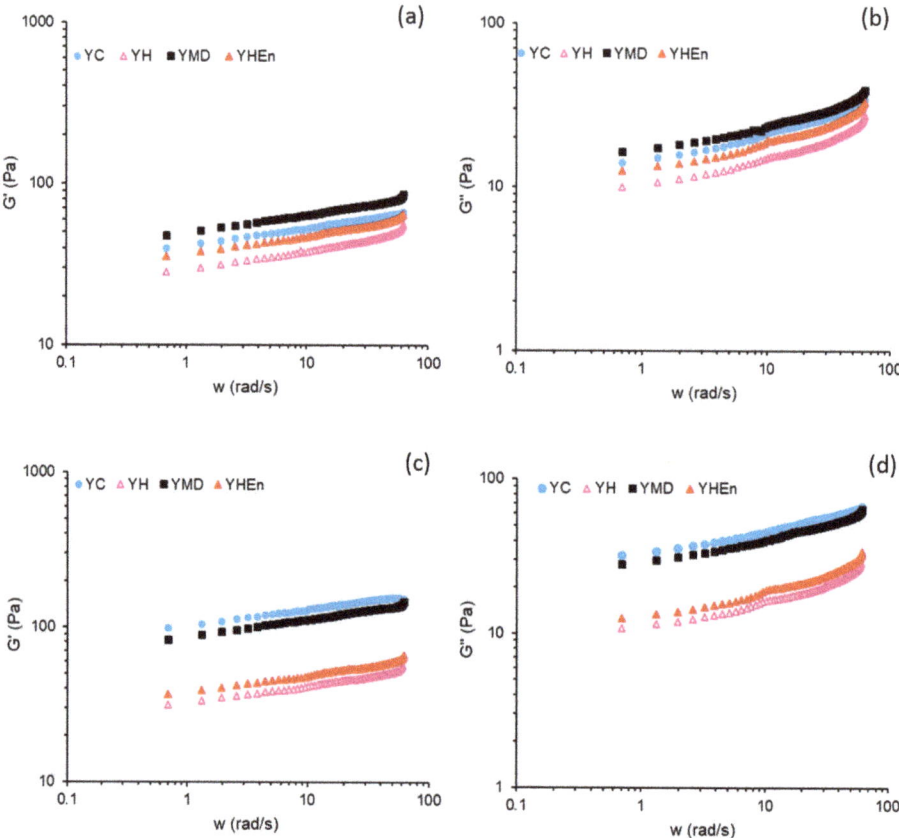

Figure 2. Mechanical spectra of yogurts during storage at 5 °C. Storage modulus $-G'$ at 1 day (**a**); loss modulus $-G''$ at 1 day (**b**); storage modulus $-G'$ at 7 days (**c**); loss modulus $-G''$ at 7 days (**d**). YC: control yogurt; YMD: yogurt with addition of maltodextrin; YH: yogurt with addition of free protein hydrolysate; YHEn: yogurt with addition of microencapsulated protein hydrolysate.

As regards n' and n'' exponents, it might be observed that for all samples, $n'' > n'$, this means that the rate of decrease of G'' with decreasing ω, was higher than that for G', resulting in a shear-induced gelation at lower frequencies (higher oscillation times) [46]. This fact shows a shear-induced increase in the energy stabilization of network bonds at lower frequencies, compatible with the decrease in the gel strength at higher oscillation times [47].

After 7 days of storage, G_0' and G_0'' values increased in YC and YMD compared with those found after day 1. This result was consistent with the observed increase in firmness from textural analysis, indicating considerable gel reinforcement during storage, which would be explained by the natural strengthening of the dipolar interactions and hydrogen bonds in the casein matrix induced by cool storage [44]. This effect was partially mitigated by the presence of maltodextrin, which would stabilize the structural rearrangements in the casein matrix during storage. In addition, both G_0' and G_0'' were scarcely modified in YH and YHEn during the storage period, showing the stabilizing role of the hydrolysate in the casein matrix, especially in microencapsulated form.

The increase in both G_0' and G_0'' after 7 days was consistent with the observed increase in L* (Table 1). L* shows the light scattered by various structural elements (casein aggregates, molecular fragments, etc.). After 7 days, the gel strength of the different networks increased moderately, so that denser matrices were formed, enhancing the diffuse reflection of electromagnetic waves and consequently increasing L* [45].

In general, n' and n'' values decreased after day 7 of storage, maintaining a similar positive difference between n'' and n' (Table 2). Therefore, the lower exponents (n' and n'') indicate an improvement in stability in the four yogurt networks over time, maintaining a similar stabilization energy at lower frequencies than on day 1.

In order to gain insight into the rheological differences among the three ingredients (H, MD and HEn) in model systems, i.e., outside the protein gel matrix environment at the same concentration, the flow behaviour and the viscoelastic characteristics of diluted aqueous solutions (1.5%, w/v) were also analyzed (Figure 3). Such a low concentration was selected to resemble the low concentration of the hydrolysate in the yogurt matrix. The dissolved maltodextrin (MD) and the microencapsulated hydrolysate (HEn) presented a similar trend (shear thinning flow) and similar viscosity values during the three-step flow, while the opposite was observed in the free hydrolysate, which exhibited the lowest viscosity and virtually Newtonian behaviour, as was evidenced by the stationary values of viscosity in the three steps (Figure 3a). The dynamic oscillatory test showed an evident fluid-like response in the three samples based on noticeably higher values for G'' compared with those for G' (Figure 3b). Both MD and HEn solutions showed considerably higher G' values compared with the H aqueous solution, while showing no significant differences between each other. This result was consistent with the decrease in viscosity in the second step, and their regeneration ability in the third step, attributed to the contribution of maltodextrin (Figure 3a). In contrast, G'' decreased in H, and more evidently in HEn, with respect to plain MD (Figure 3b). These findings reflect a lower level of intermolecular association at small oscillatory shear in HEn, and suggest that microcapsules may be less prone to interacting with each other and also with the surrounding medium, explaining why this preparation in yogurt resulted in more stable samples, which behaved differently compared with plain MD yogurts.

3.3. Microbiological Analysis

To confirm that the process of obtaining the yogurts was properly carried out, specific counts of the two major bacterial species present in yogurt were determined. The counts of *S. thermophilus* were YC: 8.72 ± 0.10 log CFU/g; YMD: 8.71 ± 0.10 log CFU/g; YH: 8.82 ± 0.04 log CFU/g; YHEn: 9.20 ± 0.01 log CFU/g ($p < 0.05$). Thus, the addition of the microencapsulated hydrolysate seems to have favored the growth of *S. thermophilus*, which could explain the lower pH and higher acidity compared with the YC sample. Regarding *L. bulgaricus*, counts were slightly higher in YH (YC: 6.42 ± 0.11; YMD: 6.92 ± 0.06; YH: 7.25 ± 0.09; YHEn: 6.94 ± 0.09 log UFC/g) ($p < 0.05$), indicating that the incorporation of the hydrolysate favored the growth of the starter culture in some way. YHEn registered higher values than the control yogurt, and similar values to those of YMD ($p < 0.05$), which indicated that maltodextrin, as a glucose supply, could also promote in some way the growth of this group of microorganisms. The viable counts of *S. thermophilus* were significantly more numerous than those of *L. bulgaricus* in all samples ($p < 0.05$). Zhao et al. [48] verified the same tendency in yogurts elaborated with α-lactalbumin hydrolysate-calcium (α-LAH-Ca) complexes.

The relatively low pH of yogurt and other fermented dairy products is a result of the fermentation of milk lactose into lactic acid, caused by the activity of lactic acid bacteria usually added as starter cultures [49,50]. In this study, as shown in Table 1, the pH of the yogurts evaluated ranged from 4.35 to 4.61. Furthermore, the yogurt starter cultures contained *Streptococcus thermophilus* and *Lactobacillus delbrueckii* ssp. *bulgaricus*, which largely out-compete other bacteria present in milk due the inhibitory effect of lactic acid and the utilization of the primary carbohydrate source (lactose), as well as the production of other inhibitory compounds [49,50].

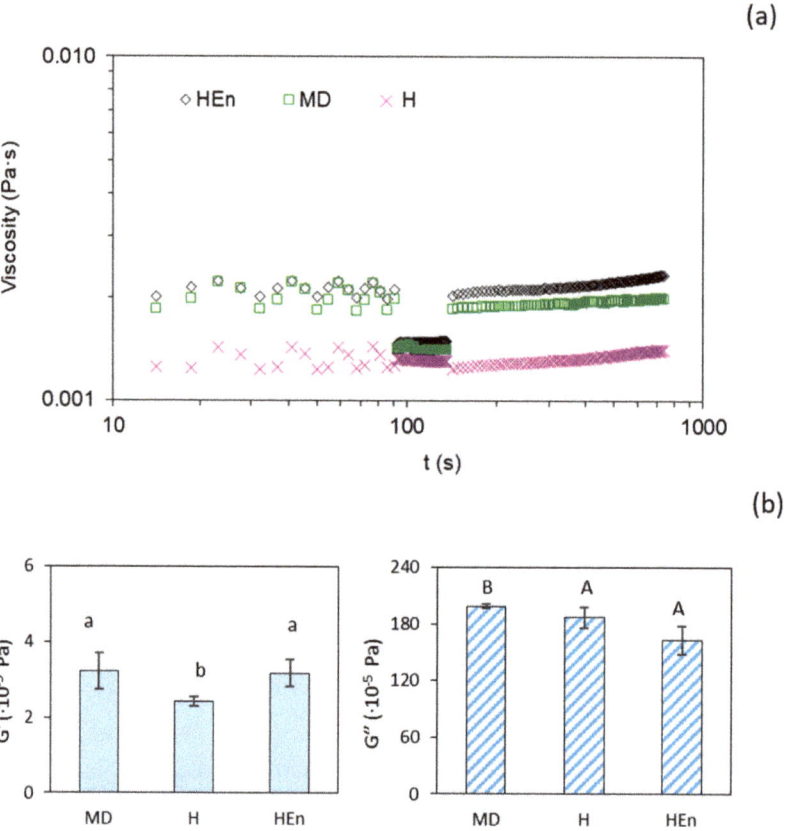

Figure 3. Evolution of viscosity values for the three-step flow (**a**) and viscoelastic parameters (**b**) for aqueous solutions at 1.5% concentration of maltodextrin (MD), hydrolysate (H) and microencapsulated hydrolysate (HEn) at 20 °C. Different letters (a,b) or (A,B) indicate a significant difference between samples ($p < 0.05$).

In this study, the presence of *Enterobacteriaceae*, molds and yeasts was not found after 1 and 7 days of storage (data not shown). There is growing evidence that the *Enterobacteriaceae* family can accurately reflect the hygienic conditions of processing milk and its derivatives, even during storage, despite the widespread use of coliforms as an indicator of hygienic–sanitary conditions. In addition, the presence of molds and yeasts in yogurt is indicative of unsatisfactory sanitary practices in manufacturing or packaging [49]. Accordingly, Barkallah et al. [4] did not observe the presence of the mentioned microorganisms in yogurt with the addition of *Spirulina platensis*.

3.4. Bioactivity

3.4.1. Antioxidant Activity

Antioxidant compounds can contribute to health promotion by eliminating free radicals, such as reactive oxygen species, as well as to increase the shelf life of foods, slowing down the process of lipid oxidation [36]. Yogurt per se, in general, may present antioxidant activity in view of its content of bioactive peptides, as reported in previous studies [6,9,36,39].

The antioxidant activity of the different yogurts after 1 and 7 days of storage was evaluated by assessing the ABTS radical scavenging activity and reducing power (Figures 4A and 4B, respectively). The addition of the free and microencapsulated hydrolysate to yogurts (YH and YHEn, respectively) increased the ABTS radicals scavenging activity ($p < 0.05$) in comparison with the YC and YMD samples, which showed a similar behavior ($p > 0.05$). The increase in antioxidant activity was probably due to peptides with antioxidant activity present in the protein hydrolysate of stripped weakfish, as previously reported by Lima et al. [25]. The activity of the hydrolysates was maintained, both in the free and microencapsulated forms in the yogurt matrix. These findings are in line with those reported by Abdel-Hamid et al. [9] and Demirci et al. [6], who found an increase in the antioxidant activity of yogurts with the addition of the fruit extracts *Siraitia grosvenorii* and rice bran, respectively. According to the authors, this was due to the composition of added phytochemicals with antioxidant activity. Additionally, Silva et al. [5] evaluated the antioxidant activity of yogurts functionalized with *Spirulina platensis* in free form and microencapsulated with MD, or with MD cross-linked with citric acid obtained by spray drying, reporting greater antioxidant activity (DPPH assay) for both functionalized yogurts compared with the control yogurt (especially in the case of the sample with MD cross-linked with citric acid). According to the authors, the addition of *Spirulina platensis* in the encapsulated form presents advantages such as masking the unpleasant fishy odor. Similarly, in the mentioned study, no differences in antioxidant activity were observed in terms of storage time (4 and 7 days).

Francisco et al. [7] reported a decrease in antioxidant activity during storage (0–7 days) for yogurts with added mushroom extracts (*Agaricus bisporus*) in free form and encapsulated by spray drying with MD cross-linked with citric acid (thermally untreated forms), while the opposite behavior was observed in thermally treated microencapsulated extracts after atomization. This decrease throughout storage time may be associated with the degradation of the extract when incorporated in free form by direct contact with the food matrix, and when thermally untreated, by the rapid release of the microencapsulated extract to the food matrix with the subsequent degradation.

In the present study there was no degradation of bioactive peptides, possibly because the antioxidant activity (ABTS and reducing power) remained stable throughout storage ($p > 0.05$). It should be noted that YH and YHEn did not differ significantly in either evaluated property ($p > 0.05$) (Figure 3). Thus, the incorporation of the microencapsulated hydrolysate may be an alternative for increasing the antioxidant activity of yogurts while possibly masking any potential changes to odor and flavor.

3.4.2. ACE Inhibitory Activity

Angiotensin-I converting enzyme (ACE) plays an important role in regulating blood pressure, as it produces the potent vasoconstrictor angiotensin II and degrades a vasodilator called bradykinin, causing an increase in blood pressure [51,52].

Fermented dairy products can provide beneficial health effects by releasing peptides with ACE inhibitory activity as a result of the proteolysis of milk proteins during fermentation [8,53]. Figure 4C shows the ACE inhibitory activity after 1 and 7 days of storage. According to these results, the activity presented by YH and YHEn was more than threefold greater compared with that of YC or YMD ($p < 0.05$). In addition, the effect of the storage time can be considered negligible in these yogurts since the ACE inhibitory activity remained stable throughout the 7 days of storage ($p > 0.05$).

The results obtained in the present study are consistent with those observed by Abdel-Hamid et al. [9] who reported a significant increase in the ACE inhibitory activity of probiotic yogurts supplemented with *Siraitia grosvenorii* fruit extract (73.36–81.39% compared to the 62.05% control). The mentioned authors correlated this activity with the degree of hydrolysis (proteolysis) of the supplemented yogurt, in which the release of small peptides occurs. In another study, Wulandani et al. [10] evaluated the ACE inhibitory activity of yogurts added with *Ficus glomerata* Roxb fruit extract (5% and 10%) during cold storage for 28 days, reporting the greatest activity on the seventh day of storage in yogurt with the addition of 10% *F. glomerata* Roxb extract (69.11 ± 0.50%) compared with yogurt without *F. glomerata* Roxb extract (53.47 ± 1.07%). Shori et al. [8] also found an increase in the ACE inhibitory activity in freshly made yogurts to which fish collagen was added compared with the control yogurt, presenting approximately 53% and 40% more inhibition, respectively. It is worth mentioning that the reported values in yogurts with the addition of fruits were lower than those observed in the present study.

Thus, the results in the present study suggest that the hydrolysate, both free and microencapsulated, is a good source of ACE inhibitor peptides, providing yogurt with increased ACE inhibitory activity after at least 7 days of storage.

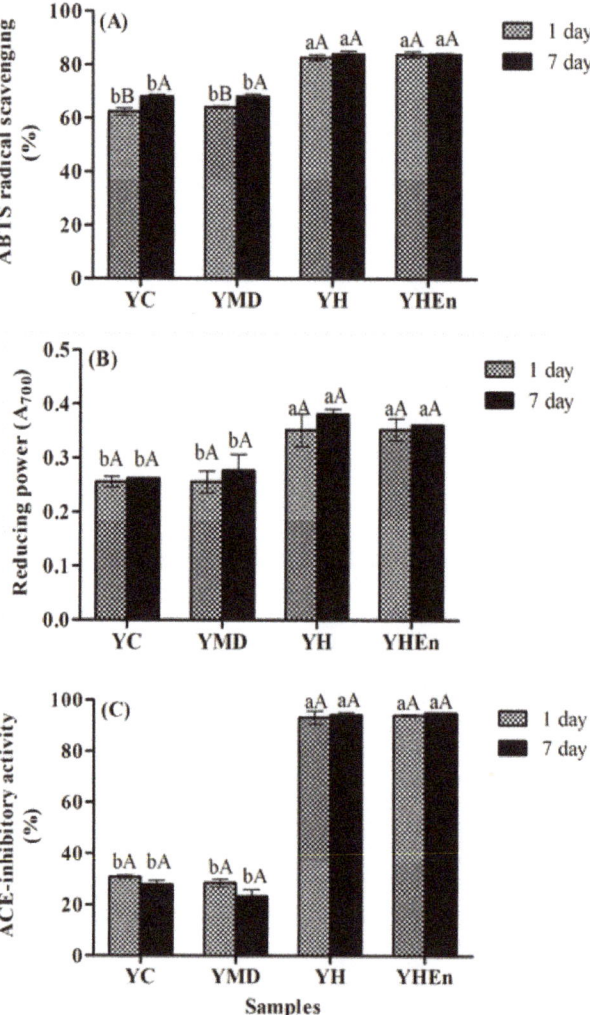

Figure 4. ABTS radical scavenging activity (**A**), reducing power (**B**) and ACE inhibitory activity (**C**) of yogurts during storage. Different lowercase letters (a,b) indicate a significant difference between samples for the same day ($p < 0.05$). Different uppercase letters (A,B) indicate a significant difference for the same sample on different days ($p < 0.05$).

3.5. Sensory Analysis

The addition of new ingredients to yogurt can cause possible interferences, mainly in regard to taste and texture. Figure 5 shows the sensory profile of the different yogurts as evaluated by the panelists. According to the results, there were no significant differences among yogurts for color and acidity ($p > 0.05$). Moreover, the presence of MD did not affect the organoleptic properties of yogurt in any of the evaluated attributes.

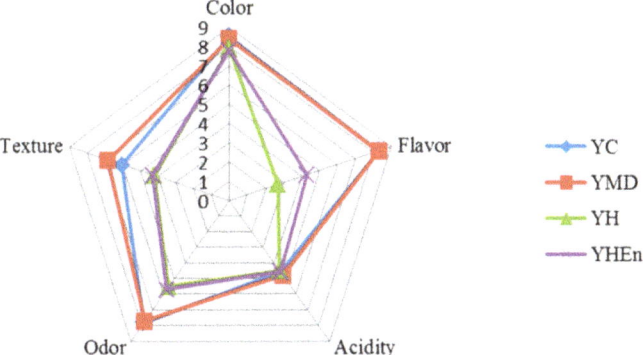

Figure 5. Sensory evaluation of different yogurts.

The addition of the free and microencapsulated protein hydrolysate (YH and YHEn) affected the texture, odor and flavor compared with the control yogurt ($p < 0.05$). Although they presented a significant difference in relation to YC, YH and YHEn showed an intermediate score in relation to texture; the panelists described the texture of YH and YHEn as slightly softer than that of the control but without losing the characteristic gel-like consistency of yogurt.

Regarding odor, a slightly fishy or peculiar smell was detected, especially in YH. Among the sensory attributes evaluated, the lowest scores obtained corresponded to flavor. However, YHEn showed a significant difference compared with YH ($p < 0.05$), presenting a higher score, and thus suggesting that encapsulation masked the fishy flavor. In the observations, the panelists indicated that the yogurts with the free and microencapsulated hydrolysates had a different flavor compared with that of a traditional yogurt (control), but were also different from each other. A more intense fishy flavor was identified in the yogurt with the free hydrolysate (YH), while YHEn was perceived as having a different flavor from that of the control, which was, however, undefined.

Taking into account the sensory attributes of a traditional yogurt, a very familiar product to consumers, it is frequently and well documented that the incorporation of ingredients (other than fruits) modifies or is detrimental to the organoleptic characteristics. Accordingly, the addition of *Allium sativum* to yogurt lowered the score for wateriness, aroma and taste (in terms of sourness) compared with the control yogurt [8]. Demirci et al. [6] reported that the addition of 1%, 2% and 3% rice bran negatively affected the appearance, texture, taste and odor of the yogurt. In the formulation of yogurt with different concentrations of jumbo squid powder (1%, 3%, 5%, 7% and 10%), significant differences in relation to color, taste, and texture were reported, while the addition of 3% did not differ statistically from the controls [35]. Barkallah et al. [4] stated that there was no significant difference in flavor between the control yogurt and that containing 0.25% *Spirulina platensis*, however, the addition of 1% had the lowest score for flavor.

In the present study, although there were differences in texture and flavor, the addition of microencapsulated hydrolysate was positive, as it masked the fish flavor while maintaining its bioactivity. There are many types of yogurt on the market; the consumer chooses among the available options, including fortified yogurt with bioactive products, which constitutes a new product. Thus, while the judges were able to differentiate the yogurts containing hydrolysates, this did not imply that the quality was worse, but rather that the yogurts were different with respect to the control; therefore, the acceptability parameter was excluded in the evaluation so as to prevent a biased judgement based on familiarity. Nevertheless, more studies should be conducted in order to improve these sensorial attributes in supplemented yogurt with fish protein hydrolysates.

4. Conclusions

Quality yogurts were made with the incorporation of a protein hydrolysate, free or microencapsulated with maltodextrin. These ingredients led to a reduction in synersis, especially for the yogurt with the added microencapsulated hydrolysate. Yogurts with hydrolysates showed a slight reduction in firmness, while they were slightly more cohesive compared with the control yogurt. Both free and microencapsulated hydrolysates caused a reduction in the viscoelastic parameters, maintaining the characteristic gel-like structure of yogurt, and provided greater rheological stability after one week of storage, especially in the microencapsulated form. In addition, the incorporation of hydrolysates in both forms resulted in yogurts with greater antioxidant and antihypertensive activities, which were maintained after 7 days of storage. The antioxidant and antihypertensive activities shown in vitro by a functional food, such as yoghurt made with probiotic bacteria and further enhanced by the incorporation of protein hydrolysates, provide such a product with a potential health effect, although in vivo studies are needed to demonstrate this effect (yogurt behavior once digested, bioavailability studies, etc.). If the health effect is proven, the inclusion of yogurt in a diet could reduce the need for medication to regulate, for example, blood pressure. Although there were differences in texture, odor and flavor, the incorporation of the hydrolysate in microencapsulated form showed advantages over the free form in relation to the masking of the fish flavor, without altering the bioactive properties, and thus a yoghurt containing this addition may constitute an acceptable fortified food.

Author Contributions: Conceptualization, K.O.L., M.E.L.-C., P.M. and C.P.; methodology, K.O.L. and A.A.; software, K.O.L. and M.d.R.; validation, M.E.L.-C., M.C.G.-G., C.A.T. and P.M.; formal analysis, K.O.L., M.E.L.-C., M.C.G.-G., C.A.T. and P.M.; investigation, K.O.L., M.d.R., A.A., C.A.T. and M.E.L.-C.; resources, K.O.L., P.M., M.C.G.-G. and C.P.; data curation, K.O.L., M.E.L.-C., P.M., M.C.G.-G. and C.A.T.; writing—original draft preparation, K.O.L. and M.E.L.-C.; writing—review and editing, K.O.L., A.A., M.E.L.-C., P.M., M.C.G.-G. and C.A.T.; visualization, M.E.L.-C., P.M. and M.C.G.-G.; supervision, M.E.L.-C., P.M., M.C.G.-G. and C.A.T.; project administration, M.E.L.-C., P.M. and M.C.G.-G.; funding acquisition, K.O.L., M.E.L.-C., M.C.G.-G., P.M. and C.P. All authors have read and agreed to the published version of the manuscript.

Funding: Rio Grande do Sul State Research Foundation (FAPERGS) (17/2551-0000916-9). This work was supported by the Spanish Ministry of Economy and Competitiveness project NANOALIVAL AGL2017-84161, cofounded with ERDF (European Regional Development Fund) and CSIC -202070E218.

Institutional Review Board Statement: Not applicable.

Informed Consent Statement: Not applicable.

Data Availability Statement: All data relevant to the study are included in the article.

Acknowledgments: This study was financed in part by the Coordenação de Aperfeiçoamento de Pessoal de Nível Superior - Brasil (CAPES) - Finance Code 001. We also thank the Rio Grande do Sul State Research Foundation (FAPERGS) (17/2551-0000916-9), the Spanish Ministry of Economy and Competitiveness project NANOALIVAL AGL2017-84161, cofounded with ERDF (European Regional Development Fund) and CSIC -202070E218.

Conflicts of Interest: The authors declare no conflict of interest.

References

1. Jaster, H.; Arend, G.D.; Rezzadori, K.; Chaves, V.C.; Reginatto, F.H.; Petrus, J.C.C. Enhancement of antioxidant activity and physicochemical properties of yogurt enriched with concentrated strawberry pulp obtained by block freeze concentration. *Food Res. Int.* **2018**, *104*, 119–125. [CrossRef]
2. Ozturkoglu-Budak, S.; Akal, C.; Yetisemiyen, A. Effect of dried nut fortification on functional, physicochemical, textural, and microbiological properties of yogurt. *J. Dairy Sci.* **2016**, *99*, 8511–8523. [CrossRef]
3. Gahruie, H.H.; Eskandari, M.H.; Mesbahi, G.; Hanifpour, M.A. Scientific and technical aspects of yogurt fortification: A review. *Food Sci. Hum. Wellness* **2015**, *4*, 1–8. [CrossRef]
4. Barkallah, M.; Dammak, M.; Louati, I.; Hentati, F.; Hadrich, B.; Mechichi, T.; Ayadi, M.A.; Fendri, M.; Attia, H.; Abdelkafi, S. Effect of *Spirulina platensis* fortification on physicochemical, textural, antioxidant and sensory properties of yogurt during fermentation and storage. *LWT—Food Sci. Technol.* **2017**, *84*, 323–330. [CrossRef]
5. Da Silva, S.C.; Fernandes, I.P.; Barros, L.; Fernandes, Â.; Alvez, M.J.; Calhelha, R.C.; Pereira, C.; Barreira, J.C.M.; Manrique, Y.; Colla, E.; et al. Spray-dried *Spirulina platensis* as an effective ingredient to improve yogurt formulations: Testing different encapsulating solutions. *J. Funct. Foods* **2019**, *60*, 103427. [CrossRef]
6. Demirci, T.; Aktaş, K.; Sözeri, D.; Öztürk, H.İ.; Akın, N. Rice bran improve probiotic viability in yoghurt and provide added antioxidative benefits. *J. Funct. Foods* **2017**, *36*, 396–403. [CrossRef]
7. Francisco, C.R.L.; Heleno, S.A.; Fernandes, I.P.M.; Barreira, J.C.M.; Calhelha, R.C.; Barros, L.; Gonçalves, O.H.; Ferreira, I.C.F.R.; Barreiro, M.F. Functionalization of yogurts with *Agaricus bisporus* extracts encapsulated in spray-dried maltodextrin crosslinked with citric acid. *Food Chem.* **2018**, *245*, 845–853. [CrossRef] [PubMed]
8. Shori, A.B.; Baba, A.S.; Chuah, P.F. The effects of fish collagen on the proteolysis of milk proteins, ACE inhibitory activity and sensory evaluation of plain- and Allium sativum-yogurt. *J. Taiwan Inst. Chem. Eng.* **2013**, *44*, 701–706. [CrossRef]
9. Abdel-Hamid, M.; Romeih, E.; Huang, Z.; Enomoto, T.; Huang, L.; Li, L. Bioactive properties of probiotic set-yogurt supplemented with *Siraitia grosvenorii* fruit extract. *Food Chem.* **2020**, *303*, 125400. [CrossRef]
10. Wulandani, B.R.D.; Marsono, Y.; Utami, T.; Rahayu, E.S. Potency of yogurt as angiotensin converting enzyme inhibitor with addition of *Ficus glomerata* Roxb fruit extract. *Int. Food Res.* **2018**, *25*, 1153–1158.
11. Hemker, A.K.; Nguyen, L.T.; Karwe, M.; Salvi, D. Effects of pressure-assisted enzymatic hydrolysis on functional and bioactive properties of tilapia (*Oreochromis niloticus*) by-product protein hydrolysates. *LWT—Food Sci. Technol.* **2020**, *122*, 109003. [CrossRef]
12. Neves, A.C.; Harnedy, P.A.; O'Keeffe, M.B.; FitzGerald, R.J. Bioactive peptides from Atlantic salmon (*Salmo salar*) with angiotensin converting enzyme and dipeptidyl peptidase IV inhibitory, and antioxidant activities. *Food Chem.* **2017**, *218*, 396–405. [CrossRef]
13. Ambigaipalan, P.; Shahidi, F. Bioactive peptides from shrimp shell processing discards: Antioxidant and biological activities. *J. Funct. Foods* **2017**, *34*, 7–17. [CrossRef]
14. Rivero-Pino, F.; Espejo-Carpio, F.J.; Guadix, E.M. Bioactive fish hydrolysates resistance to food processing. *LWT—Food Sci. Technol.* **2020**, *117*, 108670. [CrossRef]
15. Nimse, S.B.; Pal, D. Free radicals, natural antioxidants, and their reaction mechanisms. *RSC Adv.* **2015**, *5*, 27986–28006. [CrossRef]
16. Ma, Y.S.; Zhao, H.J.; Zhao, X.H. Comparison of the Effects of the Alcalase-Hydrolysates of Caseinate, and of Fish and Bovine Gelatins on the Acidification and Textural Features of Set-Style Skimmed Yogurt-Type Products. *Foods* **2019**, *8*, 501. [CrossRef] [PubMed]
17. Akbarbaglu, Z.; Jafari, S.M.; Sarabandi, K.; Mohammadi, M.; Heshmati, M.K.; Pezeshki, A. Influence of spray drying encapsulation on the retention of antioxidant properties and microstructure of flaxseed protein hydrolysates. *Colloids Surf. B Biointerfaces* **2019**, *178*, 421–429. [CrossRef] [PubMed]
18. Murthy, L.N.; Phadke, G.G.; Mohan, C.O.; Chandra, M.V.; Annamalai, J.; Visnuvinayagam, S.; Ravishankar, C.N. Characterization of Spray-Dried Hydrolyzed Proteins from Pink Perch Meat Added with Maltodextrin and Gum Arabic. *J. Aquat. Food Prod. Technol.* **2017**, *26*, 913–928. [CrossRef]
19. Mohan, A.; Rajendran, S.R.C.K.; He, Q.S.; Bazinet, L.; Udenigwe, C.C. Encapsulation of food protein hydrolysates and peptides: A review. *RSC Adv.* **2015**, *5*, 79270–79278. [CrossRef]
20. Gómez-Mascaraque, L.G.; Miralles, B.; Recio, I.; López-Rubio, A. Microencapsulation of a whey protein hydrolysate within micro-hydrogels: Impact on gastrointestinal stability and potential for functional yoghurt development. *J. Funct. Foods* **2016**, *26*, 290–300. [CrossRef]
21. Sarabandi, K.; Mahoonak, A.S.; Hamishekar, H.; Ghorbani, M.; Jafari, S.M. Microencapsulation of casein hydrolysates: Physicochemical, antioxidant and microstructure properties. *J. Food Eng.* **2018**, *237*, 86–95. [CrossRef]
22. Zhu, F. Encapsulation and delivery of food ingredients using starch based systems. *Food Chem.* **2017**, *229*, 542–552. [CrossRef]
23. Tan, S.; Zhong, C.; Langrish, T. Pre-gelation assisted spray drying of whey protein isolates (WPI) for microencapsulation and controlled release. *LWT—Food Sci. Technol.* **2020**, *117*, 108625. [CrossRef]
24. Lima, K.O.; Alemán, A.; López-Caballero, M.E.; Gómez-Guillén, M.C.; Montero, M.P.; Prentice, C.; Huisa, A.J.T.; Monserrat, J.M. Characterization, stability, and in vivo effects in Caenorhabditis elegans of microencapsulated protein hydrolysates from stripped weakfish (*Cynoscion guatucupa*) industrial byproducts. *Food Chem.* **2021**, *364*, 130380. [CrossRef]
25. Lima, K.O.; de Quadros, C.D.C.; da Rocha, M.; de Lacerda, J.T.J.G.; Juliano, M.A.; Dias, M.; Mendes, M.A.; Prentice, C. Bioactivity and bioaccessibility of protein hydrolyzates from industrial byproducts of Stripped weakfish (*Cynoscion guatucupa*). *LWT—Food Sci. Technol.* **2019**, *111*, 408–413. [CrossRef]

26. Lee, W.J.; Lucey, J.A. Formation and Physical Properties of Yogurt. *Asian Australas. J. Anim. Sci.* **2010**, *23*, 1127–1136. [CrossRef]
27. AOAC (Ed.) *Official Methods of Analysis*, 18th ed.; The Association of Official Analytical Chemists: Washington, DC, USA, 2005.
28. Santillán-Urquiza, E.; Méndez-Rojas, M.Á.; Vélez-Ruiz, J.F. Fortification of yogurt with nano and micro sized calcium, iron and zinc, effect on the physicochemical and rheological properties. *LWT—Food Sci. Technol.* **2017**, *80*, 462–469. [CrossRef]
29. Arancibia, M.Y.; López-Caballero, M.E.; Gómez-Guillén, M.C.; Montero, P. Release of volatile compounds and biodegradability of active soy protein lignin blend films with added citronella essential oil. *Food Control* **2014**, *44*, 7–15. [CrossRef]
30. Zhang, T.; Jeong, C.H.; Cheng, W.N.; Bae, H.; Seo, H.G.; Petriello, M.C.; Han, S.G. Moringa extract enhances the fermentative, textural, and bioactive properties of yogurt. *LWT—Food Sci. Technol.* **2019**, *101*, 276–284. [CrossRef]
31. Zheng, L.; Zhao, M.; Xiao, C.; Zhao, Q.; Su, G. Practical problems when using ABTS assay to assess the radical-scavenging activity of peptides: Importance of controlling reaction pH and time. *Food Chem.* **2016**, *192*, 288–294. [CrossRef] [PubMed]
32. Canabady-Rochelle, L.L.S.; Harscoat-Schiavo, C.; Kessler, V.; Aymes, A.; Fournier, F.; Girardet, J.M. Determination of reducing power and metal chelating ability of antioxidant peptides: Revisited methods. *Food Chem.* **2015**, *183*, 129–135. [CrossRef] [PubMed]
33. Alemán, A.; Giménez, B.; Pérez-Santin, E.; Gómez-Guillén, M.C.; Montero, P. Contribution of Leu and Hyp residues to antioxidant and ACE-inhibitory activities of peptide sequences isolated from squid gelatin hydrolysate. *Food Chem.* **2011**, *125*, 334–341. [CrossRef]
34. Benedetti, S.; Prudencio, E.S.; Müller, C.M.O.; Verruck, S.; Mandarino, J.M.G.; Leite, R.S.; Petrus, J.C.C. Utilization of tofu whey concentrate by nano filtration process aimed at obtaining a functional fermented lactic beverage. *J. Food Eng.* **2016**, *171*, 222–229. [CrossRef]
35. Córdova-Ramos, J.S.; Gonzales-Barron, U.; Cerrón-Mallqui, L.M. Physicochemical and sensory properties of yogurt as affected by the incorporation of jumbo squid (*Dosidicus gigas*) powder. *LWT—Food Sci. Technol.* **2018**, *93*, 506–510. [CrossRef]
36. Sah, B.N.P.; Vasiljevic, T.; Mckechnie, S.; Donkor, O.N. Effect of refrigerated storage on probiotic viability and the production and stability of antimutagenic and antioxidant peptides in yogurt supplemented with pineapple peel. *J. Dairy Sci.* **2015**, *98*, 5905–5916. [CrossRef]
37. Sah, B.N.P.; Vasiljevic, T.; McKechnie, S.; Donkor, O.N. Physicochemical, textural and rheological properties of probiotic yogurt fortified with fibre-rich pineapple peel powder during refrigerated storage. *LWT—Food Sci. Technol.* **2016**, *65*, 978–986. [CrossRef]
38. Carmona, J.C.; Robert, P.; Vergara, C.; Sáenz, C. Microparticles of yellow-orange cactus pear pulp (*Opuntia ficus-indica*) with cladode mucilage and maltodextrin as a food coloring in yogurt. *LWT—Food Sci. Technol.* **2021**, *138*, 110672. [CrossRef]
39. Öztürk, H.İ.; Aydın, S.; Sözeri, D.; Demirci, T.; Sert, D.; Akın, N. Fortification of set-type yoghurts with *Elaeagnus angustifolia* L. flours: Effects on physicochemical, textural, and microstructural characteristics. *LWT—Food Sci. Technol.* **2018**, *90*, 620–626. [CrossRef]
40. Akalin, A.S.; Unal, G.; Dinkci, N.; Hayaloglu, A.A. Microstructural, textural, and sensory characteristics of probiotic yogurts fortified with sodium calcium caseinate or whey protein concentrate. *J. Dairy Sci.* **2012**, *95*, 3617–3628. [CrossRef]
41. Caleja, C.; Barros, L.; Antonio, A.L.; Carocho, M.; Oliveira, M.B.P.P.; Ferreira, I.C.F.R. Fortification of yogurts with different antioxidant preservatives: A comparative study between natural and synthetic additives. *Food Chem.* **2016**, *210*, 262–268. [CrossRef]
42. Halim, N.R.A.; Yusof, H.M.; Sarbon, N.M. Functional and Bioactive Properties of Fish Protein Hydolysates and Peptides: A Comprehensive Review. *Trends Food Sci. Technol.* **2016**, *51*, 24–33. [CrossRef]
43. Campo-Deaño, L.; Tovar, C.A. The effect of egg albumen on the viscoelasticity of crab sticks made from Alaska pollock and Pacific whiting surimi. *Food Hydrocoll.* **2009**, *23*, 1641–1646. [CrossRef]
44. Zhang, Y.; Li, Y.; Liu, W. Dipole–dipole and H-bonding interactions significantly enhance the multifaceted mechanical properties of thermoresponsive shape memory hydrogels. *Adv. Funct. Mat.* **2015**, *25*, 471–480. [CrossRef]
45. Borderías, A.J.; Tovar, C.A.; Domínguez-Timón, F.; Díaz, M.T.; Pedrosa, M.M.; Moreno, H.M. Characterization of healthier mixed surimi gels obtained through partial substitution of myofibrillar proteins by isolate pea proteins. *Food Hydrocoll.* **2020**, *107*, 105976. [CrossRef]
46. Piñeiro-Lago, L.; Franco, M.I.; Tovar, C.A. Changes in thermoviscoelastic and biochemical properties of *Atroncau blancu* and *roxu Afuega'l Pitu* cheese (PDO) during ripening. *Food Res. Int.* **2020**, *137*, 109693. [CrossRef]
47. Moreno, H.M.; Bargiela, V.; Tovar, C.A.; Cando, D.; Borderias, A.J.; Herranz, B. High pressure applied to frozen flying fish (*Parexocoetus brachyterus*) surimi: Effect on physicochemical and rheological properties of gels. *Food Hydrocoll.* **2015**, *48*, 127–134. [CrossRef]
48. Zhao, L.L.; Wang, X.L.; Liu, Z.P.; Sun, W.H.; Dai, Z.Y.; Ren, F.Z.; Mao, X.Y. Effect of α-lactalbumin hydrolysate-calcium complexes on the fermentation process and storage properties of yogurt. *LWT—Food Sci. Technol.* **2018**, *88*, 35–42. [CrossRef]
49. Hervert, C.J.; Martin, N.H.; Boor, K.J.; Wiedmann, M. Survival and detection of coliforms, *Enterobacteriaceae*, and gram-negative bacteria in Greek yogurt. *J. Dairy Sci.* **2017**, *100*, 950–960. [CrossRef]
50. Vedamuthu, E.R. Starter cultures for yogurt and fermented milks. In *Manufacturing Yogurt and Fermented Milks*; Chandan, R.C., Ed.; Blackwell Publishing Ltd.: Ames, IA, USA, 2006.

51. Acharya, K.R.; Sturrock, E.D.; Riordan, J.F.; Ehlers, M.R.W. ACE revisited: A new target for structure-based drug design. *Nat. Rev. Drug Discov.* **2003**, *2*, 891–902. [CrossRef]
52. Hernández-Ledesma, B.; Del Mar Contreras, M.; Recio, I. Antihypertensive peptides: Production, bioavailability and incorporation into foods. *Adv. Colloid Interface Sci.* **2011**, *165*, 23–35. [CrossRef]
53. Fitzgerald, R.J.; Murray, B.A.; Walsh, D.J. Hypotensive Peptides from Milk Proteins. *J. Nutr.* **2004**, *134*, 980–988. [CrossRef] [PubMed]

 *antioxidants*

Article

Tamarillo Polyphenols Encapsulated-Cubosome: Formation, Characterization, Stability during Digestion and Application in Yoghurt

Tung Thanh Diep [1,2], Michelle Ji Yeon Yoo [1,2,*] and Elaine Rush [2,3]

1. School of Science, Faculty of Health and Environment Sciences, Auckland University of Technology, Private Bag 92006, Auckland 1142, New Zealand; tung.diep@aut.ac.nz
2. Riddet Institute, Centre of Research Excellence, Massey University, Private Bag 11222, Palmerston North 4442, New Zealand; elaine.rush@aut.ac.nz
3. School of Sport and Recreation, Faculty of Health and Environment Sciences, Auckland University of Technology, Private Bag 92006, Auckland 1142, New Zealand
* Correspondence: michelle.yoo@aut.ac.nz; Tel.: +64-9921-9999 (ext. 6456)

Abstract: Tamarillo extract is a good source of phenolic and anthocyanin compounds which are well-known for beneficial antioxidant activity, but their bioactivity maybe lost during digestion. In this study, promising prospects of tamarillo polyphenols encapsulated in cubosome nanoparticles prepared via a top-down method were explored. The prepared nanocarriers were examined for their morphology, entrapment efficiency, particle size and stability during in vitro digestion as well as potential fortification of yoghurt. Tamarillo polyphenol-loaded cubosomes showed cubic shape with a mean particle size of 322.4 ± 7.27 nm and the entrapment efficiency for most polyphenols was over 50%. The encapsulated polyphenols showed high stability during the gastric phase of in vitro digestion and were almost completely, but slowly released in the intestinal phase. Addition of encapsulated tamarillo polyphenols to yoghurt (5, 10 and 15 wt% through pre- and post-fermentation) improved the physicochemical and potential nutritional properties (polyphenols concentration, TPC) as well as antioxidant activity. The encapsulation of tamarillo polyphenols protected against pH changes and enzymatic digestion and facilitated a targeted delivery and slow release of the encapsulated compounds to the intestine. Overall, the cubosomal delivery system demonstrated the potential for encapsulation of polyphenols from tamarillo for value-added food product development with yoghurt as the vehicle.

Keywords: tamarillo extract; yoghurts; cubosome; polyphenols; encapsulation; in vitro digestion

1. Introduction

Inverse bicontinuous liquid crystalline nanoparticles, termed cubosomes, have advantageous properties that may be suitable for the delivery of bioactive compounds to the small intestine. Amphiphilic lipids such as the monoglyceride monoolein can self-assemble in water to produce dispersions of cubosomes. The basic structure of a cubosome is a honeycomb-like structure with two non-intersecting internal aqueous channels separated by lipid bilayers. The internal hydrophilic (aqueous) areas are separated by lipid bilayers that are twisted into a tightly packed three-dimensional honeycomb structure that has a high internal surface area to volume. Within this structure, encapsulation of diverse hydrophilic, hydrophobic and amphiphilic compounds of small to large molecular weights, such as proteins, peptides, amino acids and nucleic acids, is possible [1]. Within cubosomes, hydrophobic molecules can be located within the lipid bilayers, hydrophilic components in the aqueous channels or around the polar head of the lipid, and amphiphilic molecules can be located at the lipid–water interface. This structure generally maintains the efficacy—stability of actives (vitamins and proteins) without adverse effects on the recipient [2]. Both polar and non-polar compounds can be solubilized in the water channel and the bilayers, respectively; therefore, both can be loaded into these particles. According to Meikle et al. [3], cubosomes are relatively non-toxic, stable over a broad range of biologically relevant environmental conditions, and can be formulated using a wide array of lipids and stabilizers. Their size, surface charge, and bilayer structure can be tuned through adjustments in their composition. Moreover, previous studies have demonstrated the ability of cubosomes to deliver drugs to both the eukaryotic and prokaryotic cells [4–8] found in the human digestive system. Cubosome encapsulation has been used to deliver quercetin in vitro [9] and curcumin to the skin [10] and demonstrated improved solubility and availability with an entrapment of 84% and 82%, respectively. Improved anti-inflammatory/antioxidant effects and controlled diffusion of encapsulated curcumin through the skin were observed [10]. Cubosome encapsulated has also enhanced the stability and antibacterial activity of curcumin [11]. Another study had reported successful encapsulation of both piperine and curcumin in the interior of the cubosome particles [12].

Tamarillo (*Solanum betaceum* Cav.), a common fruit of New Zealand, is a good source of polyphenols compounds including anthocyanins, hydroxy benzoic acids, hydroxycinnamic acids, flavonols, flavanols and flavonol glycosides. The main polyphenols in the dried pulp of Laird's Large tamarillo cultivar were identified in our previous study [13] as delphinidin rutinoside (255 mg/100 g), pelargonidin rutinoside (201 mg/100 g DW), chlorogenic acid (66 mg/100 g), kaempferol rutinoside (50 mg/100 g) and cyanidin rutinoside (26 mg/100 g). These polyphenols are strong antioxidants possessing many potential health benefits such as preventing lipid oxidation and are associated with reduced risk for certain cancers, cardiovascular diseases and type 2 diabetes mellitus [14]. Tamarillo fruit therefore has the potential to be an ingredient in functional food products [15]. Evaluation of the stability of polyphenol compounds is important as these compounds are often degraded by oxidation during digestion, resulting in reduced biological activity [16]. For example, anthocyanins, present in high amounts in the nutrient-dense tamarillo, are oxidized into quinones, reducing the biological power of these molecules during digestion [16]. To overcome this obstacle, new generation–functional foods often use encapsulation technologies to protect polyphenols from degradation as well as maintain their bioavailability [16].

This study aimed to investigate the morphology, entrapment efficiency, and particle size of cubosome-encapsulated tamarillo extract (CUBTAM) and test the stability and antioxidant activity and release of the CUBTAM before and after in vitro digestion. In vitro digestion of yoghurt fortified with CUBTAM was similarly investigated for the stability and release of polyphenol compounds.

2. Materials and Methods

2.1. Materials

The yoghurt ingredients included standard milk (Anchor™ blue top, Fonterra) from a local supermarket (Auckland, New Zealand) and starter culture containing *Lactobacillus delbrueckii* subsp. *bulgaricus* and *Streptococcus thermophilus* (YoFlex® Express 1.1 powder) from CHR Hansen (Hoersholm, Denmark).

Dimodan® MO 90/D (monoolein) was kindly provided by Danisco (Auckland, New Zealand). Pluronic F127 (PEO_{99}–PPO_{67}–PEO_{99}) was purchased from Sigma-Aldrich (Auckland, New Zealand). The analytical grade standards of phenolics and anthocyanins were obtained from Sigma-Aldrich (Auckland, New Zealand) or Extrasynthese (Genay Cedex, France). Purite Fusion Milli-Q water purifying machine (Purite Limited, Thame, Oxon, UK) was used to produce Milli-Q water. All chemicals and reagents used were AnalaR grade or purer.

2.2. Tamarillo Extract (EXT) Preparation and Identtification of Polyphenol Components

Fresh fruits of Laird's Large (red) tamarillo cultivar were collected from growers in the Northland region of New Zealand with commercial maturity of 21–24 weeks from anthesis. The fruits were cleaned and peeled; then, the pulp was packed in polyethylene bags, sealed and frozen at −18 °C and defrosted (15 min) immediately before use. The extraction process was based on the method of Piovesana and Noreña [17] with some modifications. Aqueous citric acid (2% w/v) was used for extraction in the ratio of 1:5 (tamarillo: aqueous citric acid, w/w) and homogenised in a blender, and agitated (1000 rpm) at 55 °C in a water bath (1000 rpm) (Heidoplph, LABOROTA) for one hour after the addition of (20 µL/100 mL) pectin lyase (Novozym 33095), to improve the extraction of the bioactive compounds. Parafilm and tightly sealed lids of container and centrifuge tube were used to prevent entry of air and oxidation of polyphenols. The mixture was centrifuged at 10,000 RPM for 10 min and the supernatant (tamarillo extract) was separated and stored at −20 °C until being injected into LC-MS for identification of polyphenol component as reported in our previous study [13].

2.3. Preparation of Cubosome and Cubosome Containing Tamarillo Extract

The lyotropic liquid cubic phase was prepared and dispersed into cubosomes (CUB) as previously described in previous studies [18–20]. Briefly, molten monoolein (50 °C) was mixed with MilliQ water at a mass ratio of 60:40 to form the cubic phase. The cubic phase gel was equilibrated at room temperature for at least 48 h. Then, 25 mg of cubic phase gel was added to 5 mL of Pluronic F127 solution (2% (w/v)), and sonicated using a probe sonicator (BEM-150A, Bueno Biotech Ltd., Nanjing, China) (50% amplitude, pulsing 5 s on, 5 s off, for 7 min total run time) to obtain cubosomes (CUB). In parallel, the water was replaced by the tamarillo extract with a mass ratio of 60:40 (monoolein: tamarillo extract) to produce tamarillo polyphenol loaded-cubosomes (CUBTAM).

2.4. Polarized Light Microscopy (PLM) and Scanning Electron Microscopy (SEM)

Microstructure and morphology of CUB and CUBTAM particles were observed using a polarized light microscope (DM750, Leica, Wetzlar, Germany) equipped with a digital camera (ICC50HD, Leica, Wetzlar, Germany). Dried CUB and CUBTAM were coated by using Platinum (Pt) using a sputtering technique with a Sputter Coater (Hitachi E-1045, Hitachi, Tokyo, Japan) for 60 s at room temperature. The particle morphology was then observed by a scanning electron microscope (Hitachi SU-70 Schottky, Hitachi, Tokyo, Japan) at 25 mA and 10 kV [21].

2.5. Dynamic Light Scattering (DLS)

Each sample of the CUB and CUBTAM diluted 20× in MilliQ was measured in triplicate using disposable plastic cuvettes at 25 °C. Particle size and its distribution were determined by dynamic light scattering (DLS) using a Zetasizer Nano ZS (Malvern Instruments Ltd., Worcestershire, UK). The mean z-average diameter and polydispersity indices (PDI) were obtained from cumulative analysis using the Malvern software version 7.13 (Malvern Panalytical Ltd., Worcestershire, UK) [18].

2.6. Determination of the Entrapment Efficiency (EE)

The sample preparation procedure was applied according to Zhou et al. [22] with minor modifications. Briefly, 20 µL of CUB or CUBTAM were transferred into a 1 mL Eppendorf microcentrifuge tube and made up to volume with methanol. Then, the sample solutions were centrifuged at 5000 RPM for 5 min, and the supernatant was separated and transferred to an amber 1.8 mL glass vial, then injected into the LC-MS system.

The phenolic and anthocyanin compounds were analysed by using LC-MS according to our previous study [13] without further modification. According to Patil, Pawara, Gudewar and Tekade [23], the entrapment efficiency (EE%) is calculated as follows:

$$EE\% = \frac{A - B}{A} \times 100$$

where A is the polyphenol concentration added into cubosome and B is the polyphenol concentration present in the supernatant

2.7. Yoghurt Fermentation and Fortification with Tamarillo Polyphenol Loaded-Cubosome

CUBTAM suspensions from Section 2.3 were snap frozen by liquid nitrogen, and then lyophilized for 36 h (Alpha 1–2 LD plus Freeze Dryer, Martin Christ, New Zealand). The tamarillo bioactive loaded-cubosome (CUBTAM) powder was stored at −20 °C until use.

Kitchen yoghurt makers (Davis & Waddell, Stevens, New Zealand) were purchased to produce the yoghurt. For the control yoghurt, starter culture and milk in the ratio of 0.1:100 (w/w), were placed in the yoghurt maker. Incubation was implemented at 45 °C for 8 h until the pH of below 5.0 was obtained. The yoghurt was stored at 4 °C overnight and then homogenized at 4000 RPM using laboratory mixer (Silverson, Waterside, UK) for 2 min [24]. The yoghurt was stored at 4 °C until further analyses within 24 h.

The CUBTAM powder with concentration of 5, 10 and 15% were fortified into the yoghurt either before (PRE) or after (POS) the fermentation process. For PRE, CUBTAM powder was added to the mixture of milk and starter culture at the start of yoghurt making process, prior to fermentation. For POS, CUBTAM was added to yoghurt in the final homogenization step.

2.8. Determination of Physicochemical Properties of Fortified Yoghurts

The pH was measured with a digital pH meter to one decimal place. Syneresis index of yoghurt was identified based on the method of Wang et al. [24]. Some modifications were made from Kristo, Miao and Corredig [25] for rheological measurements, using a rheometer (RST-SST, Brookfield Ametek, Middleboro, MA, USA). The viscosity profile (viscosity, consistency coefficient, flow behavior index) was determined with a concentric cylinder with the diameters of cup and bob of 28.92 and 26.66 mm, respectively. Elastic modulus was determined with a vane spindle (SSVANE-) at a speed of 0.5 rpm for 5 min.

TA-XT plus texture analyser (Texture Technologies Corp., New York, NY, USA) was used to perform textural analysis with a backward-extrusion test based on the method of Wang et al. [24] with some modifications. The parameters of test included cylinder probe diameter of 50 mm, test speed of 1.0 mm/s, penetration distance of 25 mm and surface trigger force of 10 g.

2.9. In Vitro Digestion

The EXT, CUB, CUBTAM and yoghurt samples fortified with CUBTAM were subjected to in vitro digestion to identify the bioavailability of bioactive contents using method of Zhang et al. [26] without further modification. The sample (2 mL) was collected before digestion and after oral (5 min), gastric (120 min) and intestinal phases (180 min). The samples were snap frozen using liquid nitrogen to stop enzyme activity and centrifuged at 10,000 RPM at 4 °C for 10 min. The supernatants were collected and stored at −20 °C before being injected into the LC-MS for phenolic and anthocyanins identification as well as total phenolic content (TPC) and antioxidant activity analysis [13].

2.10. Total Phenolic Content (TPC) and Antioxidant Activity

The TPC of extracts and digests at each stage of digestion was determined using a Folin–Ciocalteu assay as described in our previous study [13]. Two different methods were used to determine the antioxidant activity, namely cupric ion reducing antioxidant capacity (CUPRAC) and ferric-reducing antioxidant power (FRAP) assays, which were mentioned in our previous study [13]. Results of TPC and antioxidant activity are presented as mg gallic acid equivalent per 100 g of tamarillo or yoghurt (mg GAE/100 g) and µmol Trolox equivalent antioxidant capacity per 100 g of tamarillo or yoghurt (µmol TEAC/100 g), respectively.

2.11. Statistical Analysis

Measurements of all the analytes were undertaken in triplicate, and the results are presented as mean ± standard deviation (SD). For comparison among different samples, one-way analysis of variance (ANOVA) was applied using SPSS 25.0 (IBM Corp., Armonk,

NY, USA). Fisher's (LSD) multiple comparison tests were used to determine the magnitude of differences between means. A p-value of <0.05 was considered statistically significant.

3. Results and Discussion

3.1. Characterization of Cubosomal Suspensions Containing Tamarillo Extract (CUBTAM)

Adapting the temperature–composition phase diagram of a monoolein/water system [27], pure monoolein cubosomes (CUB) and tamarillo polyphenols loaded-cubosomes (CUBTAM) were prepared in a top–down approach. This method allows the formation of reproducible, stable cubosomes without adding solvents. Therefore, it is unnecessary to reinvestigate phase behaviour, and there are no solvent concerns for cellular toxicity [2]. The concentration of the surfactant pluronic F127 was chosen to be 2 wt%, which yields stable cubosome dispersions [28]. The CUB dispersion appeared homogenously milky white and CUBTAM appeared semi-opaque pink (picture not shown).

Figure 1 shows the PLM and SEM photos of CUB and CUBTAM while their particle size distribution (PSD) is summarized in Figure S1 and Table S1. The addition of tamarillo extract did not significantly affect the morphology of cubosome particles. The initial cubic periodicity can be clearly visualized for both samples using PLM and SEM. Because tamarillo extracts are mainly water-soluble compounds (phenolics and anthocyanins), they should be dispersed in the water channel of the cubosome and should minimally affect the structure of the nanoparticles [19].

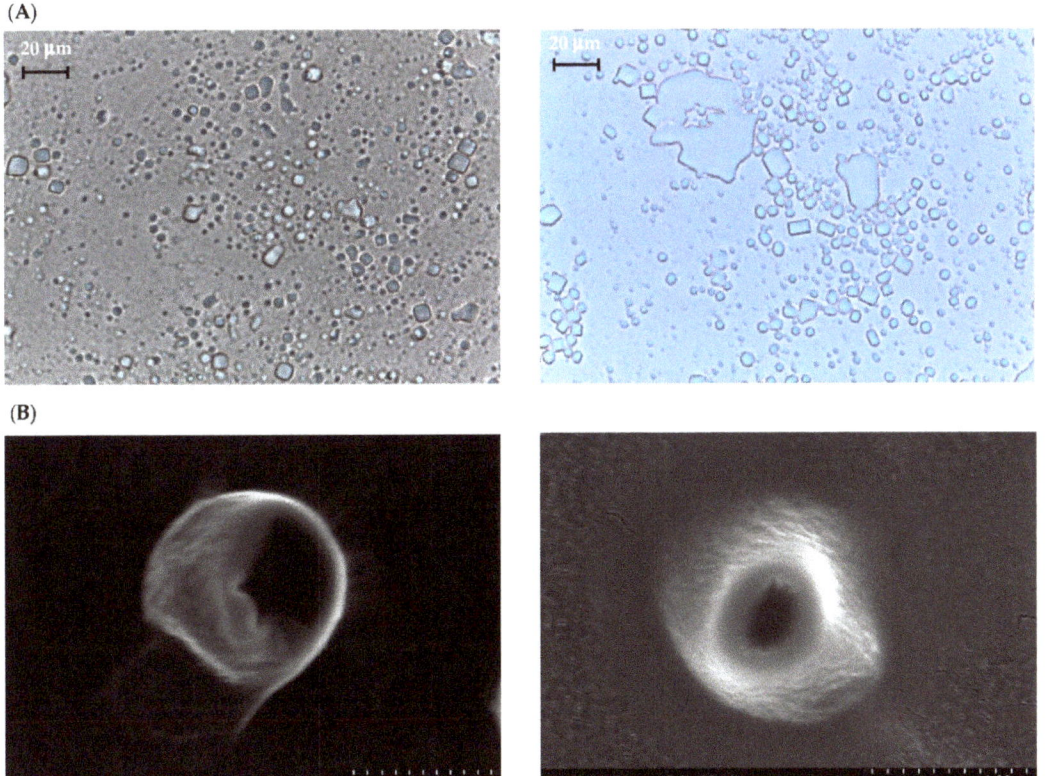

Figure 1. (**A**) PLM and (**B**) SEM micrographs of CUB (**left**) and CUBTAM (**right**).

For particle size distribution, CUBTAM had a unimodal curve and its polydispersity index (PDI) was quite small (below 0.3), as shown in Figure S1. The mean hydrodynamic diameter of liquid crystal particles increased significantly (from 270 to 327 nm) with the addition of the tamarillo extract. In general, this parameter depends on several factors such as the concentration of amphiphile (lipid and polymer), the presence of charged lipids, the ionic strength and the interactions between groups [23]. For CUBTAM, the addition of hydrophilic groups contributed to increasing electrostatic interaction as well as the coalescence between colloidal particles resulting in a bigger average particle size. However, most of the cubic particles in CUBTAM were still limited to a sub-micron range (100–1000 nm). According to Danaei et al. [29] the small particle size and the narrow size distribution (small PDI) create a large surface area that benefits cellular uptake.

Entrapment efficiency (EE%) of bioactive compounds from CUBTAM ranged from 19.8 (catechin) to 87.7% kaempferol rutinoside (Figure 2). Twelve of the fourteen tamarillo bioactive compounds had an EE of more than 50%. In addition, it is noteworthy that we show high EE% (>69%) for the major polyphenols in tamarillo (chlorogenic acid, kaempferol rutinoside, delphinidin rutinoside, cyanidin rutinoside and pelargonidin rutinoside). The high EE in the CUBTAM could be attributed to the fact that polyphenols in tamarillo extract are water-soluble compounds which embed in the water channels. The EE difference between polyphenols encapsulated by cubosome might also depend on the number of -OH groups in molecular structure. For example, the hydroxycinnamic acids chlorogenic acid, caffeic acid and ferulic acid (with >2 -OH groups) showed higher EE than p-coumaric acid (which has only 1 -OH group). More -OH groups will more easily attach in the aqueous channel of cubosome particles. Furthermore, different polyphenol classes showed different EE. For instance, the hydroxycinnamic acids (chlorogenic acid, caffeic acid and ferulic acid), hydroxybenzoic acid (gallic acid), flavonol glycosides (rutin, kaempferol rutinoside, isorhamnetin rutinoside) and anthocyanins (delphinidin rutinoside, cyanidin rutinosid, pelargonidin rutinoside) showed higher EE than flavanols (catechin, epicatechin). According to Patil et al. [23], the EE is dependent on particle size rather than the amount of poloxamer (pluronic F127) used to stabilise the cubosome. The larger the nanoparticles, the higher entrapment efficiency for the polyphenols. This is because surface area to volume ratio of large particles is less than that of smaller particles and exposure to water of active compounds also decreased. Thus, the active compound loss due to diffusion also decreased in larger particles.

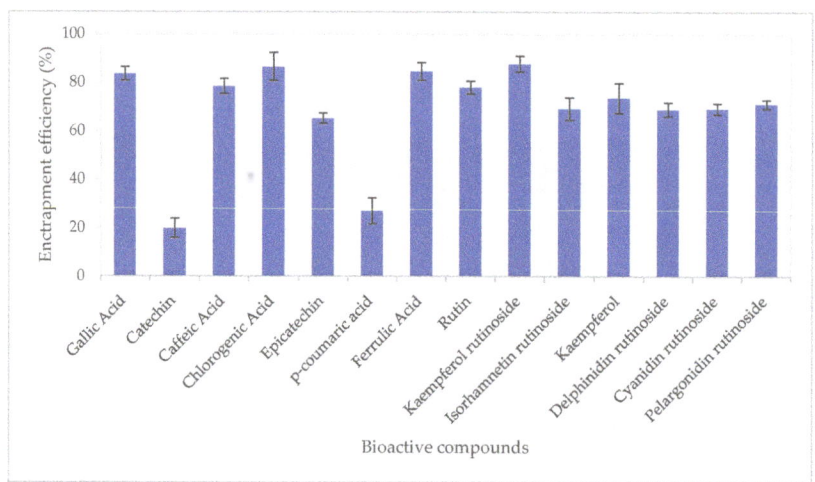

Figure 2. Entrapment efficiency of polyphenols from tamarillo extract using lyotropic liquid crystalline nanoparticles. Data are presented as mean and error bar (standard deviation) (n = 3).

3.2. TPC and Antioxidant Activities of Encapsulated and Non-Encapsulated Extracts during In Vitro Digestion

The impact of digestion on TPC and antioxidant activities of EXT and CUBTAM is shown in Figure 3. The TPC and antioxidant activities decreased significantly after digestion for both non-encapsulated and encapsulated in comparison to the undigested samples.

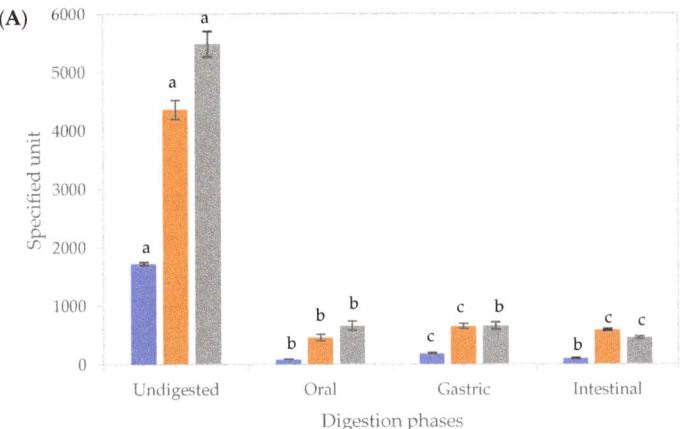

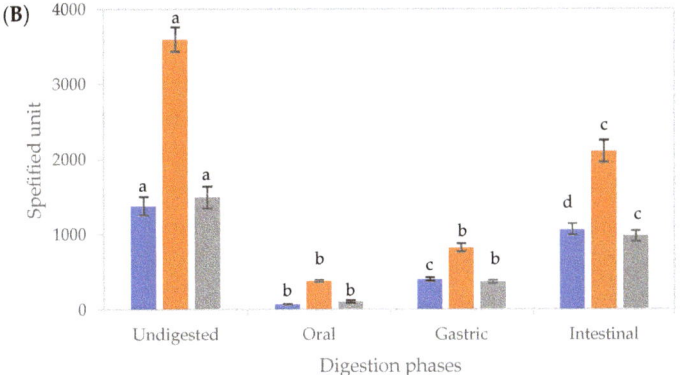

Figure 3. Changes in the total phenolic content and antioxidant activities of tamarillo extract (**A**) and tamarillo polyphenol loaded-cubosomes (**B**) before and after in vitro digestion. The units of TPC (■), CUPRAC (■) and FRAP (■) were mg GAE/100 g tamarillo, μmol TEAC/100 g tamarillo and μmol TEAC/100 g tamarillo, respectively. Data are presented as mean and error bar (standard deviation) (n = 3). Different alphabets indicate statistical difference ($p < 0.05$) for each assay.

For both EXT and CUBTAM, there were significant differences ($p < 0.05$) between the amounts of TPC in the supernatant after each stage of digestion. Gastric digests recorded the highest TPC for EXT, while, in the oral and intestinal phases, no significant differences were observed (Figure 3A). Low values of TPC in the supernatant of the oral digest (after 2 min of digestion) are related to the short time for diffusion and low solubility of polyphenols. The loss of polyphenols during digestion could be explained by physicochemical transformations, such as oxidation or the presence of yoghurt molecules (fats and proteins) in the digestion mixture. Furthermore, the decrease of bioactive content could arise from precipitation of several compounds with proteins or enzymes in the digest [30].

However, for the CUBTAM, the release of polyphenols increased during the digestion and was greater in the intestinal phase than the gastric phase (Figure 3B), which was not seen for the non-encapsulated extract, demonstrating a protective effect of the cubosome encapsulation technique against digestive enzymes and pH changes during gastric digestion. Results obtained from LC-MS (Table 1) further support the protective effect of the cubosomes on polyphenols. Similar findings have been reported on release properties of encapsulated blueberry extract [31] and carob pulp extract [32] during simulated gastrointestinal digestion. Both studies showed that TPC in the supernatant increased throughout gastric to intestinal digestion. The materials used for lipid bilayer and stabilisation of cubosomes determine susceptibility of polyphenols to digestive enzymes as well as pH at each stage [33]. The reduction in TPC of tamarillo extract during in vitro digestion might be associated to sensitivity of phenolic compounds to higher pH (>6), since, at that pH, monomers obtained by hydrolysis from larger molecules might be less stable [34]. The increase in the TPC of CUBTAM could be related to the release of complexed bioactive compounds as a result of the digestive process [35].

Table 1. Phenolics and anthocyanins in tamarillo extract and tamarillo polyphenol loaded-cubosome released during three stages of in vitro digestion.

Bioactive/Phases	Tamarillo Extract			Tamarillo Polyphenol Loaded-Cubosome		
	Oral	Gastric	Intestinal	Oral	Gastric	Intestinal
Phenolics						
Gallic Acid	3.86 ± 0.84 [a]	46.5 ± 6.02 [b]	20.4 ± 1.86 [c]	4.51 ± 0.79 [a]	8.57 ± 0.85 [a]	52.9 ± 11.7 [d]
Catechin	24.5 ± 5.01 [a]	27.2 ± 7.93 [a]	25.4 ± 6.09 [a]	6.37 ± 0.52 [b]	23.5 ± 4.09 [a]	16.3 ± 0.39
Caffeic Acid	24.1 ± 4.55 [a]	31.4 ± 3.90 [b]	21.9 ± 7.06 [a]	81.2 ± 15.2 [c]	2.73 ± 0.91 [d]	8.03 ± 0.35 [e]
Chlorogenic Acid	9.40 ± 1.24 [a]	67.7 ± 12.3 [b]	4.82 ± 1.12 [c]	4.99 ± 0.44 [c]	9.33 ± 0.43 [a]	28.8 ± 4.02 [d]
Epicatechin	24.2 ± 6.22 [a]	55.2 ± 10.3 [b]	13.9 ± 1.10 [c]	9.42 ± 0.49 [d]	16.6 ± 2.62 [c]	16.3 ± 3.66 [ac]
p-coumaric acid	38.8 ± 8.36 [a]	39.4 ± 6.29 [a]	4.14 ± 0.36 [b]	31.1 ± 1.16 [c]	16.2 ± 0.97 [d]	23.2 ± 4.20 [e]
Ferulic Acid	5.55 ± 0.98 [a]	13.4 ± 2.62 [b]	17.8 ± 2.64 [b]	54.4 ± 8.75 [c]	3.42 ± 0.98 [a]	32.3 ± 7.54 [d]
Rutin	31.0 ± 6.09 [a]	38.7 ± 7.40 [a]	23.0 ± 5.96 [b]	16.5 ± 1.39 [bc]	14.7 ± 2.69 [c]	24.0 ± 5.79 [b]
Kaempferol rutinoside	32.1 ± 8.67 [a]	37.6 ± 5.87 [a]	15.5 ± 2.00 [b]	21.0 ± 3.76 [c]	21.5 ± 4.49 [c]	29.1 ± 4.72 [ac]
Isorhamnetin rutinoside	8.21 ± 1.04 [a]	10.3 ± 1.08 [b]	9.81 ± 1.08 [ab]	3.76 ± 0.52 [c]	10.2 ± 2.20 [b]	9.43 ± 1.89 [ab]
Kaempferol	43.8 ± 12.1 [a]	20.4 ± 5.01 [b]	5.83 ± 0.70 [c]	29.6 ± 4.51 [b]	20.2 ± 5.21 [b]	33.1 ± 3.25 [d]
Anthocyanins						
Delphinidin rutinoside	10.1 ± 0.76 [a]	24.6 ± 5.97 [b]	7.87 ± 0.40 [c]	5.30 ± 0.32 [d]	10.3 ± 1.16 [a]	19.4 ± 1.14 [b]
Cyanidin rutinoside	14.3 ± 1.26 [a]	35.9 ± 6.17 [b]	4.31 ± 0.73 [c]	8.68 ± 0.55 [d]	17.1 ± 6.90 [a]	25.2 ± 0.09 [e]
Pelargonidin rutinoside	20.6 ± 1.49 [a]	48.7 ± 5.26 [b]	6.16 ± 0.31 [c]	10.3 ± 1.46 [d]	14.6 ± 2.80 [d]	27.8 ± 1.36 [e]

Results are expressed as % with respect to the initial concentration. Data are expressed as Mean ± SD (n = 3). Different letters of the alphabet superscripts indicate statistical difference ($p < 0.05$) across each row.

The antioxidant activity of tamarillo fruit phenolic extracts is mainly linked to their anthocyanins, chlorogenic acid and kaempferol rutinoside compounds. However, due to the chemical transformations from different mechanisms, the antioxidant properties of these compounds might change during digestion. Thus, the influence of digestion on the antioxidant capacity of tamarillo pulp extracts in non-encapsulation and encapsulation form was assessed by using CUPRAC and FRAP assays (Figure 3).

All activities tested significantly ($p < 0.05$) decreased after digestion in comparison to the raw material, which coincide with the decrease in bioactive compounds, mainly polyphenols, after digestion (Table 1). There were substantial and significant differences in CUPRAC values between the non-encapsulated and encapsulated extracts ($p < 0.05$) throughout the process of in vitro digestion. Tamarillo extract had the highest CUPRAC value in the gastric phase, whereas, for CUBTAM, the highest supernatant activity was noted in the intestinal medium. EXT and CUBTAM presented significant differences ($p < 0.05$) in FRAP values during the digestion process. The highest FRAP value was observed in the oral phase for non-encapsulated extract, whereas, for encapsulated samples, this activity increased with the progress of digestion with the highest FRAP activity of the supernatant was recorded at the end of the intestinal phase. In other studies, an increase in the FRAP with digestion was most pronounced at the intestinal phase for both encapsulated blueberry extract [31] and carob pulp extract [32].

The difference in FRAP and CUPRAC activities, in the oral and intestinal phases, respectively, may not be due to the content of phenolics and anthocyanins, but rather to the diversity and characteristics of the polyphenols present. However, the highest activities (FRAP and CUPRAC) after the gastric phase might be due to the higher release of phenolics and anthocyanins content and the quenching and reducing properties of the acidic medium of the sample. The effect of the pH could also be different among various polyphenols. At neutral pH, some polyphenols have exhibited pro-oxidant activities, whereas, at lower pH, others have demonstrated antioxidant activities [32]. Furthermore, the difference might be related to in vitro digestion conditions used and/or change of polyphenol availability related to the release of matrix associated compounds [36]. In fact, free polyphenols have shown higher antioxidant activity than iron-phenol chelates. Together with the enzymatic action, the pH influence within the gastrointestinal digestion and the presence of compounds that were not analysed in this study (e.g., peptides or complex polyphenols) enhance antioxidant activity [35]. The increase in antioxidant power of the supernatant between the acidic gastric phase and alkaline intestinal phase environments, as seen in this study, can be partially explained by the deprotonation of the hydroxyl groups on the aromatic rings of the polyphenols [36].

Under the intestinal conditions, the decrease in antioxidant activity (CUPRAC and FRAP) for the non-encapsulated extract would be related to the lower TPC alongside transformation of some polyphenols into conformations related to the neutral pH (Figure 3A). Meanwhile, the highest antioxidant activities for CUBTAM supernatant, in the intestinal phase (Figure 3B), could be explained by their release from the microcapsules as they are degraded in the neutral pH. The weak activities recorded in oral and gastric phases of digestion might be due to a small amount of polyphenol release from the microcapsule surface and/or via the penetration of salivary and gastric fluids into the microcapsules through their surface pores.

3.3. Release of Tamarillo Polypehnols from Cubosomes during Digestion

In order to evaluate the stability of individual polyphenol compounds during digestion, a total of fourteen compounds were evaluated by LC-MS (Table 1). The CUB was also analysed as a control. The results showed that 11 phenolic compounds and three anthocyanins were released from the microcapsules after the digestion process, demonstrating that these phytochemical compounds were well encapsulated by the cubosomes.

The phenolics presented different behaviours during the simulated digestion (Table 1). Analysis of phenolics released from EXT during digestion showed a significant instability for the major phenolic acids (gallic acid, chlorogenic acid and p-coumaric acid), other phenolics (epicatechin, rutin and kaempferol rutinoside) as well as all anthocyanins after oral and gastric phases. For gallic acid, the concentrations in oral phase remained stable and only a significant ($p < 0.05$) increase was observed in the gastric phase (46.49%) when compared to the initial undigested EXT. Then, the concentration of this acid dropped down to 20.4%. Tagliazucchi, Verzelloni, Bertolini and Conte [37] reported the degradation (43%) of pure gallic acid after gastrointestinal digestion, while the total degradation for gallic acid from grape extract and carob pulp extract had been explored by Jara-Palacios et al. [38] and Ydjedd et al. [32], respectively. Meanwhile, caffeic acid showed insignificant changes during the digestion (24.13% at the oral phase, 31.42% after the gastric phase and 21.94% at the end of the intestinal phase). According to Wojtunik-Kulesza et al. [39], the remaining percentage of caffeic acid decreased to 75% and 78% after oral and gastric phases, respectively. Some studies have reported that the gastric phase has increased the bioaccessibility of some phenolic acids, while, during the intestinal phase, their levels could be decreased. This behaviour has been closely related to the stability and structural changes that each type of polyphenolic acid undergoes [37]. Due to its low molecular weight, gallic acid has been better absorbed in humans compared to other phenolic acids, which makes it highly bioaccessible [36]. For chlorogenic acid, the highest concentration was detected in the gastric phase; then, the concentration of this compound reduced by 63% in the intestinal phase (Table 1). According to Tagliazucchi, Helal, Verzelloni and Conte [40], the degradation of chlorogenic acid during gastro-pancreatic digestion might be due to the oxidation and polymerization to form quinone in an alkaline environment. Significant reductions of free phenolic acids (gallic, chlorogenic, caffeic, p-coumaric acids) during in vitro digestion have been reported in previous studies [41–43]. These decreases in phenolic acids could be related to changes in pH and the presence of bile salts in the intestinal phase, which may lead to the formation of precipitates [42], which may explain the reductions observed at the end the intestinal phase of this study.

The concentrations of kaempferol rutinoside in EXT remained stable after the oral and gastric phases but decreased significantly ($p < 0.05$) at the end of the intestinal phase (Table 1). A similar trend for kaempferol rutinoside during in vitro digestion of the Cactus Cladodes plant had been observed [44]. Hydrolysis of glycoside flavonoids starts in the mouth by means of β-glycosidase action, but the degree of hydrolysis depends on the types of sugars present in the flavonoid compounds. For example, polyphenol compounds with more hydrophobic properties often interact more strongly with proteins [39]. Degradation of polyphenols with high molecular weights (such as kaempferol rutinoside) may be related to their strong affinities with human salivary proline- and histidine-rich proteins to form non-covalent and covalent associations [39].

All of the anthocyanins, especially delphinidin rutinoside and pelargonidin rutinoside, showed the same releasing behavior during in vitro digestion (Table 1). For these main anthocyanins in tamarillo extract, a significantly ($p < 0.05$) higher proportion of anthocyanins (43 to 76%) was released after the intestinal phase when compared to the undigested samples. The instability of anthocyanins at neutral or slightly basic pH has been observed for polyphenols from grape and chokeberry [37,45]. The instability can be explained by the formation of a colourless chalcone pseudo-base, resulting in the destruction of the anthocyanin chromophore [46]. The current results support these previous findings, suggesting that anthocyanins are stable in the acidic conditions of the gastric phase but are degraded in the alkaline/neutral conditions of the intestinal phase. The reduction of anthocyanins may also be related to the fact that, in aqueous solution in response to changes in pH, anthocyanins undergo structural rearrangements, change colour, may form complexes with proteins in food and digestate and be degraded to phenolic acids [42].

The quantity of individual bioactive compounds from the CUBTAM at the end of each digestive phase varied by compound (Table 1). Catechin, epicatechin, isorhamnetin rutinoside and all anthocyanins (delphinidin rutinoside, cyanidin rutinoside and pelargonidin rutinoside) were released after the gastric phase in acidic medium; gallic acid, caffeic acid, chlorogenic acid, p-coumaric acid, ferulic acid, rutin, kaempferol and kaempferol rutinoside were released into the neutral medium after oral and intestinal phases. It is worth noting that the percentage of free polyphenols was lower in CUBTAM (encapsulated) than in EXT (non-encapsulated) ones, and remained fairly constant along different in vitro digestion phases. These results were also expected because the initial amount of polyphenols in the encapsulated sample was lower owing to the encapsulated efficiency (over 50%). According to Ydjedd et al. [32], the properties of encapsulating material play a significant role in enhancing the entrapment efficiency and controlled release of the core compounds. They reported a slow release of some phenolics (gallic acid, p-coumaric acid, and kaempferol) from the microcapsules and a period of more than 3 h in the intestinal phase (neutral medium) has been necessary for complete release of these compounds, when the encapsulating material was completely degraded [32].

The present study is the first to report the proportion of cubosome encapsulated polyphenols released after each phase of in vitro digestion, demonstrating the potential of cubosomes to protect bioactive compounds in their matrix. Similarly, reduction of the degradation in cubosome encapsulated bioactive antimicrobial peptide has been reported, showing resistance towards the enzymatic degradation [18]. Cubosomes have a high viscosity which hinders the diffusion of polyphenols into the release medium and slows the entry of water, which sustains the slow release profile [47]. The rate of release controlled by the structure also depends both on the partition coefficient and on the diffusion of the drug through the hydrocarbon tail region [48].

3.4. Physicochemical Properties of Yoghurt Fortified with CUBTAM

The addition of 5%, 10% and 15% CUBTAM to yoghurt was associated with a small but statistically significant fall in pH and reduced syneresis (Table 2). This can be explained by the use of the freeze-drying treatment to prepare the powder for the cubosome, which would result in an increase in total dry solids, which in turn would increase the water holding capacity, reduce porosity and reduce the syneresis.

Table 2. Physicochemical properties of yoghurt fortified with CUBTAM (5, 10 and 15%) in PRE and POS. Control yoghurt contained no CUBTAM.

Parameters/Samples	Control	POS5	POS10	POS15	PRE5	PRE10	PRE15
pH	4.35 ± 0.03 [a]	4.27 ± 0.02 [b]	4.14 ± 0.01 [c]	4.09 ± 0.02 [d]	4.30 ± 0.00 [a]	4.18 ± 0.01 [c]	4.10 ± 0.02 [cd]
Syneresis (%)	29.3 ± 0.91 [a]	28.8 ± 1.48 [a]	27.1 ± 1.53 [ab]	26.6 ± 1.91 [b]	28.7 ± 1.36 [a]	27.9 ± 1.70 [b]	26.9 ± 1.52 [b]
Textural parameters							
Firmness (N)	1.027 ± 0.005 [a]	1.031 ± 0.004 [a]	1.034 ± 0.002 [ab]	1.042 ± 0.008 [b]	1.033 ± 0.004 [a]	1.035 ± 0.003 [ab]	1.045 ± 0.006 [b]
Consistency (N.sec)	16.08 ± 0.004 [a]	16.12 ± 0.029 [a]	16.21 ± 0.021 [b]	17.19 ± 0.068 [c]	16.09 ± 0.102 [a]	16.23 ± 0.039 [b]	17.22 ± 0.065 [d]
Cohesiveness (N)	−0.005 ± 0.003 [a]	−0.007 ± 0.002 [a]	−0.011 ± 0.004 [ab]	−0.014 ± 0.005 [b]	−0.008 ± 0.003 [a]	−0.010 ± 0.001 [ab]	−0.016 ± 0.007 [b]
Rheological parameters							
Consistency coefficient (K, Pa.s)	<0.005 [a]	0.009 ± 0.003 [b]	0.010 ± 0.003 [b]	0.014 ± 0.007 [b]	0.008 ± 0.02 [b]	0.012 ± 0.005 [b]	0.015 ± 0.008 [b]
Flow behaviours index (n)	0.781 ± 0.012 [a]	0.729 ± 0.022 [a]	0.658 ± 0.010 [b]	0.604 ± 0.018 [c]	0.735 ± 0.014 [a]	0.678 ± 0.020 [b]	0.642 ± 0.015 [b]
Viscosity at 350 s^{-1} (Pa.s)	0.026 ± 0.000 [a]	0.030 ± 0.000 [b]	0.033 ± 0.001 [c]	0.041 ± 0.001 [d]	0.032 ± 0.000 [b]	0.037 ± 0.002 [cd]	0.045 ± 0.003 [d]
Elastic modulus (Pa)	N/A	0.003 ± 0.000 [a]	0.005 ± 0.000 [b]	0.005 ± 0.001 [b]	0.003 ± 0.000 [a]	0.004 ± 0.001 [b]	0.006 ± 0.002 [b]

N/A: not applicable. Data are expressed as Mean ± SD (n = 3). Different alphabetic superscripts indicate statistical difference ($p < 0.05$) across each row. CUBTAM: tamarillo polyphenols loaded-cubosomes. POS5, POS10, POS15: addition of 5%, 10%, 15% of CUBTAM post to fermentation process, respectively. PRE5, PRE10 and PRE15: addition of 5%, 10%, 15% of CUBTAM prior to fermentation process, respectively.

Viscosity of yoghurt increased with the increase of the concentration of CUBTAM, showing significant ($p < 0.05$) differences across the yoghurt samples (Table 2). Within the same % CUBTAM fortification, there was no significant ($p > 0.05$) difference in viscosity between PRE and POS. Based on the Oswald–de Waele power law model, yoghurts fortified with CUBTAM made from both fermentation processes can be considered as non-Newtonian fluids with shear-thinning behaviour due to the flow behaviour index (n) below 1. The breakage of bonds between the protein aggregates as a consequence of shear stress led to the pseudoplastic behaviour of the yoghurt samples [49]. The consistency index (K) and flow behaviour index (n) of yoghurts were not significantly influenced by the fermentation process, whereas the increase of encapsulated powder concentration led to the increase of K and decrease of n values. The increase of K value might be attributed to the water holding capacity, caused by the addition of powder.

The elastic modulus of all yoghurts was very low, indicating the same relatively weak structure with or without CUBTAM (Table 2).

3.5. Total Phenol Content, Antioxidant Activity and Release of Polyphenol Compounds in Yoghurt Fortified with CUBTAM during Digestion

In a yoghurt matrix, catalase and super oxidase enzymes, casein as well as lactic acid bacteria which have antioxidant properties are present [50]. Without digestion, as expected, the addition of CUBTAM led to a dose-dependent increase in TPC and total antioxidant capacity; i.e., as expected, a higher level of fortification led to a higher TPC as well as antioxidant activity ($p < 0.05$). Furthermore, fortification in PRE resulted in higher TPC and antioxidant activity than in POS at the same concentration (Figure 4A).

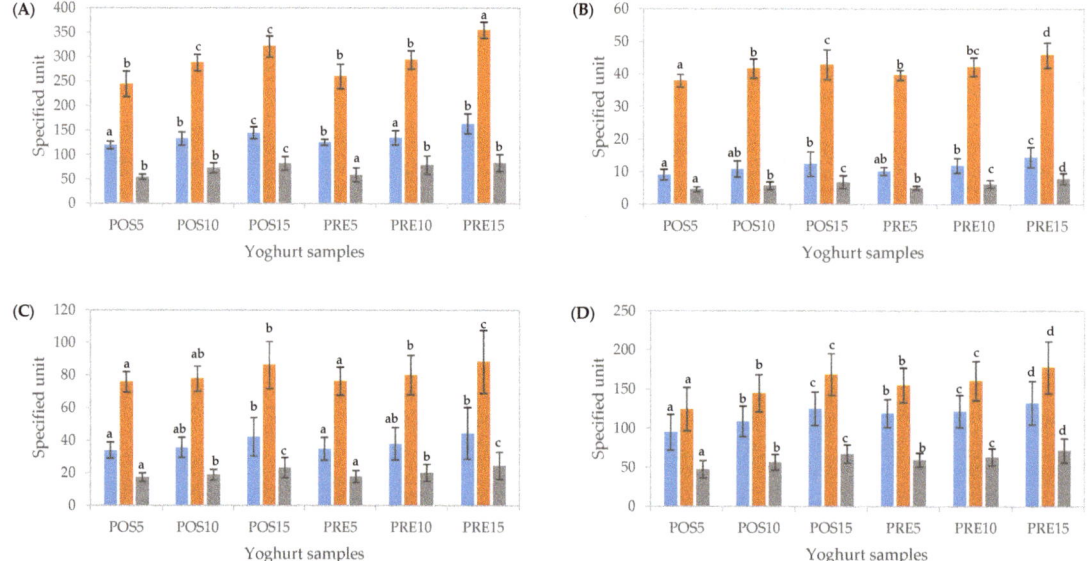

Figure 4. Total phenolic content (TPC) and antioxidant activity (CUPRAC and FRAP assay) of yoghurts fortified with cubosome containing tamarillo extract before digestion (**A**), after oral (**B**), gastric (**C**) and intestinal (**D**) phases of in vitro digestion. The units of TPC (■), CUPRAC (■) and FRAP (■) were mg GAE/100 g yoghurt, µmol TEAC/100 g yoghurt and µmol TEAC/100 g yoghurt, respectively. Data are presented as mean and error bar (standard deviation) (n = 3). Different alphabets indicate statistical difference ($p < 0.05$) for each assay.

After the oral phase, TPC and total antioxidant capacity determined using CUPRAC and FRAP assays in both POS and PRE samples at all fortified concentrations reduced by 7.6–8.9%, 12.9–15.5% and 7.8–9.4%, respectively, compared to undigested samples (Figure 4B). After the gastric phase, significant increases in TPC and total antioxidant capacity were obtained compared to those of the oral phase (up to 3-, 2- and 3-fold, respectively), ($p < 0.05$) (Figure 4C). After simulated intestinal digestion, further significant increases in TPC (64–71%) were observed ($p < 0.05$). Furthermore, total antioxidant activity resulted in an additional 39–50% and 63–70% increase for CUPRAC and FRAP, respectively (Figure 4D). These results were in line with the measures of TPC and antioxidant activity of encapsulated tamarillo extracts (CUBTAM) (Figure 3) as well as polyphenol concentration determined by LC-MS (Table 1), in which the TPC and antioxidant activity increased greatly during the gastric and intestinal phases of in vitro digestion. Previous research showed that the antioxidant activity of the yoghurt samples containing encapsulated phenolics was increased due to the controlled release of the phenolic components from the encapsulation network [50]. Some researchers have considered the Folin–Ciocalteau assay as an antioxidant capacity test since this assay not only measures phenolic compounds but also the total reducing capacity of a sample and hence [42]. This may explain the difference in polyphenols content measured by LC-MS and by the colorimetric tests. Considering the pH conditions of the total antioxidant capacity assays performed in this study, the CUPRAC assay (pH 7.0) could be more appropriate for evaluating total antioxidant activity after the oral and intestinal phases, while the FRAP assay (pH 3.6) could be more suitable to assess total antioxidant activity after the gastric phase.

The data in Table 3 are described as a percentage of each digestion phase in the supernatant compared to the content of the undigested sample. The TPC and antioxidant activities measured by CUPRAC and FRAP between CUBTAM and yoghurt fortified with CUBTAM were relatively similar at each phase of the digestion simulation (Table 3), demonstrating that yoghurt was a suitable carrier of cubosome without significant interference. For the oral phase, the TPC released and antioxidant activities in CUBTAM fortified yoghurt were higher than that for the CUBTAM itself by 1.5–1.7, 1.2–1.5 and 1.1–1.4 times, respectively. For the gastric phase, the amount of TPC released for CUBTAM was slightly higher than the fortified yoghurts, whereas the opposite trend was observed for the antioxidant activities. For the intestinal phase, the TPC and FRAP of CUBTAM were lower than that for CUBTAM fortified yoghurts, whereas, for the CUPRAC, the CUBTAM showed a similar value to PRE5, which were both much higher compared to the rest of the samples. Overall, cubosomes containing tamarillo extract showed effective protection for polyphenols in the oral and gastric phases.

The differences in the LC-MS profiles between PRE and POS were significant ($p < 0.05$). Major polyphenols in tamarillo were also detected in CUBTAM fortified yoghurts (both from PRE and POS) before digestion (Table S2). Thus, the yoghurt matrix as well as encapsulation had helped to retain these individual polyphenols during processing. Encapsulation of bioactive compounds promoted lower loss of polyphenols under refrigerated conditions [51]. At the same concentration of CUBTAM yoghurt, addition of tamarillo polyphenol loaded-cubosomes prior to fermentation was associated with a higher concentration of major polyphenols than addition post fermentation. The concentrations of chlorogenic acid, kaempferol rutinoside and delphinidin rutinoside (accounted for over 65% of the total polyphenol content in yoghurts) were higher for pre-fermentation versus the post fermentation approach (Table S2). Delphinidin rutinoside was the dominant anthocyanin in fortified yoghurts that was in agreement with the main anthocyanin in Laird's Large tamarillo pulp reported in our previous study [13]. The results showed that the fermentation process appeared to have little impact on anthocyanins present in both yoghurts, which might be due to the encapsulation of anthocyanins.

The yields of polyphenols were associated with the extractability of polyphenols from the original tamarillo extract. According to Sun-Waterhouse, Zhou and Wadhwa [52], during fermentation, the yoghurt starter cultures could transform the polyphenols into other forms/types of compounds, e.g., via flavonoid glycosides hydrolysis or C-ring cleavage. Such a conversion may result in the deactivation of bioactive compounds or activation of previously inactive compounds, e.g., polyphenol glycosides are hydrolyzed into their aglycones of higher free radical scavenging ability, and procyanidins break down to flavan-3-ols or to smaller molecular phenolic acids. Acidity of yoghurt may have induced acid hydrolysis of polyphenols, hence this could explain an increased amount of hydroxycinnamates such as caffeic acid, chlorogenic acid, ferulic acid and p-coumaric acid in the fortified yoghurts. Sun-Waterhouse, Zhou and Wadhwa [53] stated that the yields of hydroxycinnamic acids and flavonols detected might be dependent on the extractability of these polyphenols from different product matrices.

Table 3. Measures of total phenolic content and antioxidant activity of supernatant of cubosome encapsulated tamarillo and yoghurt fortified with encapsulated tamarillo during in vitro digestion.

Phases	Samples	TPC	CUPRAC	FRAP
Oral	CUBTAM	5.07	10.55	6.97
	POS5	7.64	15.50	8.38
	POS10	8.19	14.47	7.83
	POS15	8.59	13.34	8.33
	PRE5	8.14	15.25	8.67
	PRE10	8.86	14.36	7.93
	PRE15	8.87	12.89	9.45
Gastric	CUBTAM	29.13	22.77	24.70
	POS5	28.34	31.04	31.84
	POS10	26.79	27.05	25.88
	POS15	29.20	26.86	28.24
	PRE5	27.75	29.41	30.18
	PRE10	28.18	27.29	25.39
	PRE15	27.19	24.84	29.25
Intestinal	CUBTAM	77.03	58.53	65.27
	POS5	79.20	50.83	87.22
	POS10	81.72	50.27	77.47
	POS15	86.30	52.49	81.92
	PRE5	94.90	59.64	101.19
	PRE10	90.41	54.74	79.94
	PRE15	81.08	49.99	85.62

Data are presented as % of the sample before digestion. CUBTAM: tamarillo polyphenols loaded-cubosomes. POS5, POS10, POS15: addition of 5%, 10%, 15% of CUBTAM post to fermentation process, respectively. PRE5, PRE10 and PRE15: addition of 5%, 10%, 15% of CUBTAM prior to fermentation process, respectively.

Despite the importance of the recovery in each digestion phase, bioactive compounds will need to be released from their food matrix and reach the intestine where they can be absorbed and be metabolised [54]. The release rate during in vitro digestion has been considered as an indicator to assess the effectiveness of compound carriers [55]. In general, non-encapsulated phenolic compounds in drinking yoghurt were highly degraded after digestion [56] while microencapsulated formulation showed the ability to preserve the antioxidant activity of extract in yoghurt when compared with the free form [51,57]. The amount of individual polyphenol in yoghurt samples was significantly different ($p < 0.05$) after each phase of in vitro digestion. The quantity of polyphenols released at each stage is dependent on the time at each phase, pH and the concentration of CUBTAM (Table S3). Most polyphenol components were detected in both PRE and POS in each digestion phase. The oral digestion lasts for a few minutes; the encapsulated polyphenol in yoghurts release from oral digestion was significantly lower than the gastric (post to 2 h) and intestinal (post to 3 h) simulated digests. These data are in line with the findings from Section 3.3 that encapsulating polyphenol in cubosome particles could effectively protect the bioactive compounds from the gastric enzymes and facilitate the utilization of polyphenols in the human body. Studies on the metabolism of bioactive compounds in the humans have shown that bioactives are mainly metabolised by a large number of small and large intestinal bacteria, and the metabolites are absorbed into the human blood [58]. Therefore, for the polyphenols to be absorbed by the human body or to be active in the microbiome of the small and large intestines, the polyphenols should be protected in encapsulated form until completely released in the intestinal tract. When the polyphenol capsules were present in the simulated oral phase, the concentration of all bioactive compounds was low, indicating good retention in the cubosomes. Furthermore, compared to undigested samples, a significantly ($p < 0.05$) lower amount of most polyphenols (percentage loss < 10%) were released during the oral phase (except for epicatechin, rutin and kaempferol rutinoside). The findings provide evidence that cubosomes protect bioactive compounds from interaction with, for example, milk proteins and digestive enzymes, reducing the risk of polyphenol degradation.

To our knowledge, most of reports of the loads of cubosomes are for proteins or small molecules such as drugs while research about encapsulation of hydrophilic polyphenols, mainly presented in tamarillo, using the cubosome is limited. Effects of the digestion process on properties and stability of cubosome as well as application of cubosome in food are still scarce. Hence, a strength of this study is it is the first attempt to encapsulate polyphenols from tamarillo and has demonstrated the proof of principle that this technique can be used to fortify yoghurt with a fruit extract. Entrapment efficiency was greater than 50% and, together with enhancement of antioxidant effects, stability and bioavailability of polyphenols in vivo and in vitro [2], the results showed the potential of cubosome to minimize degradation of polyphenols and contribute to controlled release of these and other bioactives during digestion. Application of encapsulated polyphenols into yoghurt did not significantly change texture and rheology of yoghurts when compared to the control, except for 15% fortification, where higher texture and rheology values were observed. However, there are still challenges about applications of cubosomes, including a deeper understanding of the stabilizer and possible cytotoxicity. In the future interaction between yoghurt components (mainly protein), starter culture and encapsulated bioactives should be evaluated with longer intestinal digestion and possible effects on the microbiota explored. According to Wei et al. [21], monoolein has been easily hydrolysed to free oleic acid due to the presence of pancreatic lipase and bile salt; therefore, it can be assumed that cubosome particles (with 60% of monoolein in components) can only be degraded in intestinal phase. However, degradation or stability of cubosome after each phase of digestion should be evaluated to ensure the safety, effectiveness and acceptability to the consumer of this approach. It is a limitation that we did not validate the correct encapsulation of the tamarillo polyphenols in the cubosome. In future work, we could confirm the encapsulation with confocal Raman/FTIR microscopy.

4. Conclusions

This study demonstrated the proof-of-principle that tamarillo polyphenols could be effectively encapsulated by cubosome nanoparticles with relatively high loading efficiency and preservation of high antioxidant activity. Compared to the unencapsulated extract, cubosomal encapsulation provided a protective effect to the tamarillo polyphenols under simulated gastrointestinal conditions, exhibiting good free polyphenol concentrations at the end of the intestinal phase. A cubosomal system was employed for the delivery of tamarillo polyphenols via yoghurt, and the addition of encapsulated bioactive improved the physicochemical and nutritional properties of yoghurt. The addition of CUBTAM at increasing concentrations successfully increased the concentration of polyphenols, TPC and antioxidant activity of yoghurts, with controlled stability during digestion, suggesting that polyphenols with enhanced bioavailability could be delivered in a dose-controlled manner. This research informs application of cubosome encapsulation to fortification of food products, for example both water-soluble and lipid-soluble vitamins and carotenoids (β-carotene). However, although the components of cubosomes (monoolein and Pluronic F127) are listed as "generally recognised as safe" (GRAS) by the FDA and approved in principle, further investigations should be carried out before sensory testing or consumption by humans as a food.

Supplementary Materials: The following supporting information can be downloaded at: https://www.mdpi.com/article/10.3390/antiox11030520/s1, Table S1: Particle size and polydispersity index of CUB and CUBTAM, Table S2: Concentrations (µg/g FW) of individual polyphenols in yoghurts fortified with CUBTAM before in vitro digestion, Table S3: Concentrations (µg/g FW) of individual polyphenols in yoghurts fortified with CUBTAM after each step of in vitro digestion, Figure S1: Size distribution of CUB (♦) and CUBTAM (●), Figure S2: Concentration of polyphenols added into cubosome ((■) and concentration of polyphenols present in the supernatant (■).

Author Contributions: Conceptualization, M.J.Y.Y., T.T.D.; Methodology, T.T.D., M.J.Y.Y.; Software, T.T.D.; Validation, T.T.D., M.J.Y.Y.; Formal Analysis, T.T.D.; Investigation, T.T.D.; Resources, M.J.Y.Y.,

E.R.; Data Curation, T.T.D., M.J.Y.Y.; Writing—Original Draft Preparation, T.T.D.; Writing—Review and Editing, M.J.Y.Y., E.R.; Visualization, T.T.D., M.J.Y.Y.; Supervision, M.J.Y.Y., E.R.; Project Administration, M.J.Y.Y., E.R.; Funding Acquisition, M.J.Y.Y., E.R. All authors have read and agreed to the published version of the manuscript.

Funding: This research received a Performance Based Research Fund and internal postgraduate research project funds from the Auckland University of Technology. A PhD scholarship for the first author and internal project fund from Riddet Institute was also received.

Institutional Review Board Statement: Not applicable.

Informed Consent Statement: Not applicable.

Data Availability Statement: All data are contained within the article and supplementary material.

Acknowledgments: The authors would like to acknowledge the Auckland University of Technology for the Performance Based Research Funding and postgraduate research project funds received for the project. The authors would like to thank the Riddet Institute for the Doctoral Scholarship provided to the first author, and the internal project fund to support the research. We also thank DuPont Nutrition and Bioscience (Danisco Ltd.) for supplying Dimodan®. We thank Azelis New Zealand for supplying Novozym to our research.

Conflicts of Interest: The authors declare no conflict of interest.

References

1. Mezzenga, R.; Seddon, J.M.; Drummond, C.J.; Boyd, B.J.; Schröder-Turk, G.E.; Sagalowicz, L. Nature-Inspired design and application of lipidic lyotropic liquid crystals. *Adv. Mater.* **2019**, *31*, 1900818. [CrossRef] [PubMed]
2. Barriga, H.M.; Holme, M.N.; Stevens, M.M. Cubosomes: The next generation of smart lipid nanoparticles? *Angew. Chem. Int. Ed.* **2019**, *58*, 2958–2978. [CrossRef] [PubMed]
3. Meikle, T.G.; Dharmadana, D.; Hoffmann, S.V.; Jones, N.C.; Drummond, C.J.; Conn, C.E. Analysis of the structure, loading and activity of six antimicrobial peptides encapsulated in cubic phase lipid nanoparticles. *J. Colloid Interface Sci.* **2021**, *587*, 90–100. [CrossRef] [PubMed]
4. Boge, L.; Browning, K.L.; Nordström, R.; Campana, M.; Damgaard, L.S.; Seth Caous, J.; Hellsing, M.; Ringstad, L.; Andersson, M. Peptide-loaded cubosomes functioning as an antimicrobial unit against *Escherichia coli*. *ACS Appl. Mater. Interfaces* **2019**, *11*, 21314–21322. [CrossRef] [PubMed]
5. Kwon, T.K.; Hong, S.K.; Kim, J.-C. In vitro skin permeation of cubosomes containing triclosan. *J. Ind. Eng. Chem.* **2012**, *18*, 563–567. [CrossRef]
6. Luo, Q.; Lin, T.; Zhang, C.Y.; Zhu, T.; Wang, L.; Ji, Z.; Jia, B.; Ge, T.; Peng, D.; Chen, W. A novel glyceryl monoolein-bearing cubosomes for gambogenic acid: Preparation, cytotoxicity and intracellular uptake. *Int. J. Pharm.* **2015**, *493*, 30–39. [CrossRef]
7. Saber, M.M.; Al-Mahallawi, A.M.; Nassar, N.N.; Stork, B.; Shouman, S.A. Targeting colorectal cancer cell metabolism through development of cisplatin and metformin nano-cubosomes. *BMC Cancer* **2018**, *18*, 1–11. [CrossRef]
8. Zabara, M.; Senturk, B.; Gontsarik, M.; Ren, Q.; Rottmar, M.; Maniura-Weber, K.; Mezzenga, R.; Bolisetty, S.; Salentinig, S. Multifunctional nano-biointerfaces: Cytocompatible antimicrobial nanocarriers from stabilizer-free cubosomes. *Adv. Funct. Mater.* **2019**, *29*, 1904007. [CrossRef]
9. Cortesi, R.; Cappellozza, E.; Drechsler, M.; Contado, C.; Baldisserotto, A.; Mariani, P.; Carducci, F.; Pecorelli, A.; Esposito, E.; Valacchi, G. Monoolein aqueous dispersions as a delivery system for quercetin. *Biomed. Microdevices* **2017**, *19*, 41. [CrossRef]
10. Puglia, C.; Cardile, V.; Panico, A.M.; Crascì, L.; Offerta, A.; Caggia, S.; Drechsler, M.; Mariani, P.; Cortesi, R.; Esposito, E. Evaluation of monoolein aqueous dispersions as tools for topical administration of curcumin: Characterization, in vitro and ex-vivo studies. *J. Pharm. Sci.* **2013**, *102*, 2349–2361. [CrossRef]
11. Archana, A.; Vijayasri, K.; Madhurim, M.; Kumar, C. Curcumin loaded nano cubosomal hydrogel: Preparation, in vitro characterization and antibacterial activity. *Chem. Sci. Trans.* **2015**, *4*, 75–80.
12. Tu, Y.; Fu, J.; Sun, D.; Zhang, J.; Yao, N.; Huang, D.; Shi, Z. Preparation, characterisation and evaluation of curcumin with piperine-loaded cubosome nanoparticles. *J. Microencaps.* **2014**, *31*, 551–559. [CrossRef] [PubMed]
13. Diep, T.; Pook, C.; Yoo, M. Phenolic and Anthocyanin Compounds and Antioxidant Activity of Tamarillo (*Solanum betaceum* Cav.). *Antioxidants* **2020**, *9*, 169. [CrossRef] [PubMed]
14. Cory, H.; Passarelli, S.; Szeto, J.; Tamez, M.; Mattei, J. The role of polyphenols in human health and food systems: A mini-review. *Front. Nutr.* **2018**, *5*, 87. [CrossRef] [PubMed]
15. Diep, T.T.; Rush, E.C.; Yoo, M.J.Y. Tamarillo (*Solanum betaceum* Cav.): A Review of Physicochemical and Bioactive Properties and Potential Applications. *Food Rev. Int.* **2020**, 1–25. [CrossRef]
16. Martínez-Ballesta, M.; Gil-Izquierdo, Á.; García-Viguera, C.; Domínguez-Perles, R. Nanoparticles and controlled delivery for bioactive compounds: Outlining challenges for new "smart-foods" for health. *Foods* **2018**, *7*, 72. [CrossRef]

17. Piovesana, A.; Noreña, C.P.Z. Microencapsulation of bioactive compounds from hibiscus calyces using different encapsulating materials. *Int. J. Food Eng.* **2018**, *14*. [CrossRef]
18. Boge, L.; Hallstensson, K.; Ringstad, L.; Johansson, J.; Andersson, T.; Davoudi, M.; Larsson, P.T.; Mahlapuu, M.; Håkansson, J.; Andersson, M. Cubosomes for topical delivery of the antimicrobial peptide LL-37. *Eur. J. Pharm. Biopharm.* **2019**, *134*, 60–67. [CrossRef]
19. Park, S.H.; Kim, J.-C. In vitro anti-inflammatory efficacy of Bambusae Caulis in Taeniam extract loaded in monoolein cubosomes. *J. Ind. Eng. Chem.* **2019**, *77*, 189–197. [CrossRef]
20. Seo, S.R.; Kang, G.; Ha, J.W.; Kim, J.-C. In vivo hair growth-promoting efficacies of herbal extracts and their cubosomal suspensions. *J. Ind. Eng. Chem.* **2013**, *19*, 1331–1339. [CrossRef]
21. Wei, Y.; Zhang, J.; Zheng, Y.; Gong, Y.; Fu, M.; Liu, C.; Xu, L.; Sun, C.C.; Gao, Y.; Qian, S. Cubosomes with surface cross-linked chitosan exhibit sustained release and bioavailability enhancement for vinpocetine. *RSC Adv.* **2019**, *9*, 6287–6298. [CrossRef]
22. Zhou, Y.; Guo, C.; Chen, H.; Zhang, Y.; Peng, X.; Zhu, P. Determination of sinomenine in cubosome nanoparticles by HPLC technique. *J. Anal. Methods Chem.* **2015**, *2015*, 931687. [CrossRef] [PubMed]
23. Patil, R.P.; Pawara, D.D.; Gudewar, C.S.; Tekade, A.R. Nanostructured cubosomes in an in situ nasal gel system: An alternative approach for the controlled delivery of donepezil HCl to brain. *J. Liposome Res.* **2019**, *29*, 264–273. [CrossRef] [PubMed]
24. Wang, X.; Kristo, E.; LaPointe, G. Adding apple pomace as a functional ingredient in stirred-type yogurt and yogurt drinks. *Food Hydrocoll.* **2020**, *100*, 105453. [CrossRef]
25. Kristo, E.; Miao, Z.; Corredig, M. The role of exopolysaccharide produced by Lactococcus lactis subsp. cremoris in structure formation and recovery of acid milk gels. *Int. Dairy J.* **2011**, *21*, 656–662. [CrossRef]
26. Zhang, R.; Yoo, M.J.; Gathercole, J.; Reis, M.G.; Farouk, M.M. Effect of animal age on the nutritional and physicochemical qualities of ovine bresaola. *Food Chem.* **2018**, *254*, 317–325. [CrossRef]
27. Caffrey, M.; Cherezov, V. Crystallizing membrane proteins using lipidic mesophases. *Nat. Protoc.* **2009**, *4*, 706–731. [CrossRef]
28. Flak, D.K.; Adamski, V.; Nowaczyk, G.; Szutkowski, K.; Synowitz, M.; Jurga, S.; Held-Feindt, J. AT101-Loaded Cubosomes as an Alternative for Improved Glioblastoma Therapy. *Int. J. Nanomed.* **2020**, *15*, 7415. [CrossRef]
29. Danaei, M.; Dehghankhold, M.; Ataei, S.; Hasanzadeh Davarani, F.; Javanmard, R.; Dokhani, A.; Khorasani, S.; Mozafari, M. Impact of particle size and polydispersity index on the clinical applications of lipidic nanocarrier systems. *Pharmaceutics* **2018**, *10*, 57. [CrossRef]
30. Ortega, N.; Macià, A.; Romero, M.-P.; Reguant, J.; Motilva, M.-J. Matrix composition effect on the digestibility of carob flour phenols by an in-vitro digestion model. *Food Chem.* **2011**, *124*, 65–71. [CrossRef]
31. Flores, F.P.; Singh, R.K.; Kerr, W.L.; Phillips, D.R.; Kong, F. In vitro release properties of encapsulated blueberry (Vaccinium ashei) extracts. *Food Chem.* **2015**, *168*, 225–232. [CrossRef] [PubMed]
32. Ydjedd, S.; Bouriche, S.; López-Nicolás, R.N.; Sánchez-Moya, T.; Frontela-Saseta, C.; Ros-Berruezo, G.; Rezgui, F.; Louaileche, H.; Kati, D.-E. Effect of in vitro gastrointestinal digestion on encapsulated and nonencapsulated phenolic compounds of carob (Ceratonia siliqua L.) pulp extracts and their antioxidant capacity. *J. Agric. Food Chem.* **2017**, *65*, 827–835. [CrossRef] [PubMed]
33. Saura-Calixto, F.; Serrano, J.; Goñi, I. Intake and bioaccessibility of total polyphenols in a whole diet. *Food Chem.* **2007**, *101*, 492–501. [CrossRef]
34. Pavan, V.; Sancho, R.A.S.; Pastore, G.M. The effect of in vitro digestion on the antioxidant activity of fruit extracts (Carica papaya, Artocarpus heterophillus and Annona marcgravii). *LWT-Food Sci. Technol.* **2014**, *59*, 1247–1251. [CrossRef]
35. da Silva Haas, I.C.; Toaldo, I.M.; Gomes, T.M.; Luna, A.S.; de Gois, J.S.; Bordignon-Luiz, M.T. Polyphenolic profile, macro-and microelements in bioaccessible fractions of grape juice sediment using in vitro gastrointestinal simulation. *Food Biosci.* **2019**, *27*, 66–74. [CrossRef]
36. Ferreyra, S.; Torres-Palazzolo, C.; Bottini, R.; Camargo, A.; Fontana, A. Assessment of in-vitro bioaccessibility and antioxidant capacity of phenolic compounds extracts recovered from grapevine bunch stem and cane by-products. *Food Chem.* **2021**, *348*, 129063. [CrossRef]
37. Tagliazucchi, D.; Verzelloni, E.; Bertolini, D.; Conte, A. In vitro bio-accessibility and antioxidant activity of grape polyphenols. *Food Chem.* **2010**, *120*, 599–606. [CrossRef]
38. Jara-Palacios, M.J.; Gonçalves, S.; Hernanz, D.; Heredia, F.J.; Romano, A. Effects of in vitro gastrointestinal digestion on phenolic compounds and antioxidant activity of different white winemaking byproducts extracts. *Food Res. Int.* **2018**, *109*, 433–439. [CrossRef]
39. Wojtunik-Kulesza, K.; Oniszczuk, A.; Oniszczuk, T.; Combrzyński, M.; Nowakowska, D.; Matwijczuk, A. Influence of in vitro digestion on composition, bioaccessibility and antioxidant activity of food polyphenols—A non-systematic review. *Nutrients* **2020**, *12*, 1401. [CrossRef]
40. Tagliazucchi, D.; Helal, A.; Verzelloni, E.; Conte, A. The type and concentration of milk increase the in vitro bioaccessibility of coffee chlorogenic acids. *J. Agric. Food Chem.* **2012**, *60*, 11056–11064. [CrossRef]
41. Celep, E.; Charehsaz, M.; Akyüz, S.; Acar, E.T.; Yesilada, E. Effect of in vitro gastrointestinal digestion on the bioavailability of phenolic components and the antioxidant potentials of some Turkish fruit wines. *Food Res. Int.* **2015**, *78*, 209–215. [CrossRef] [PubMed]

42. Kamiloglu, S.; Ozkan, G.; Isik, H.; Horoz, O.; Van Camp, J.; Capanoglu, E. Black carrot pomace as a source of polyphenols for enhancing the nutritional value of cake: An in vitro digestion study with a standardized static model. *LWT* **2017**, *77*, 475–481. [CrossRef]
43. Sengul, H.; Surek, E.; Nilufer-Erdil, D. Investigating the effects of food matrix and food components on bioaccessibility of pomegranate (Punica granatum) phenolics and anthocyanins using an in-vitro gastrointestinal digestion model. *Food Res. Int.* **2014**, *62*, 1069–1079. [CrossRef]
44. De Santiago, E.; Pereira-Caro, G.; Moreno-Rojas, J.M.; Cid, C.N.; De Pena, M.-P. Digestibility of (poly) phenols and antioxidant activity in raw and cooked cactus cladodes (Opuntia ficus-indica). *J. Agric. Food Chem.* **2018**, *66*, 5832–5844. [CrossRef] [PubMed]
45. Bermúdez-Soto, M.-J.; Tomás-Barberán, F.-A.; García-Conesa, M.-T. Stability of polyphenols in chokeberry (Aronia melanocarpa) subjected to in vitro gastric and pancreatic digestion. *Food Chem.* **2007**, *102*, 865–874. [CrossRef]
46. McDougall, G.; Fyffe, S.; Dobson, P.; Stewart, D. Anthocyanins from red wine–their stability under simulated gastrointestinal digestion. *Phytochemistry* **2005**, *66*, 2540–2548. [CrossRef]
47. Zhang, X.; Xiao, Y.; Huang, Z.; Chen, J.; Cui, Y.; Niu, B.; Huang, Y.; Pan, X.; Wu, C. Smart phase transformation system based on lyotropic liquid crystalline@ hard capsules for sustained release of hydrophilic and hydrophobic drugs. *Drug Deliv.* **2020**, *27*, 449–459. [CrossRef]
48. Martiel, I.; Baumann, N.; Vallooran, J.J.; Bergfreund, J.; Sagalowicz, L.; Mezzenga, R. Oil and drug control the release rate from lyotropic liquid crystals. *J. Control. Release* **2015**, *204*, 78–84. [CrossRef]
49. Cui, B.; Lu, Y.-m.; Tan, C.-p.; Wang, G.-q.; Li, G.-H. Effect of cross-linked acetylated starch content on the structure and stability of set yoghurt. *Food Hydrocoll.* **2014**, *35*, 576–582. [CrossRef]
50. Tavakoli, H.; Hosseini, O.; Jafari, S.M.; Katouzian, I. Evaluation of physicochemical and antioxidant properties of yogurt enriched by olive leaf phenolics within nanoliposomes. *J. Agric. Food Chem.* **2018**, *66*, 9231–9240. [CrossRef]
51. De Moura, S.C.; Schettini, G.N.; Garcia, A.O.; Gallina, D.A.; Alvim, I.D.; Hubinger, M.D. Stability of hibiscus extract encapsulated by ionic gelation incorporated in yogurt. *Food Bioprocess Technol.* **2019**, *12*, 1500–1515. [CrossRef]
52. Sun-Waterhouse, D.; Zhou, J.; Wadhwa, S.S. Effects of adding apple polyphenols before and after fermentation on the properties of drinking yoghurt. *Food Bioprocess Technol.* **2012**, *5*, 2674–2686. [CrossRef]
53. Sun-Waterhouse, D.; Zhou, J.; Wadhwa, S.S. Drinking yoghurts with berry polyphenols added before and after fermentation. *Food Control* **2013**, *32*, 450–460. [CrossRef]
54. Ribeiro, T.B.; Bonifácio-Lopes, T.; Morais, P.; Miranda, A.; Nunes, J.; Vicente, A.A.; Pintado, M. Incorporation of olive pomace ingredients into yoghurts as a source of fibre and hydroxytyrosol: Antioxidant activity and stability throughout gastrointestinal digestion. *J. Food Eng.* **2021**, *297*, 110476. [CrossRef]
55. Remanan, M.K.; Zhu, F. Encapsulation of rutin using quinoa and maize starch nanoparticles. *Food Chem.* **2021**, *353*, 128534. [CrossRef] [PubMed]
56. Altin, G.; Gültekin-Özgüven, M.; Ozcelik, B. Liposomal dispersion and powder systems for delivery of cocoa hull waste phenolics via Ayran (drinking yoghurt): Comparative studies on in-vitro bioaccessibility and antioxidant capacity. *Food Hydrocoll.* **2018**, *81*, 364–370. [CrossRef]
57. Martins, A.; Barros, L.; Carvalho, A.M.; Santos-Buelga, C.; Fernandes, I.P.; Barreiro, F.; Ferreira, I.C. Phenolic extracts of Rubus ulmifolius Schott flowers: Characterization, microencapsulation and incorporation into yogurts as nutraceutical sources. *Food Funct.* **2014**, *5*, 1091–1100. [CrossRef]
58. Aravind, S.M.; Wichienchot, S.; Tsao, R.; Ramakrishnan, S.; Chakkaravarthi, S. Role of dietary polyphenols on gut microbiota, their metabolites and health benefits. *Food Res. Int.* **2021**, *142*, 110189.s. [CrossRef]

MDPI
St. Alban-Anlage 66
4052 Basel
Switzerland
Tel. +41 61 683 77 34
Fax +41 61 302 89 18
www.mdpi.com

Antioxidants Editorial Office
E-mail: antioxidants@mdpi.com
www.mdpi.com/journal/antioxidants

www.ingramcontent.com/pod-product-compliance
Lightning Source LLC
LaVergne TN
LVHW070449100526
838202LV00014B/1690